AF181205

Growth Hormone

ENDOCRINE UPDATES

Shlomo Melmed, M.D., Series Editor

---

1. E.R. Levin and J.L. Nadler (eds.): Endocrinology of Cardiovascular Function. 1998. ISBN: 0-7923-8217-X
2. J.A. Fagin (ed.): Thyroid Cancer. 1998. ISBN: 0-7923-8326-5
3. J.S. Adams and B.P. Lukert (eds.): Osteoporosis: Genetics, Prevention and Treatment. 1998. ISBN: 0-7923-8366-4.
4. B.-Å. Bengtsson (ed.): Growth Hormone. 1999. ISBN: 0-7923-8478-4

Growth Hormone

edited by

Bengt-Åke Bengtsson, M.D.
Sahlgrenska University Hospital
Göteborg, Sweden

Springer Science+Business Media, LLC

Library of Congress Cataloging-in-Publication Data

Growth Hormone / edited by Bengt-Åke Bengtsson.
p. cm. -- (Endocrine updates ; 4)
Includes bibliographical references and index.
ISBN 978-1-4613-7351-3 ISBN 978-1-4615-5163-8 (eBook)
DOI 10.1007/978-1-4615-5163-8
1. Dwarfism, Pituitary--Hormone therapy. 2. Somatotropin-Therapeutic Use. I. Bengtsson, Bengt-Åke. II. Series.
[DNLM: 1.Hormone Replacement Therapy. 2. Somatotropin-deficiency. 3. Somatotropin-metabolism. 4. Somatotropin-therapeutic use. WK 550 G88397 1999]
RC658.7.G76 1999
616.4'7--dc21
DNLM/DLC 99-19993
for Library of Congress CIP

Copyright © 1999 by Springer Science+Business Media New York
Originally published by Kluwer Academic Publishers in 1999
Softcover reprint of the hardcover 1st edition 1999

All rights reserved. No part of this publication may be reproduced, stored in a retrieval system or transmitted in any form or by any means, mechanical, photo-copying, recording, or otherwise, without the prior written permission of the publisher, Springer Science+Business Media, LLC.

Printed on acid-free paper.

List of contributors

Christoph J. Auernhammer, Klinikum Grosshadern, Ludwig-Maximilians-Universität, Munich, Germany

Gerhard Baumann, Center for Endocrinology, Metabolism, and Molecular Medicine, Northwestern University Medical School, Chicago, Illinois, USA

Bengt-Åke Bengtsson, Research Centre for Endocrinology and Metabolism, Sahlgrenska University Hospital, Göteborg, Sweden

Martin Bidlingmaier, Klinikum Innerstadt, Ludwig-Maximilians-Universität Munich, Germany

Cyril Y. Bowers, Tulane Medical School, New Orleans, USA

P. V. Carroll, Division of Medicine, St Thomas' Hospital, London, UK

Jens Sandahl Christiansen, Department of Endocrinology and Diabetes M, Aarhus University Hospital, Aarhus, Denmark

Antonio Cittadini, Federico II University, Institute of Internal Medicine, Naples, Italy

David R. Clemmons, Division of Endocrinology, Department of Medicine, The University of North Carolina, Chapel Hill, NC, USA

E. R. Christ, Division of Medicine, St Thomas' Hospital, London, UK

Rolf Dall, Medical Department M, Aarhus University Hospital, Aarhus, Denmark

Suzanne L. Dickson, Department of Physiology, University of Cambridge, Cambridge, UK

Serfino Fazio, Federico II University, Institute of Internal Medicine, Naples, Italy

Janet K. Fernihough, University Department of Medicine, Bristol Royal Infirmary Bristol, UK

Ken K. Y. Ho, The Garvan Institute of Medical Research, St Vincent's Hospital Sydney, Australia

Jeff M. P. Holly, University Department of Medicine, Bristol Royal Infirmary, Bristol, UK

David M. Hoffman,The Garvan Institute of Medical Research, St Vincent's Hospital, Sydney, Australia

Simon J. Howell, Department of Endocrinology, Christie Hospital, Manchester, UK

Jörgen Isgaard, Research Centre for Endocrinology and Metabolism, Sahlgrenska University Hospital, Göteborg, Sweden

John-Olov Jansson, Research Centre for Endocrinology and Metabolism, Sahlgrenska University Hospital, Göteborg, Sweden

Gudmundur Johannsson, Research Centre for Endocrinology and Metabolism, Sahlgrenska University Hospital, Göteborg, Sweden

Jan-Ove Johansson, Research Centre for Endocrinology and Metabolism, Sahlgrenska University Hospital, Göteborg, Sweden

Troels Krarup Hansen, Medical Department M, Aarhus University Hospital, Aarhus, Denmark

Kerstin Landin-Wilhelmsen, Research Centre for Endocrinology and Metabolism, Sahlgrenska University Hospital, Göteborg, Sweden

Peter Lobie, Institute of Molecular and Cell Biology, The National University of Singapore, Singapore

Jens Otto Lunde Jörgensen, Medical Department M, Aarhus University Hospital, Aarhus, Denmark

John P. Monson, Department of Endocrinology, St Bartholomew's Hospital, London, UK

Steen Nielsen, Medical Department M, Aarhus University Hospital, Aarhus, Denmark

Helene Nörrelund, Medical Department M, Aarhus University Hospital, Aarhus, Denmark

Claes Ohlsson, Research Centre for Endocrinology and Metabolism, Sahlgrenska University Hospital, Göteborg, Sweden

A. Rahim, Department of Endocrinology, Christie Hospital, Manchester, UK

Thord Rosén, Research Centre for Endocrinology and Metabolism, Sahlgrenska University Hospital, Göteborg, Sweden

David Russell-Jones, Division of Medicine, St Thomas' Hospital, London, UK

Luigi Saccà, MD, Federico II University, Institute of Internal Medicine, Naples, Italy

S. M. Shalet, MD, Department of Endocrinology, Christie Hospital, Manchester, UK

Christian J. Strasburger, Klinikum Innerstadt, Ludwig-Maximilians-Universität, Munich, Germany

P. H. Sönksen, Division of Medicine, St Thomas' Hospital, London, UK

A. Margot Umpleby, Division of Medicine, St Thomas' Hospital, London, UK

Nina Vahl, Medical Department M, Aarhus University Hospital, Aarhus, Denmark

Douglas Wilmore, Brigham and Women's Medical School, Harvard Medical School, Boston, USA

Preface - Bengt-Åke Bengtsson

Ten years ago, many endocrinologists were still skeptical that growth hormone (GH) played an active role in adult metabolism. This is, perhaps, surprising given that GH deficiency (GHD) in adulthood had been "recognized" as long ago as in the 1960's. At that time Falkheden reported on patients who were treated with hypophysectomy (1). After the operation, the patients displayed symptoms now considered typical of GHD, including reductions in basal metabolic rate, cardiac output, heart rate, stroke volume, glomerular filtration rate and red cell volume. In a case history, reported one year earlier, Raben described a 35-year old female patient who was treated for hypopituitarism for eight years with standard hormone replacement therapy, and then given, additionally, human GH, 3 mg three times per week (2). After 2 months of GH treatment, the patient noticed "increased vigor, ambition and sense of well-being". Later investigations into quality of life after GH replacement therapy in adults were to confirm these earlier findings.

During the last 10 years studies have shown that GHD in adulthood is a far-reaching syndrome associated with abnormal body composition, reduced bone mineral density with an increased fracture rate, increased cardiovascular morbidity, and impaired cardiac function, as well as reduced exercise performance and decreased psychological well-being. Importantly, it is also clear, from placebo-controlled trials, that GH replacement therapy can normalize body composition, increase bone mineral content and improve the cardiovascular risk factor profile. GH replacement therapy also improves cardiac performance and exercise capacity, increases muscle strength and improves well-being and quality of life.

Some of the signs and symptoms of GHD are reminiscent of other familiar disease processes, and the known effects of GH replacement therapy suggest that GH could be indicated for conditions other than GHD. In this book some of the potential uses of GH outside classical endocrinology are being discussed. In this new, exciting development it is important that endocrinologists take leadership in exploring the potential role of GH in these conditions.

The aim of this book is to provide a critical update of current knowledge about adult GHD and the future role of GH/IGF-1 in adult medicine. For this purpose, contributions from a number of research groups have been generated. It is my hope that the book may serve not only as an introduction to the field, but also stimulate further research within this exciting area.

References

1. Falkheden T. Pathophysiological studies following hypophysectomy in man. Theses, University of Göteborg, 1963.
2. Raben MS. Clinical use of human growth hormone. N. Engl J. Med. 1962;266:82-86.

Growth Hormone

Growth hormone secretion and mechanisms of action

1 NEUROENDOCRINE CONTROL OF GROWTH HORMONE SECRETION

John-Olov Jansson
Suzanne L. Dickson

Research Centre for Endocrinology and Metabolism (RCEM), Sahlgrenska University Hospital, Göteborg, Sweden and Dept. Physiology, University of Cambridge, Downing Street, Cambridge UK.

GROWTH HORMONE SECRETION IS PULSATILE

Growth hormone (GH), like many other pituitary hormones, is released from the pituitary in a highly episodic manner. In the conscious male rat, the pulsatile pattern of GH secretion is particularly striking, with high amplitude pulses occurring regularly, once every 3-4 h (Tannenbaum & Martin, 1976). The pattern of GH secretion reflects a changing balance in the output of two neuroendocrine systems: the GH-releasing hormone (GRF) neurones, which stimulate GH secretion (Guillemin et al. 1982; Rivier et al. 1982) and the somatostatin neurones, which inhibit GH secretion (Brazeau et al. 1973). GRF and somatostatin are released into portal blood at the median eminence and are then transported to the anterior pituitary where they bind to different receptors on the somatotroph cells. Passive immunization studies have established that GRF and somatostatin control different aspects of the GH secretory pattern: GRF controls GH pulse amplitude and its release is a prerequisite for pulsatile GH secretion whereas somatostatin is important for regulating the period between pulses (see Jansson et al. 1985). Somatostatin also appears to be involved in the GH pulse generating mechanism, since a pulsatile pattern of GH secretion can be generated by an intermittent intravenous infusion of somatostatin, in which GH pulses occurred when delivery of somatostatin was interrupted for a short period every 3 hours (Clark & Robinson 1988).

Distribution of GRF- and somatostatin-containing cells

The somatostatin cells projecting to the median eminence are located within the rostral periventricular nucleus (Ishikawa et al. 1987) while the hypophysiotropic GRF neurones are confined to the ventrolateral part of the arcuate nucleus (Jacobowitz et al. 1983; Sawchenko et al. 1985). Additional populations of somatostatin cell bodies can be found in other regions of the hypothalamus, including the arcuate, the suprachiasmatic and the dorsomedial nuclei (Johansson et al. 1984). By contrast, the GRF cells are located only in the arcuate nucleus (where they constitute a major cell population) and in the area around the ventrolateral aspects of the ventromedial nucleus (peri-VMN). These peri-VMN GRF cells do not project to the median eminence and the extent to which they are involved in GH regulation is not known.

Hypothalamic interactions between GRF and somatostatin neurones

There is anatomical evidence that GRF and somatostatin neurones interact bi-directionally within the hypothalamus, in addition to having direct effects on GH secretion at the pituitary level. Within the arcuate nucleus, direct synaptic associations between somatostatin fibres and GRF-containing neurones have been demonstrated and high affinity somatostatin receptors have been found to co-localise on GRF cell bodies (Epelbaum et al. 1989). At least 5 somatostatin receptor sub-types have been identified. Both SSTR1 and SSTR2 are present on sub-populations of GRF neurones (Tannenbaum et al. 1998), and it has been suggested that these are likely to be the principal sub-types involved in the control of GH secretion at the hypothalamic level (Beaudet et al. 1995). Neuroanatomical evidence for control of periventricular somatostatin neurones by the GRF system is suggested from studies demonstrating that GRF fibres closely approach and synapse with somatostatin cell bodies in the periventricular nucleus (Horvath et al. 1989). Studies investigating the origin of the arcuate somatostatin and the periventricular GRF fibres are fairly inconclusive. Fodor et al (1994) found that 3% of the GRF cells in the arcuate nucleus were retrogradely-labelled following injection of wheat germ agglutinin-apohorseradish peroxidase conjugate-gold complex into the periventricular nucleus, providing confirmation that GRF neurones do project to the periventricular nucleus, although this is clearly not a major projection. Willoughby and colleagues (1989) reported that very few of the periventricular somatostatin cells were retrogradely labelled following Fluorogold injection into the arcuate nucleus, and suggested that the innervation of GRF neurones by somatostatin fibres may be derived from local arcuate somatostatin neurones. That these arcuate somatostatin cells are involved in GH regulation has yet to be established.

Functionally, there is also strong evidence for reciprocal bidirectional CNS control of GH secretion by GRF and somatostatin. Intracerebroventricular injections of GRF inhibits GH secretion via a mechanism which includes increased somatostatin release into portal blood (Katakami et al.1986) whereas intracerebroventricular somatostatin stimulates GH secretion (Abe et al. 1983) by stimulating GRF release (Murakami et al. 1987). Intravenous infusion of somatostatin greatly suppresses GH release, and this is characteristically followed by a large rebound release of GH once the infusion is stopped, the so-called 'somatostatin withdrawal' response (Clark et al. 1988). Some of this GH rebound is generated at the level of the pituitary, but the rebound is much larger in vivo than in vitro, suggesting the involvement of an active hypothalamic component. Since the administration of GRF antiserum attenuates the GH rebound in vivo, it has been suggested that GRF neurones are activated during the withdrawal period (Clark et al. 1988). Okada et al (1991) suggested that the somatostatinergic control is likely to be mediated by the periventricular somatostatin cells since GH release is inhibited by electrical stimulation of the periventricular nucleus and this gave way to a rebound release once the stimulation was stopped.

In electrophysiological studies we demonstrated that electrical stimulation of the periventricular nucleus exerts a similar response in a sub-population of putative neurosecretory arcuate neurones which fulfilled multiple additional criteria for identification as GRF neurones (Dickson et al. 1993; 1994). These cells were inhibited during periventricular nucleus stimulation and many of these cells exhibited a rebound hyperactivation following the end of stimulation. Thus, our electrophysiological studies provide evidence for the existence of a direct or indirect pathway from the periventricular nucleus to the arcuate nucleus, which influences the activity of cells in this region, including putative GRF neurones.

Therefore these electrophysiological data provide neurophysiological evidence for previously presumed regulation of GRF neurones by the periventricular somatostatin system.

What role might an endogenous GH secretagogue play in the control of GH secretion?

Recent interest in the neuroendocrine control of GH secretion has focused on the possible site and mechanism of action of a groups of synthetic compounds, the GH secretagogues, of which GH-releasing peptide (GHRP-6) was the first identified member (Bowers et al. 1984). These substances stimulate GH secretion by a direct pituitary action (Bowers et al. 1984) and also activate a sub-population of cells in the hypothalamic arcuate nucleus (Dickson et al. 1993; 1995). The recent cloning of the GH secretagogue receptor (Howard et al. 1996), shown to be present at pituitary and hypothalamic sites, has suggested the possible existence of an endogenous GH secretagogue. The precise role that such a ligand might play in the hypothalamic orchestration of pulsatile GH secretion is currently being elucidated. Certainly, GRF neurones appear to be one target population for GH secretagogue action within the CNS since approximately 40% of cells expressing GRF mRNA could be identified, on the consecutive section, as express c-fos mRNA following systemic GHRP-6 injection (Dickson & Luckman 1997). It is unlikely that the sole action of GH secretagogues in the central nervous system is to activate GRF neurones, since this would not explain the large synergy between GHRP-6 and GRF for GH secretion (Bowers et al. 1984). To explain this synergy, CY Bowers proposed that the GH secretagogues stimulate the release into the portal blood of an unknown factor that acts together with GRF to stimulate GH secretion. Indeed, it remains to be determined whether the putative endogneous GH secretagogue ligand is also an adenohypophysiotrophic factor, produced by the hypothalamus and released into the portal circulation to control GH secretion.

The activation of cells in the hypothalamus by GH secretagogues, shown both in electrophysiological studies and in studies detecting Fos protein (the protein product of the c-fos gene) is attenuated by systemic or central injection of Sandostatin, a long acting somatostatin analogue (Dickson et al. 1998; Zheng et al. 1997). This suggests that the central actions of the GH secretagogues are subject to inhibition by central somatostatin pathways, possibly reflecting an inhibitory effect of somatostatin on GRF producing cells as discussed above. Interesting this was not seen in mice with disrupted somatostatin type 2 receptor, indicating that the inhibitory effect of central somatostatin pathways on GH secretagogue induced Fos protein expression is mediated by type 2 receptors (Zheng et al. 1997).

Effects of gender and gonadal steroids on GH secretion

In the rat, pulsatility in the GH secretory pattern is very pronounced in the male, as already mentioned, but in the female GH is secreted in a more continuous manner, comprising low amplitude, frequent GH pulses (see Jansson et al. 1985). Subsequent experiments demonstrated that the plasma GH levels are higher during the night in female rats (Clark et al. 1987).

This sexual dimorphism is likely to reflect differences in the patterning of the GRF-somatostatin GH pulse generator. Certainly, the effects of regular injections of GRF on GH secretion are sexually dimorphic: female rats display a GH response following each GRF injection, while males show a regular cycle of responsiveness (Clark & Robinson, 1985). The most likely explanation is that portal concentrations of somatostatin determine whether GRF injection can induce a GH pulse and that the pattern of somatostatin secretion is likely to be more continuous in the female rat than in the male. This is supported by the finding that intermittent somatostatin treatment masculinizes the continuous GH secretory pattern normally seen in female rats (Clark & Robinson1988), as described above.

Both male and female rats show an increase in plasma GH levels immediately before the onset of puberty (see Jansson et al. 1985). This prepubertal rise in GH does not appear to be dependent on the presence of gonadal steroids since it occurs in neonatally gonadectomized rats (Jansson et al.1984) and it precedes the rise in plasma sex steroid levels observed at the onset of puberty (see Jansson et al. 1985). In humans, the results of several studies indicate that the GH secretion is more continuous with higher baseline levels in young women than in young men Winer et al. 1990; Ho et al. 1987; Jaffe et al. 1998).

In most mammals, following puberty, there is a sex difference in the rate of body growth as well as the pattern of hormone secretion (Howard, 1989) and this difference is very striking in the rat (Edén, 1979; Jansson et al.1985). Several lines of evidence suggest that sex-related differences in the control of GH secretion depend on gonadal steroids. Exposure to or deprivation of specific sex steroid hormones during the neonatal period causes permanent alterations of the adult GH secretory pattern (Jansson et al. 1984; 1985; Jansson & Frohman 1987). Neonatal or prepubertal gonadectomy of male rats causes an increase in basal GH secretion and neonatal gonadectomy also reduces pulse height to levels similar to those of normal adult female rats. In female rats, neonatal gonadectomy causes an increase in the magnitude of GH pulses during puberty and a reduction in baseline levels post-pubertally (Jansson et al. 1984). Estrogen treatment feminizes the GH-secretory pattern in male rats, such that basal GH release is elevated, and GH pulses are more frequent with a reduced pulse height (Painson et al. 1992). Similarly, the administration of testosterone to ovariectomized adult rats induces a GH-secretory pattern which is almost identical to that in normal adult male rats (Akira et al. 1988). These data suggest that gonadal androgens maintain a high GH pulse amplitude and low basal GH levels, while gonadal oestrogens reduce the pulse height, increase the pulse frequency and elevate baseline GH levels.

The mechanism of action of gonadal steriods for maintaining a sexually dimorphic GH secretory pattern is not completely elucidated, but seems to reflect differences in the secretory pattern of somatostatin and/or GRF into portal blood. Oestrogen not only feminizes the pattern of GH secretion in castrated adult rats but also feminizes the pattern of GH responsiveness to GRF in these rats (Painson et al. 1992). Since the sexual dimorphism in the GH responsiveness to GRF probably reflects a sexually dimorphic pattern of somatostatin release into portal blood (Tannenbaum & Ling 1984), it is conceivable that oestrogens are involved in maintaining a steady non-cyclic pattern of somatostatin release.

More direct evidence that gonadal steroids regulate GH secretion via GRF and somatostatin is provided by experiments in which androgens (either testosterone or dihydrotestosterone) increases the expression of both somatostatin mRNA in the periventricular nucleus (Argente et al. 1990) and GRF mRNA in the arcuate nucleus (Argente et al. 1991) of castrated rats. Nuclear androgen receptors are located in both arcuate and periventricular nuclei (Roselli et al. 1989) and more androgen receptors are present in the periventricular nucleus of male rats than female rats (Roselli, 1991).

Masculinization of the GH secretory pattern in ovariectomized rats by androgen treatment has been shown to be accompanied by increased GRF and somatostatin mRNA levels in the hypothalamus (Hasegawa et al. 1992). Thus, gonadal steroids appear to influence GH secretion by a hypothalamic mechanism that includes the regulation of GRF and/or somatostatin gene transcription.

Growth hormone feedback

Spontaneous GH secretion in conscious rats is effectively inhibited by infusing exogenous GH (Tannenbaum, 1980; Clark et al. 1988). GH does not inhibit its own release in vitro from pituitary cells, and it is therefore unlikely that GH directly feeds back at the level of the pituitary (Richman et al. 1981). A hypothalamic site of GH feedback is suggested from experiments in which an i.c.v. injection of human GH suppresses spontaneous GH secretion in conscious male rats (Tannenbaum, 1980; Conway et al. 1985).

One paradox of GH induced feedback effects, is that it is a large peptide (Niall, 1971) and may not have direct access to the brain, with the possible exception of those areas where the blood-brain barrier is leaky, such as the median eminence and most ventromedial portions of the arcuate nucleus (Merchenthaler 1991; Banks & Broadwell 1994, Dickson et al, 1996). GH receptor mRNA has been colocalized in hypothalamic somatostatin neurones (Burton et al. 1992) and GRF neurones (Burton et al. 1995). GH feedback may be mediated (at least partially) by insulin-like growth factors (IGF)-I and -II, which are produced in many tissues in response to GH. Receptors for both IGFs have been localized on anterior pituitary cells (Goodyer et al. 1984) and in the brain (Gammeltoft et al. 1985). At the level of the pituitary, IGF-I and -II inhibits basal (Goodyer et al. 1984) and GRF-induced (Ceda et al. 1987) GH release. However, concomitant i.c.v. administration of IGF-I and -II has been shown to greatly inhibit GH secretion, providing evidence for a central inhibitory action of these substances on GH release (Harel & Tannenbaum, 1992).This effect was probably exerted at the level of the hypothalamus, where IGF-I may act either to stimulate somatostatin release (Aquila, 1991) or to inhibit GRF release (Shibasaki et al. 1986).

Growth hormone feedback and sex differentiation of GH secretion

Studies on male rats have shown that exogenous i.v. GH pulses given at 3 h intervals can synchronize endogenous GH so occur at the same time as the exogenous pulses (Carlsson & Jansson 1990). This study together with data obtained by Chihara and coworkers using a different experimental approach (Sato et al. 1989), indicate that each GH pulse in male rats causes the low plasma GH levels during the subsequent through period via negative feedback.. This effect is probably mediated by enhanced somatostin secretion, since studies on male rats demonstrated that an intracerebroventricular injection of GH stimulates somatostatin release into portal blood (Chihara et al. 1981).

In female rats, GH feedback may be mediated by an inhibition of GRF release, rather than a stimulation of somatostatin release, since the infusion of exogenous human GH does not block the regular GH responses to GRF in conscious female rats (Clark et al. 1988). Furthermore, in the male rat, exogenous GH reduces the effects of serial injections of GRF and disturbs the regular cycle of GRF responsiveness usually found in male rats (Carlsson et al. 1990). The assumption that the GH effect on somatostatin release is gender specific is supported by the finding that prolonged exposure to GH from a tumour induces more hypothalamic somatostatin in male rats than in females (Maiter & Martin 1990). Hence, the sexual dimorphic patterns of GH secretion may be determined by a difference in autofeedback regulation. In females, GH feedback may be primarily controlled by an inhibition of GRF release, whereas in males, the feedback may reflect increased somatostatin release. Interestingly, the results of a recent study indicate that negative on GH secretion by IGF-1 is gender specific in humans. The feed back effect in women seemed to be due in part to suppressed GRF secretion (Jaffe et al. 1998).

Minami and colleagues (1992) have found that systemic administration of GH induces expression of the immediate early gene, c-fos, in the arcuate and periventricular nuclei of hypophysectomised male rats. These authors also reported that the distribution of cells expressing c-fos is identical to the distribution of cells expressing somatostatin mRNA suggesting that GH feeds back on somatostatin neurones of the hypothalamic periventricular and arcuate nuclei. The results of a recent study suggest that GH can inhibit GHRP responsiveness in the arcuate nucleus via somatostatin acting on somatostatin type 2 receptors (Zheng et al. 1997). It remains to be determined whether this reflects a suppressive effect on the activity of GRF producing cells.

Possible roles of leptin and Neuropeptide Y (NPY) in the regulation of GH secretion

Pulsatile GH secretion is modulated by a number of peripheral and central mechanisms, some of which reflect changing physiological conditions of the animal. For example, pulsatile GH secretion is abolished in adverse metabolic situations such as severe food restriction (Tannenbaum et al. 1979) or diabetes (Tannenbaum, 1981). The block of pulsatile GH secretion in fasted rats can be reversed by administering leptin (Carro et al. 1997, Vuagnat et al. 1998), a hormone produced by adipose tissue. This effect could be due to leptin-induced reversal of the somatostatin release reported to be responsible for starvation-induced suppression of GH secretion in rats (see Jansson et al. 1985). The results of a recent study suggest that leptin also may suppress somatostatin release in fed male rats, thereby enhancing basal GH secretion Tannenbaum 1998). On the other hand, there are indications that Zucker rats, which have defective leptin receptor and decreased GH pulse height, do not have markedly increased somatostatin release. Since they respond well to exogenous GRF, the suppressed GH pulses might be due to decreased endogenous GRF release rather than increased somatostatin (Tannenbaum, 1990). It has recently been reported that some of the GRF containing neurons in the arcuate nucleus have leptin receptors (Meister 1997).

Leptin has access to the hypothalamic arcuate nucleus, where its actions include the suppression of the NPY cells (Mercer et al, 1997). This suppression of NPY may be of importance to restore GH secretion, since central administration of NPY inhibits GH secretion (Pierroz et al, 1996). Moreover, the suppression of serum IGF-1 and body length that is observed in leptin-deficient ob/ob mice is reversed by NPY gene depletion (Erickson et al 1996). Thus, it could be speculated that NPY neurones in the arcuate nucleus play a role for suppression of GH secretion in the fasted animal. In this context it may seem paradoxical that the GH secretagogue GHRP-6, in addition to activating the GRF neurones also activates a sub-population of NPY containing neurons in the arcuate nucleus of the rat in vivo (Dickson & Luckman 1997), an effect that also has been seen in human tumour cells (Jansson et al 1998). It seems likely that the effects of GHRP-6 on NPY containing neurones are of importance for the reported GHRP-6 induction of food intake and/or activation of the hypothalamo-pituitary-adrenal axis (Pierroz et al, 1996) rather than its effect on GH secretion.

At present, the effects of leptin and NPY on GH secretion in humans is not completely understood. Children with defective leptin receptors have suppressed GH secretion and delayed body growth (Clement et al. 1998), although leptin deficient children appear to grow normally (Montague et al. 1997).

REFERENCES

Abe H, Molitch ME, Van-Wyk JJ, Underwood LE. Human growth hormone and somatomedin C suppress the spontaneous release of growth hormone in unanesthetized rats. Endocrinology 1983; 113: 1319-1324

Akira S, Wakabayashi I, Sugihara H, Minami S, Takahashi F, Motoyama A. Effect of testosterone on growth hormone secretion in female rats during a continuous infusion of growth hormone releasing factor. Neuroendocrinology 1988; 47: 116-124

Aquila MC. Insulin-like growth factor I (IGF-I) regulates somatostatin (SRIF) release, messenger ribonucleic acid levels in the rat hypothalamus. Proceedings of the 73rd Annual Meeting of the Endocrine Society, 1991, P351

Argente J, Chowen-Breed JA, Steiner RA, Clifton DK. Somatostatin messenger RNA in hypothalamic neurons is increased by testosterone through activation of androgen receptors and not by aromatization to estradiol. Neuroendocrinology 1990; 52: 342-349

Argente J, Chowen JA, Zeitler P, Clifton DK, Steiner RA. Sexual dimorphism of growth hormone-releasing hormone and somatostatin gene expression in the hypothalamus of the rat during development. Endocrinology 1991; 128: 2369-2375

Banks WA, Broadwell RD. Blood to brain and brain to blood passage of native horseradish peroxidase, wheat germ agglutinin, and albumin: pharmacokinetic and morphological assessments. J Neurochem 1994; 62: 2404-2419

Beaudet A, Tannenbaum GS. Expression of SSTR1 and SSTR2 in rat hypothalamus: correlation with receptor binding and distribution of growth hormone regulatory peptides. Ciba Found Symp 1995; 190: 142-152

Bowers CY, Momany FA, Reynolds GA, Hong A. On the in vitro and in vivo activity of a new synthetic hexapeptide that acts on the pituitary to specifically release growth hormone. Endocrinology 1984; 114: 1537-1545

Brazeau P, Vale W, Burgus R, Ling N, Butcher M, Rivier J, Guillemin R. Hypothalamic polypeptide that inhibits the secretion of immunoreactive pituitary growth hormone. Science 1973; 179: 77-79

Burton KA, Kabigting EB, Clifton DK, Steiner RA. Growth hormone receptor messenger ribonucleic acid distribution in the adult male rat brain and its colocalization in hypothalamic somatostatin neurons. Endocrinology 1992; 131: 958-963

Burton KA, Kabigting EB, Steiner RA, Clifton DK. Identification of target cells for growth hormone's action in the arcuate nucleus. Am J Physiol 1995; 269: E716-E722

Carlsson LMS, Clark RG, Robinson ICAF. Sex difference in growth hormone feedback in the rat. J Endocrinol 1990; 126: 27-35

Carlsson LMS, Jansson J-O. Endogenous growth hormone (GH) secretion in male rats is synchronized to pulsatile GH infusions given at 3-hour intervals. Endocrinology 1990; 126: 6-10

Carro E, Señaris R, Considine RV, Casanueva FF, Dieguez C. Regulation of in vivo growth hormone secretion by leptin. Endocrinology 1997; 138: 2203-2206

Ceda GP, Davis RG, Rosenfeld RG, Hoffman AR. The growth hormone (GH)-releasing hormone (GHRH)-GH-somatomedin axis: evidence for rapid inhibition of GHRH-elicited GH release by insulin-like growth factors I and II. Endocrinology 1987; 120: 1658-1662

Chihara K, Minamitani N, Kaji H, Arimura A, Fujita T. Intraventricularly injected growth hormone stimulates somatostatin release into rat hypophysial portal blood. Endocrinology 1981; 109: 2279-2281

Clark RG, Robinson ICAF. Growth hormone responses to multiple injections of a fragment of human growth hormone-releasing factor in conscious male and female rats. J Endocrinol 1985; 106: 281-289

Clark RG, Carlsson LMS, Robinson ICAF. Growth hormone secretory profiles in conscious female rats. J Endocrinol 1987; 114: 399-407

Clark RG, Carlsson LMS, Rafferty B, Robinson ICAF. The rebound release of growth hormone (GH) following somatostatin infusion in rats involves hypothalamic GH-releasing factor release. J Endocrinol 1988; 119: 397-404

Clark RG, Robinson ICAF. Paradoxical growth-promoting effects induced by patterned infusions of somatostatin in female rats. Endocrinology 1988; 122: 2675-2682

Clark RG, Carlsson LMS, Robinson ICAF. Growth hormone (GH) secretion in the conscious rat: negative feedback of GH on its own release. J Endocrinol 1988; 119: 201-209

Clement K, Vaisse C, Lahlou N, Cabrol S, Pelloux V, Cassuto D, Gourmelen M, Dina C, Chambaz J, Lacorte JM, Basdevant A, Bougneres P, Lebouc Y, Froguel P, Guy GB. A mutation in the human leptin receptor gene causes obesity and pituitary dysfunction. Nature 1998; 392: 398-401

Conway S, McCann SM, Krulich L. On the mechanism of growth hormone autofeedback regulation: possible role of somatostatin and growth hormone-releasing factor. Endocrinology 1985; 117: 2284-2292

Dickson SL, Leng G, Robinson ICAF. Systemic administration of growth hormone-releasing peptide activates hypothalamic arcuate neurons. Neuroscience 1993; 53: 303-306

Dickson SL, Leng G, Robinson ICAF. Electrical stimulation of the rat periventricular nucleus influences the activity of hypothalamic arcuate neurones. J Neuroendocrinol 1994; 6: 359-367

Dickson SL, Leng G, Dyball RE, Smith RG. Central actions of peptide and non-peptide growth hormone secretagogues in the rat. Neuroendocrinology 1995; 61: 36-43

Dickson SL, Doutrelant-Viltart O, Dyball RE, Leng G. Retrogradely labelled neurosecretory neurones of the rat hypothalamic arcuate nucleus express Fos protein following systemic injection of GH-releasing peptide-6. J Endocrinol 1996; 151: 323-331

Dickson SL, Viltart O, Bailey ART, Leng G. Attenuation of the growth hormone secretagogue induction of Fos protein in the rat arcuate nucleus by central somatostatin action. Neuroendocrinology 1997; 66: 188-194

Dickson SL, Luckman SM. Induction of c-fos messenger ribonucleic acid in neuropeptide Y and growth hormone (GH)-releasing factor neurons in the rat arcuate nucleus following systemic injection of the GH secretagogue, GH-releasing peptide-6. Endocrinology 1997; 138: 771-777

Edén S. Age- and sex-related differences in episodic growth hormone secretion in the rat. Endocrinology 1979; 105: 555-560

Epelbaum J, Moyse E, Tannenbaum GS, Kordon C, Beaudet A. Combined autoradiographic and immunohistochemical evidence for an association of somatostatin binding sites with growth hormone-releasing factor nerve cell bodies in the rat arcuate nucleus. J Neuroendocrinol 1989; 1: 109-115

Erickson JC, Clegg KE, Palmiter RD. Sensitivity to leptin and susceptibility to seizures of mice lacking neuropeptide Y. Nature 1996; 381: 415-421

Fodor M, Csaba Z, Kordon C, Epelbaum J. Growth hormone-releasing hormone, somatostatin, galanin and beta-endorphin afferents to the hypothalamic periventricular nucleus. J Chem Neuroanat 1994; 8: 61-73

Gammeltoft S, Haselbacher GK, Humbel RE, Fehlmann M, Van-Obberghen E. Two types of receptor for insulin-like growth factors in mammalian brain. EMBO J 1985; 4: 3407-3412

Goodyer CG, De-Stephano L, Guyda HJ, Posner BI. Effects of insulin-like growth factors on adult male rat pituitary function in tissue culture. Endocrinology 1984; 115: 1568-1576

Guillemin R, Brazeau P, Bohlen P, Esch F, Ling N, Wehrenberg WB. Growth hormone-releasing factor from a human pancreatic tumor that caused acromegaly. Science 1982; 218: 585-587

Hakansson ML, Brown H, Ghilardi N, Skoda RC, Meister B. Leptin receptor immunoreactivity in chemically defined target neurons of the hypothalamus. J Neurosci 1998; 18: 559-572

Harel Z, Tannenbaum GS. Centrally administered insulin-like growth factor II fails to alter pulsatile growth hormone secretion or food intake. Neuroendocrinology 1992; 56: 161-168

Hasegawa O, Sugihara H, Minami S, Wakabayashi I. Masculinization of growth hormone (GH) secretory pattern by dihydrotestosterone is associated with augmentation of hypothalamic somatostatin and GH-releasing hormone mRNA levels in ovariectomized adult rats. Peptides 1992; 13: 475-481

Ho KY, Evans WS, Blizzard RM, Veldhuis JD, Merriam GR, Samojlik E, Furlanetto R, Rogol AD, Kaiser DL, Thorner MO. Effects of sex and age on the 24-hour profile of growth hormone secretion in man: importance of endogenous estradiol concentrations. J Clin Endocrinol Metab 1987; 64:51-58

Horvath S, Palkovits M, Gorcs T, Arimura A. Electron microscopic immunocytochemical evidence for the existence of bidirectional synaptic connections between growth hormone-releasing hormone- and somatostatin-containing neurons in the hypothalamus of the rat. Brain Res 1989; 481: 8-15

Howard AD, Feighner SD, Cully DF, Arena JP, Liberator PA, Rosenblum CI, Hamelin M, Hreniuk DL, Palyha OC, Anderson J, Paress PS, Diaz C, Chou M, Liu KK, McKee KK, Pong SS, Chaung LY, Elbrecht A, Dashkevicz M, Heavens R, Rigby M, Sirinathsinghji DJS, Dean DC, Melillo DG, Van-der-Ploeg LH, Smith RG. A receptor in pituitary and hypothalamus that functions in growth hormone release. Science 1996; 273: 974-977

Howard CP. "Endocrinology of puberty. ". In: "Adolescent medicine", Appleton & Lange, 1989. pp17-20.

Ishikawa K, Taniguchi Y, Kurosumi K, Suzuki M, Shinoda M. Immunohistochemical identification of somatostatin-containing neurons projecting to the median eminence of the rat. Endocrinology 1987; 121:94-97

Jacobowitz DM, Schulte H, Chrousos GP, Loriaux DL. Localization of GRF-like immunoreactive neurons in the rat brain. Peptides 1983; 4: 521-524

Jaffe CA, Ocampo-Lim B, Guo W, Krueger K, Sugahara I, DeMott-Friberg R, Barkan AL. Regulatory mechanisms of growth hormone secretion are sexually dimorphic. J Clin Invest 1998; 102: 153-164

Jansson J-O, Ekberg S, Isaksson OG, Edén S. Influence of gonadal steroids on age- and sex-related secretory patterns of growth hormone in the rat. Endocrinology 1984; 114: 1287-1294

Jansson J-O, Edén S, Isaksson O. Sexual dimorphism in the control of growth hormone secretion. Endocr Rev 1985; 6: 128-150

Jansson J-O, Frohman LA. Differential effects of neonatal and adult androgen exposure on the growth hormone secretory pattern in male rats. Endocrinology 1987; 120: 1551-1557

Jansson J-O, Svensson J, Bengtsson B-Å, Frohman LA, Ahlman H, Wangberg B, Nilsson O, Nilsson M. Acromegaly and Cushing's syndrome due to ectopic production of GHRH and ACTH by a thymic carcinoid tumour: in vitro responses to GHRH and GHRP-6. Clin Endocrinol Oxf 1998; 48: 243-250

Johansson O, Hökfelt T, Elde RP. Immunohistochemical distribution of somatostatin-like immunoreactivity in the central nervous system of the adult rat. Neuroscience 1984; 13: 265-339

Katakami H, Arimura A, Frohman LA. Growth hormone (GH)-releasing factor stimulates hypothalamic somatostatin release: an inhibitory feedback effect on GH secretion. Endocrinology 1986; 118: 1872-1877

Maiter DM, Gabriel SM, Koenig JI, Russell WE, Martin JB. Sexual differentiation of growth hormone feedback effects on hypothalamic growth hormone-releasing hormone and somatostatin. Neuroendocrinology 1990; 51:174-180

Mercer JG, Moar KM, Rayner DV, Trayhurn P, Hoggard N. Regulation of leptin receptor and NPY gene expression in hypothalamus of leptin-treated obese (ob/ob) and cold-exposed lean mice. FEBS Lett 1997; 402: 185-188

Merchenthaler I. Neurons with access to the general circulation in the central nervous system of the rat: a retrograde tracing study with fluoro-gold. Neuroscience 1991; 44: 655-662

Minami S, Kamegai J, Sugihara H, Hasegawa O, Wakabayashi I. Systemic administration of recombinant human growth hormone induces expression of the c-fos gene in the hypothalamic arcuate and periventricular nuclei in hypophysectomized rats. Endocrinology 1992; 131: 247-253

Montague CT, Farooqi IS, Whitehead JP, Soos MA, Rau H, Wareham NJ, Sewter CP, Digby JE, Mohammed SN, Hurst JA, Cheetham CH, Earley AR, Barnett AH, Prins JB, O'Rahilly S. Congenital leptin deficiency is associated with severe early-onset obesity in humans. Nature 1997; 387: 903-908

Murakami Y, Kato Y, Kabayama Y, Inoue T, Koshiyama H, Imura H. Involvement of hypothalamic growth hormone (GH)-releasing factor in GH secretion induced by intracerebroventricular injection of somatostatin in rats. Endocrinology 1987; 120: 311-316

Niall HD. Revised primary structure for human growth hormone. Nat New Biol 1971; 230: 90-91

Okada K, Wakabayashi I, Sugihara H, Minami S, Kitamura T, Yamada J. Electrical stimulation of hypothalamic periventricular nucleus is followed by a large rebound secretion of growth hormone in unanesthetized rats. Neuroendocrinology 1991; 53: 306-312

Painson JC, Thorner MO, Krieg RJ, Tannenbaum GS. Short-term adult exposure to estradiol feminizes the male pattern of spontaneous and growth hormone-releasing factor-stimulated growth hormone secretion in the rat. Endocrinology 1992; 130: 511-519

Pierroz DD, Catzeflis C, Aebi AC, Rivier JE, Aubert ML. Chronic administration of neuropeptide Y into the lateral ventricle inhibits both the pituitary-testicular axis and growth hormone and insulin-like growth factor I secretion in intact adult male rats. Endocrinology 1996; 137: 3-12

Richman RA, Weiss JP, Hochberg Z, Florini JR. Regulation of growth hormone release: evidence against negative feedback in rat pituitary cells. Endocrinology 1981; 108: 2287-2292

Rivier J, Spiess J, Thorner M, Vale W. Characterization of a growth hormone-releasing factor from a human pancreatic islet tumour. Nature 1982; 300: 276-278

Roselli CE, Handa RJ, Resko JA. Quantitative distribution of nuclear androgen receptors in microdissected areas of the rat brain. Neuroendocrinology 1989; 49: 449-453

Roselli CE. Sex differences in androgen receptors and aromatase activity in microdissected regions of the rat brain. Endocrinology 1991; 128: 1310-1316

Sato M, Chihara K, Kita T, Kashio Y, Okimura Y, Kitajima N, Fujita T. Physiological role of somatostatin-mediated autofeedback regulation for growth hormone: importance of growth hormone in triggering somatostatin release during a trough period of pulsatile growth hormone release in conscious male rats. Neuroendocrinology 1989; 50: 139-151

Sawchenko PE, Swanson LW, Rivier J, Vale WW. The distribution of growth-hormone-releasing factor (GRF) immunoreactivity in the central nervous system of the rat: an immunohistochemical study using antisera directed against rat hypothalamic GRF. J Comp Neurol 1985; 237: 100-115

Shibasaki T, Yamauchi N, Hotta M, Masuda A, Imaki T, Demura H, Ling N, Shizume K. In vitro release of growth hormone-releasing factor from rat hypothalamus: effect of insulin-like growth factor-1. Regul Pept 1986; 15: 47-53

Tannenbaum GS, Martin JB. Evidence for an endogenous ultradian rhythm governing growth hormone secretion in the rat. Endocrinology 1976; 98:562-570

Tannenbaum GS, Rorstad O, Brazeau P. Effects of prolonged food deprivation on the ultradian growth hormone rhythm and immunoreactive somatostatin tissue levels in the rat. Endocrinology 1979; 104: 1733-1738

Tannenbaum GS. Evidence for autoregulation of growth hormone secretion via the central nervous system. Endocrinology 1980; 107: 2117-2120

Tannenbaum GS. Growth hormone secretory dynamics in streptozotocin diabetes: evidence of a role for endogenous circulating somatostatin. Endocrinology 1981; 108: 76-82

Tannenbaum GS, Ling N. The interrelationship of growth hormone (GH)-releasing factor and somatostatin in generation of the ultradian rhythm of GH secretion. Endocrinology 1984; 115: 1952-1957

Tannenbaum GS. Interrelationship of somatostatin and growth hormone-releasing hormone in the genesis of the rhythmic secretion of growth hormone. Acta Paediatr Scand Suppl 1990; 367: 76-80

Tannenbaum GS, Zhang WH, Lapointe M, Zeitler P, Beaudet A. Growth hormone-releasing hormone neurons in the arcuate nucleus express both SST1 and SST2 somatostatin receptor genes. Endocrinology 1998; 139: 1450-1453

Vuagnat BAM, Pierroz DD, Lalaoui M, Englaro P, Pralong FP, Blum WF, Aubert ML. Evidence for a leptin-neuropeptide Y axis for the regulation of growth hormone secretion in the rat. Neuroendocrinology (in press).

Willoughby JO, Brogan M, Kapoor R. Hypothalamic interconnections of somatostatin and growth hormone releasing factor neurons. Neuroendocrinology 1989; 50: 584-591

Winer LM, Shaw MA, Baumann G. Basal plasma growth hormone levels in man: new evidence for rhythmicity of growth hormone secretion. J Clin Endocrinol Metab 1990; 70: 1678-1686

Zheng H, Bailey ART, Jiang MH, Honda K, Chen HY, Trumbauer ME, Van-der-Ploeg LH, Schaeffer JM, Leng G, Smith RG. Somatostatin receptor subtype 2 knockout mice are refractory to growth hormone-negative feedback on arcuate neurons. Mol Endocrinol 1997; 11: 1709-1717

Acknowledgements

The authors are supported by grants from the Swedish Medical Research Council (No 9894 and 11501 for JOJ) and the British Medical Research Council (No G9521630 for SLD).

2 SIGNAL TRANSDUCTION THROUGH THE GROWTH HORMONE RECEPTOR

Peter E. Lobie

Institute of Molecular and Cell Biology and Defence Medical Research Institute, National University of Singapore, 10 Kent Ridge Crescent 119260, Singapore, Republic of Singapore

I. INTRODUCTION

The importance of the cloned growth hormone (GH) receptor in the regulation of postnatal somatic growth is evidenced by the demonstration of point mutations and deletions in the gene encoding the GH receptor in Laron type dwarfism (1,2) and sex linked dwarfism in the chicken (3,4). Further, GH receptor gene deletion in mice results in a growth retarded phenotype (5). The mechanism by which the GH receptor mediates the general pleiotropic and specific somatic responses to its ligand have only recently begun to be understood. This review provides a brief discourse on the signal transduction pathways which have been demonstrated to be utilized by GH. The identification of such pathways, at least, provides a basis for understanding the pleiotropic actions of GH.

II. THE GROWTH HORMONE RECEPTOR

II.1. Cytokine Receptor Superfamily

The GH receptor was the first identified member of the cytokine receptor superfamily (6-8). Other members of the class 1 cytokine receptor superfamily include prolactin (PRL), erythropoietin (EPO), granulocyte colony stimulating factor (G-CSF), granulocyte-macrophage colony stimulating factor (GM-CSF), ciliary neurotrophic factor (CNTF), thrombopoietin, leptin, interleukins (IL) 2-7, IL-9, IL-11 and IL-12. The superfamily also includes receptor subunits that interact with more than one cytokine receptor. Such receptor subunits include gp130 (shared by LIF, oncostatin M, IL-6, IL11, IL-12) , the β-subunit of the leukaemia inhibitory factor (LIF) receptor, the γ-chain of the IL-2 receptor (shared by IL-2, IL-4, IL-7, IL-9 and IL-15) and the IL-3 receptor common β chain. A second class of more distantly related cytokine receptors has been classified and includes the α and β subnits of the interferon (IFN)- α/β receptor, both the α and β subunits of the IFN-γ receptor and the IL-10 receptor (7). Characteristics which group members of the class 1 cytokine receptor superfamily include the following (9,10) (a) possession of a single putative membrane spanning domain ; (b) limited amino acid homology (14-44 %) in a region spanning approximately 210 amino acids in the extracellular domain (corresponding to two fibronectin III domains of similar size) ; (c) conserved pairs of cysteine residues in the extracellular domain and a conserved tryptophan residue adjacent to the second cysteine in the N-terminal fibronectin domain ; (d) a WSXWS (trp,ser,any,trp,ser)-like motif in the C-terminal fibronectin domain (YXXFS in the mammalian GH receptor) (11) ; (e) the absence of a canonical tyrosine kinase consensus sequence and (f) two short homologous (proline rich) domains in the intracellular domain. Box 1 is 8 residues in length and is located within 20 residues of the transmembrane domain. Box 1 displays an alΨPXalPXP or ΨXXXalPXP consensus sequence where al is any aliphatic residue, Ψ is any hydrophobic residue and P is proline. In the mammalian GH receptor the Box 1 sequence is ILPPVPVP (11). Box 1 is one site of association of Janus Kinase 2 (JAK2) with the GH receptor and is critical for most GH stimulated cellular functions (12-15). Box 2 is less well defined and comprises approximately 15 amino acid residues situated 30 residues distal to Box 1 (11). It consists of a cluster of hydrophobic and acidic residues ending with one or two basic residues. Deletion or mutation of these boxes abrogates JAK activation and subsequent signal transduction leading to proliferation (15,16).

II.2. Receptor Dimerization

With use of recombinant hGH and the extracellular domain of the receptor, the Genentech group has defined the physical interaction between these two molecules (17). Use of X-ray crystallography has enabled the visualization of the three dimensional structure of the receptor extracellular domain (GHBP) and regions that interact with the ligand (18). One hGH molecule interacts with the extracellular domains of two receptors leading to receptor homodimerization (19). Accordingly a model was proposed whereby hGH possesses two receptor binding sites ; a high affinity site "site 1" and a lower affinity site "site 2" that sequentially interact with binding 'pockets' in two discrete receptor molecules (20). Contact with hGH at site 2 and a dimerization interface of approximately 500 Å between the two extracellular domains of the receptor stabilize the binding of the second receptor to the complex (17,21). Interestingly, in the GHBP/receptor the two ligand binding sites are largely composed of the same amino acid residues and possess similar overall shape and structure. The amino acid residues that form sites one and two in the GHBP are located in six clusters between residues 43 and 218 (17).

The resulting receptor dimerization presumably provides the signal to generate the biological response (17,20). Evidence in support of this hypothesis includes the following ; (a) self antagonism for GH receptor activation by GH at high concentration (at high ligand concentration the higher affinity receptor site 1 will be saturated with ligand thereby precluding receptor dimerization via site 2) ; (b) activation of myeloid cell replication with bivalent but not univalent monoclonal antibodies directed against the GH receptor ; (c) disruption of interaction with site 2 by point mutation in the hormone (eg. hGH-G120R) generates GH antagonistic activity *in vitro* (20) and *in vivo* (22) ; (d) Use of a monoclonal antibody that binds to the dimerization interface between the two receptor molecules, and would therefore be expected to prevent dimerization, antagonizes GH dependent cellular proliferation (19) and (e) naturally occurring receptor isoforms lacking the intracellular domain inhibit the function of the full length receptor presumably by the formation of non-productive dimers (23). However, some data suggests that simple dimerization is not sufficient and not all the effects of GH are antagonistic at high GH concentration (24). Receptor dimerization has similarly been demonstrated to be involved in signal transduction by other cytokine receptors (7,25). With use of such data on dimerization of the receptor, it may be possible to design small molecule agonists or antagonists for the GH receptor, as has been described for the related EPO receptor (26).

III. SIGNAL TRANSDUCTION PATHWAYS UTILIZED BY GH

III.1. Janus Kinases

Recently it has become apparent that the receptors of the cytokine receptor superfamily share common mechanistic features for signal transduction although such features were not readily discernable from the primary sequence (8, 25). The GH receptor is tyrosine phosphorylated upon ligand binding presumably by one or more members of the Janus family of tyrosine kinases (27-29). Members of the JAK family of tyrosine kinases to date include JAK1, JAK2, JAK3 and Tyk2 (8). Phosphorylation of the receptor by, and signalling through, members of the JAK family is not limited to GH. Most if not all members of the cytokine receptor superfamily utilize members of the JAK family for signal transduction (8, 25). Some specificity is evident as the individual cytokine molecules activate different combinations of JAKs and the activation is of both different extent and duration (30). The predominant JAK utilized by the GH receptor is JAK2 (27) although GH has been reported to induce the tyrosine phosphorylation of JAK1 (31) and JAK3 (32) as well. It is not known whether GH also causes tyrosine phosphorylation and activation of Tyk2, although GH dependent phosphorylation of Tyk2 was not detectable in IM-9 lymphocytes (30).

Tyrosine phosphorylation of JAK proteins results in the association of many adaptor molecules and enzymes to the signalling complex. Such recruitment includes SHP-1 (33-35), SHP-2 (36), focal adhesion kinase (FAK) (Zhu, Goh, Lobie, submitted), SHC (37,38), Tec tyrosine kinase (and therefore phosphatidylinositol 3'-kinase (PI-3 kinase) (40)), p59fyn (41) and the signal transducer and activator of transcription (STAT) molecules (for review see 8). Of these GH has been reported to stimulate the association of SHP-1 (34,36), SHP-2 (34), FAK (Zhu, Goh, Lobie,submitted), SHC (37), and the STAT molecules (31,42-44) to JAK. GH also stimulates the tyrosine phosphorylation of p59fyn (45) although it is not known if GH stimulates the association of fyn and JAK2.

SHP-1 has been reported to be phosphorylated by GH (34,35) and is responsible for the dephosphorylation of JAK2 mediated via the EPO receptor, requiring a specific tyrosine residue in the intracellular domain of the EPO receptor for JAK2 dephosphorylation (33). Interestingly, SHP-1 does not associate with the GH receptor as might be expected from the EPO receptor data but rather associates directly with JAK2 (34) as does SHP-2 (see below). However, a region of the GH receptor between amino acids 520 and 540 in the cytoplasmic domain has been identified to be required for attenuation of GH stimulated JAK-STAT signalling (35). In any case, SHP-1 deficient mice (motheaten mice) display prolonged GH stimulated tyrosine phosphorylation of JAK2 compared to normal littermates (35). It has also been postulated that proteasome mediated protein degradation modulates protein tyrosine phosphatase activity that negatively regulates the JAK-STAT signalling pathways (46). Presumably GH will also utilize the newly described JAB proteins as one method to downregulate the JAK response. The expression of JAB is stimulated by the cytokine itself and the expressed JAB binds to JAK and inhibits JAK activation (47).

III.2. Focal Adhesion Kinase

Other diverse actions of GH include the stimulation of chemotaxis and migration of monocytic cells (48). Concordantly, we have demonstrated that GH stimulates the re-organization of the actin cytoskeleton in cells with fibroblastic morphology (49). Focal adhesion kinase (p125 FAK) has been postulated to play a central role in the response of the cell to the ECM (for review see 50) and in cell morphology and motility (for review see 51). We have demonstrated that GH activates focal adhesion kinase (FAK) and this activation results in the tyrosine phosphorylation of two FAK substrates, namely paxillin and tensin (Zhu, Goh and Lobie, submitted). FAK and paxillin are constitutively associated in the unstimulated state, remain associated during the stimulation phase and recruit tyrosine phosphorylated tensin to the complex after GH stimulation. Half of the carboxylterminal region of the GH receptor is dispensable for FAK activation, but FAK activation does require the proline rich box 1 region of the GH receptor indicative that FAK is downstream of JAK2. FAK associates with JAK2 but not JAK1 after GH stimulation of cells. Using FAK replete and FAK deficient cells, we have also shown that FAK is not required for STAT mediated transcriptional activation (Zhu, Goh and Lobie, submitted).

The use of FAK by GH for signal transduction permits the GH signal to be propogated through multiple alternate transduction pathways. Tyrosine phosphorylation of the p85 subunit of PI-3 kinase is regulated by cell adhesion *in vivo* and it can be phosphorylated by FAK *in vitro* (52) suggestive that PI-3 kinase may be a substrate of FAK *in vivo*. The association of FAK and PI-3 kinase is direct and dependent on FAK autophophorylation (52,53). It is therefore possible that GH may utilize the FAK-PI-3 kinase pathway to increase phosphatidylinositol 3,4,5-triphosphate levels within the cell as well as the IRS-PI-3 kinase pathway (see below). Such potential utilization of two alternate pathways to activate the same kinase may permit the use of PI-3 kinase for distinct cellular purposes. For example, the activation of PI-3 kinase via IRS may be involved in GH stimulation of metabolic events such as lipogenesis (54) whereas activation of PI-3 kinase via FAK may regulate GH stimulated cytoskeletal re-organization (49). Such alternate use of pathways would allow the cell to respond precisely to hormonal stimuli dependent on cell type and differentiation status. GH has also been demonstrated to stimulate mitogen activated protein kinase (MAP kinase) activity (see below) and the association of SHC and Grb2 with JAK2 (37). JAK2 is required for GH stimulation of MAP kinase activity (14,55) and also for Ras and Raf activation which mediate GH activation of MAP kinase (55). Tyr-925 of FAK is phosphorylated by c-src and serves as a binding site for the Grb2-Sos complex both *in vivo* and *in vitro* (53). Since both FAK (Zhu, Goh, Lobie, submitted) and Grb2 (37) associate with JAK2 it is possible that the Grb2 association to JAK2 is mediated by FAK and that FAK is an upstream intermediary in the GH stimulation of the MAP kinase pathway. Studies using FAK deficient cells stably transfected with GH receptor cDNA should be useful to delineate the role of FAK in the GH activation of various signal transduction pathways.

The FAK-paxillin signalling complex also contains other proteins which have not been described here (for review see 56). An early step in FAK activation involves targetting of Src family kinases to FAK, and the Src kinases (c-src and c-fyn) are responsible for further tyrosine phosphorylation of FAK. GH has previously been reported to stimulate the tyrosine phosphorylation of c-fyn (45). FAK also associates with p130 Cas, and Cas and paxillin recruit other SH2 domain-containing signalling proteins such as c-Crk that couple to guanine nucleotide exchangers (Sos and C3G) for Ras or Rap1 (56). Further analysis of the components of the GH stimulated FAK signalling complex is currently in progress. We have preliminary evidence to suggest that at least Cas and Crk are tyrosine phosphorylated in response to cellular stimulation by GH (Zhu and Lobie, unpublished observations). In any case, FAK is likely to be pivotal to mediate many of the pleiotropic cellular effects attributed to GH.

III.3. MAP kinase

Mitogen activated protein kinases are serine/threonine kinases that require phosphorylation on both tyrosine and threonine residues for activation (57). MAP kinases play a pivotal role in the mammalian cell including regulation of gene transcription, cellular proliferation and prevention of apoptosis. To date, ten mammalian MAP kinase family members have been identified (57). MAP kinase 1 (ERK1) and MAP kinase 2 (ERK2) are predominantly activated by phorbol esters and polypeptide growth factors (57) including GH (58-60). The other 8 MAP kinase family members are predominantly activated by adverse stimuli such as heat, UV irradiation, osmotic stress and also by pro-inflammatory cytokines and are therefore termed stress activated protein kinases (SAP Kinase) (57). To date, GH has not been reported to phosphorylate or activate the SAP kinases. It is likely that GH does utilize some of the SAP kinases as SAP kinase 1 has been reported to be downstream effector of c-Crk leading to JUN phosphorylation (61). GH stimulates the association of c-Crk to p130 Cas and c-Crk phosphorylation (Zhu and Lobie, unpublished). One pathway for the activation of MAP kinases is by increasing the activity of the Ras GTPase via SHC, Grb-2 and Sos. Ras-Raf coupling results in the activation of MAP kinase kinase (MEK) 1 or 2 with subsequent activation of the MAP kinases (57). Indeed, ras and raf have been demonstrated to be required for GH activation of the MAP kinases using dominant negative forms of H-Ras and Raf-1 (55). GH has also been reported to promote the rapid phosphorylation of 46, 52 and 66 kDa splice variants of SHC and the association of SHC to the GH receptor-JAK2 complex (37). JAK2 has been demonstrated to be necessary for the GH stimulated tyrosine phosphorylation of SHC (62). The proline rich Box 1 of the GH receptor has also been demonstrated to be required for GH activation of the MAP kinases (14). Direct association of SHC and JAK2 has been reported after EPO activation of JAK2 through the EPO receptor (38). Grb2 has also been demonstrated to associate with SHC after cellular GH stimulation (37). Recently other components of the classical MAP kinase pathway have also been demonstrated to be utilized by GH (63). Interestingly, PI-3 kinase activity is required for full GH activation of the MAP kinases (64) thus lending credence to the possibility of multiple alternate mechanisms for GH activation of different signalling pathways.

MAP kinase have been reported to phosphorylate and/or activate further downstream proteins such as $p70^{s6k}$, $p90^{rsk}$, phospholipase A_2 (PLA_2), c-Raf-1, c-jun, ternary complex factor (p62TCF/Elk1) and STAT molecules (57,65, references in 66). GH has been demonstrated to utilize all of these downstream proteins in its signal transduction pathways (55, 64, 67-76). $p90^{rsk}$ phosphorylates the serum response factor which binds to the serum response element (SRE) of the c-fos promoter (77). Binding of both SRF and Elk-1 to the SRE contributes to the induction of c-fos gene transcription by GH (70,78). Elk-1 has been demonstrated to be phosphorylated by GH and both the GH stimulated phosphorylation of Elk-1 and GH stimulated SRE mediated transcriptional activation (and c-fos expression) were inhibited by the MEK inhibitor PD098059 (70). Activation of PLA_2 by GH increases the level of arachidonic acid and subsequent formation of arachidonic acid metabolites (68). This GH stimulated PLA_2 dependent formation of arachidonic acid metabolites has been implicated in the Ca^{2+} dependent GH stimulation of P4502C12 gene transcription (68). MAP kinase has also been demonstrated to regulate STAT mediated transcription by association with and phosphorylation of STAT molecules on serine residues (79). Indeed, the MEK kinase inhibitor, PD98059, decreases the GH stimulated transcription mediated by STAT5 (66) without affecting the cytoplasmic to nuclear translocation of STAT5. Phosphorylation by MAP kinase can also affect protein stability which has obvious consequences for transcription factor activity (80). Phosphorylation of FOS by MAP kinase and/or $p90^{rsk}$ stabilizes the protein and is required for the transcriptional repression of the c-fos promoter by c-fos (80). Whether GH also utilizes such mechanisms for transcriptional regulation via the MAP kinase pathway remains to be determined.

III.4. IRS-1, IRS-2 and PI-3 kinase

GH and insulin share many cellular effects in common including the stimulation of amino acid transport, protein synthesis, glucose transport, lipogenesis, gene expression (81), mitogenesis, prevention of apoptosis (82) and re-organization of cytoskeletal architecture (49). Thus, GH and insulin can be expected to utilize in common, components of signal transduction pathways. Consistent with this is the ability of GH to stimulate the tyrosine phosphorylation of both insulin receptor substrate-1 (IRS-1) and IRS-2 (83-86). Whether GH also stimulates the tyrosine phosphorylation of the newly described IRS-3 and IRS-4 (87) has not been investigated. Other cytokine molecules have also been demonstrated to stimulate the tyrosine phosphorylation of IRS-1 and/or IRS-2 (including IL-2, IL-4, IL-7, IL-9, IL-13, IL-15, LIF, OSM, IFN-γ and IFN-α) indicative that the IRS molecules may also be a common component of the cytokine receptor superfamily signal transduction pathway (88). IRS proteins bind directly to several enzymes and adaptor proteins and many of these interactions require tyrosine phosphorylation of IRS proteins. Such IRS associated molecules include PI-3 kinase, SHP-2, Fyn, Grb-2, nck and crk (for review see 88). GH has been demonstrated to activate or tyrosine phosphorylate PI-3 kinase (83-85), SHP-2 (34), fyn (45), Grb-2 (37) and crk (Zhu and Lobie, unpublished observations).

GH has been reported to promote the association of the p85 subunit of PI-3 kinase with both IRS-1 (83,85) and IRS-2 (84) and to increase the PI-3 kinase activity associated with IRS-1, IRS-2 (84,85) and JAK2 (64). Interestingly, a Tec tyrosine kinase has been demonstrated to link JAK1 and JAK2 to PI-3 kinase upon cytokine stimulation (40). PI-3 kinase is pivotal in many cellular processes including cell proliferation and survival, cytoskeletal re-organization and cellular metabolism (88). Many of the PI-3 kinase dependent effects are further mediated by downstream enzymes including $p70^{s6k}$ and protein kinase C (PKC)-ζ of which $p70^{s6k}$ is also activated by GH (67). Insulin stimulation of $p70^{s6k}$ and PKC-ζ requires IRS proteins indicative that the PI-3 kinase activity required for $p70^{s6k}$ and PKC-ζ activation is IRS associated (88). GH activation of $p70^{s6k}$ has been demonstrated to be PI-3 kinase dependent (64). PKC-ζ has been reported to stimulate general protein synthesis in response to insulin while $p70^{s6k}$ is required for synthesis of a subset of proteins that are essential for the control of cell proliferation (references in 88). $p70^{s6k}$ mediates this growth regulated protein synthesis through two translation factors PHAS-1 and elF4E. These factors are phosphorylated by insulin in a IRS/PI-3 kinase dependent pathway (88). To date at least GH stimulated lipid synthesis (54) and GH stimulated actin cytoskeletal re-organization (49) has been demonstrated to be PI-3 kinase dependent. PI-3 kinase has also been reported to function upstream of both ras and raf (89) and wortmannin prevents the activation of MAP kinase by other growth factors and GH (64). Recently PI-3 kinase was also shown to be a serine/threonine kinase that phosphorylates both IRS-1 (90) and STAT 3 (91) although the biological significance of this activity is not understood. Assocation of p85 with STAT3 may also constitute an alternate pathway of PI-3 kinase activation and we have demonstrated that GH stimulates the association of STAT3 with p85 (Zhu and Lobie, unpublished observations).

SHP-2 is a phosphotyrosine specific phosphatase (references in 36). SHP-2 has been reported to be tyrosine phosphorylated in response to cellular GH treatment but apparently does not require IRS proteins to be phosphorylated (34). A SHP-2 binding site on the GH receptor which requires tyrosine phosphorylation has been reported to lie between amino acids 389 and 620 in the rabbit GH receptor (34) and SHP-2 will also bind to JAK2 (34,36). A region between amino acids 232 and 272 in SHP-2 is required for the association to the amino terminal domain of JAK2 (36). Thus, there is some apparent divergence in IRS signalling mediated by GH and insulin which may account for some of the specificity encountered in the cellular response to the two hormones. Differential tyrosine phosphorylation of the IRS proteins by JAK kinases and the insulin receptor kinase may be responsible. Such is the case with insulin and IL-4 which engage a different set of SH2 domains in IRS-1 (for review see 88). Thus, the proteins recruited to the IRS complex will reflect the selection of specific IRS tyrosine phosphorylation sites by the respective kinases.

As mentioned above IRS proteins also recruit several adaptor proteins to the complex including crk, nck, Shc and Grb-2 (88). Crk (Zhu and Lobie, unpublished data) and Shc and Grb-2 (37) are both tyrosine phosphorylated in response to GH although it has not been demonstrated that GH also induces the association of these two molecules to IRS proteins. Crk was initially identified as a proto-oncogene and overexpression of crk has been demonstrated to potentiate IGF-1 induced mitogenesis (92).

The GH receptor does not contain the NPXY consensus sequence required for the association of IRS-proteins with receptors such as insulin, IGF-1 and IL-4 (88) and no specific tyrosine residues in the GH receptor are required for for IRS-1 or IRS-2 phosphorylation (83,84). Although Box 1 of the receptor (and therefore presumably JAK2) is required for GH stimulated phosphorylation of IRS-1 and IRS-2 (83,84), JAK2 also does not possess the NPXY consensus sequence. The mechanism of the GH stimulated phosphorylation of the IRS proteins is therefore fundamentally different to the mechanism utilized by insulin (88). It is possible that the IRS proteins may associate with JAK2 via an adaptor molecule and Grb-2 may be a suitable candidate (37,88).

III.5. Protein Kinase C

Several GH stimulated events within the cell are reduced by chronic treatment of cells with phorbol 12-myristate 13 acetate which depletes PKC and also by inhibitors of PKC activity indicative that PKC is required. Such cellular functions include lipogenesis (93), the expression of the c-fos gene (69,94,95), the increase in intracellular Ca^{2+} concentration (96) and the ability of nuclear proteins to bind to the C/EBP consensus sequence (97). Protein kinase C is a family of 12 isoforms of serine/threonine kinases that phosphorylate and activate multiple intracellular proteins (98). GH has been reported to increase PKC activity in pre-adipocyte cells (99). In these cells, GH activated PKC-ε in contrast to insulin which activated PKC-β,-γ, and -ζ but not PKC-ε (99). The majority of PKC isoforms require 1,2-diacylglycerol (DAG) for their activation and translocation from the cytosol to the plasma membrane. GH has also been demonstrated to elicit rapid but transient increases in cellular DAG levels (95, 100-103) through the utilization of variant pathways in different cells. The GH dependent DAG increase in renal proximal tubule cells is accompanied by a rapid transient increase in IP3 suggestive of GH stimulated DAG production from phosphatidylinositol (102). In contrast, in Ob1771 preadipocytes, GH stimulates the formation of DAG associated with a phospholipase C mediated hydrolysis of phosphatidylcholine (100). In rat adipocytes, the GH dependent DAG formation is inhibited by wortmannin suggestive that in some cell types GH activation of PKC may involve PI-3 kinase (103). GH induced DAG formation is inhibited by a tyrosine kinase inhibitor which also inhibits JAK2, suggestive that PKC is downstream of JAK2 (104).

GH has also been reported to increase the intracellular free calcium concentration in a variety of cell types (96, 105-108). It is suggested that GH activates L-type calcium channels since the increase in intracellular [Ca^{2+}] is dependent on extracellular Ca^{2+} and is blocked by verapamil and nimodipine (108). The GH activated increase in intracellular [Ca^{2+}] is apparently PKC dependent since the increase is blocked by the PKC inhibitor calphostin C and is also mimicked by the addition of DAG (96). Verapamil has been demonstrated to prevent the GH stimulated STAT5 mediated transcription (108), arachidonic acid production (68), expression of P4502C12 mRNA (68) and the GH induced re-organization of the actin cytoskeleton (49). Interestingly, the GH induced increase in intracellular [Ca^{2+}] does not require the box 1 region of the GH receptor (108) indicative that this is not a JAK mediated function.

III.6. STATs

STAT proteins are latent cytoplasmic transcription factors. Upon tyrosine phosphorylation, the STAT molecules either homo- or heterodimerize, translocate to the nucleus, bind to their appropriate DNA response element and stimulate transcription. Many excellent detailed reviews exist on STAT signal transduction and I refer the reader to such reviews (8,25,109). Suffice to say, GH has been demonstrated to utilize STATs 1, 3 (71-73) and 5 (74-76) for the regulation of a variety of genes. GH has been demonstrated to stimulate the formation of STAT1 homodimers, STAT3 homodimers and STAT1/STAT3 heterodimers all binding to the SIE of the c-fos gene (71,72). Interestingly, STAT5 has been demonstrated to be involved in the sexually dimorphic responses of cytochrome P450 genes to male (pulsatile) and female (continuous) modes of GH secretion (75). In this regard STAT5 has also been demonstrated to be responsible for the male specific growth pattern in mice (109).

GH activation of STATs 1,3 and 5 requires the presence and activity of JAK2 (62,63) although the mechanism differs for STATs 1 and 3 and STAT5 (31, 42-44). STATs 1 and 3 are activated by the GH activation of JAK2 and do not require tyrosine residues in the carboxyterminal intracellular domain of the GH receptor as does STAT5. This is consistent with the fact that the GH receptor does not possess either the STAT1 or STAT3 association motifs such as exist in the IFN-γ receptor or gp 130 proteins (110). The activation of JAK2 by GH may obviate such a requirement for receptor binding of STATs 1 and 3 since JAK2 itself has a STAT1 like association motif and 2 STAT3 association motifs. There is some controversy as to whether the membrane proximal tyrosine residues (333 and 338 in the rat GH receptor) are required for the STAT signalling response (31, 111, 112). Apart from JAK2, STAT5 also utilizes tyrosine residues in the intracellular domain of the GH receptor for its full activation (113).

Activation of STAT molecules has also been demonstrated to require serine phosphorylation in addition to tyrosine phosphorylation (79) with such serine phosphorylation provided by MAP kinase (79) or PI-3 kinase (91). At least STAT3 has been demonstrated to associate directly with PI-3 kinase upon IFN-α stimulation (91) and GH stimulation (Zhu and Lobie, unpublished observation). Inhibition of MEK has also been demonstrated to prevent GH stimulated transcription mediated by STAT5 (66) but not nuclear translocation of STAT5 itself. It has also been suggested that STAT3 phosphorylated by GH stimulation behaves as if it is also serine phosphorylated on sodium dodecyl sulphate polyacrylamide gel electrophoresis (104).

The GH activation of STAT molecules may be cell type specific despite the presence of functional receptors. For example, GH was not observed to stimulate STAT1 or STAT3 activity in IM-9 lymphocytes despite the ability of IFN-γ to activate these same STATs in the same cell line (114,115). Some of this specificity may be provided by the interaction of different transcription factors binding to multiple response elements of the particular gene. One such example is the inhibition of transcription by STAT5 binding to the IRF-1 promoter (116). Another is the regulation of STAT5 mediated transcription by a direct association with the glucocorticoid receptor (117).

IV. SUBCELLULAR LOCATION OF GH SIGNAL TRANSDUCTION

GH signal transduction into the cell is likely to occur in specific cellular compartments. Caveolae in the plasma membrane are one such cellular compartment where signal transducing molecules such as JAK2 are concentrated (118). We have also demonstrated that the GH receptor is concentrated in caveolae after ligand stimulation and caveolae mediate partial cellular internalization of GH, the remainder being internalized via the clathrin coated pathway (Lobie, Sadir, Graichen, Mertani, Morel, submitted). Disruption of the structural integrity of caveolae by sterol binding agents prevents the GH stimulated tyrosine phosphorylation of JAK2 indicative that GH signal transduction at the cell surface is transferred into the cell via receptors concentrated in caveolae.

Another cellular subcompartment where GH signal transduction via the GH receptor is mediated is the nucleus. Various components of the GH signal transduction cascade either translocate to the nucleus upon GH stimulation (GH (119), GH receptor (120), STAT5 (121)) or are already located in the nucleus in the unstimulated state (JAK1, JAK2 (122), STAT1 (123), STAT5 (121)). GH has been reported to stimulate the tyrosine phosphorylation of the nuclear localized JAK2 (122) and also to stimulate the association of JAK2 and STAT3 in the nucleus (124). It remains to be determined what functions of GH are mediated entirely in the cytoplasmic compartment, entirely in the nuclear compartment or by nucleocytoplasmic co-operation.

V. CONCLUSION

Rapid progress has been made recently in the definition of GH receptor signal transduction pathways. It is now apparent that many cytokines, including GH, share identical or similar signalling components to exert their cellular effects. Although many of these cellular effects are pleiotropic and/or universal to the cytokine receptor superfamily, some level of specificity does exist. The mechanisms by which the specific cellular effects of GH are achieved remain to be elucidated.

REFERENCES

1 Godowski PJ, Leung DW, Meachem LR, Galgani JP, Helmiss R, Keret R, Rotwein P, Parks JS, Laron Z, Wood WI (1989) Characterization of the human growth hormone receptor gene and demonstration of a partial gene deletion in two patients with Laron type dwarfism. Proc. Natl. Acad. Sci. USA 86: 8083-8087

2 Amselem S, Duquesnoy P, Attree O, Novelli G, Bousnina S, Postel-Vinay MC, Goosens M (1989) Laron dwarfism and mutations of the growth hormone receptor gene. N. Engl. J. Med. 321: 989-995

3 Duriez B, Sobrier ML, Duquesnoy P, Tixier-Boichard M, Decuypere E, Coquerelle C, Zeman M, Goosens M, Amselem S (1993) A naturally occurring growth hormone receptor mutation: in vivo and in vitro evidence for the functional importance of the WS motif common to all members of the cytokine receptor superfamily. Mol. Endocrinol. 7: 806-814

4 Huang N, Cogburn LA, Agarwal SK, Marks HL, Burnside J (1993) Overexpression of a truncated growth hormone receptor in the sex linked dwarf chicken. Mol. Endocrinol. 7: 1391-1398

5 Zhou Y, Xu BC, Maheshwari H, He L, Reed M, Lozykowski M, Chen N, Knapp JR, Cataldo LA, Okada S, Wagner TE, Baumann G, Kopchick JJ (1997) A mouse model for Laron syndrome produced by targeted disruption of the growth hormone receptor/binding protein gene. Proc 79th Ann. Meeting Endo. Soc. (Abstract p341)

6 Cosman D., Lyman SD., Idzerda RL., Beckmann MP., Park LS., Goodwin RG., March CJ (1990) A new cytokine receptor superfamily. TIBS 15: 265-270

7 Kitamura T, Ogorochi T, Miyajima A (1994) Multimeric cytokine receptors. TEM. 5: 8-13

8 Ihle JN, Witthuhn BA, Quelle FW, Yamamoto K, Silvennoinen O (1995) Signalling through the haemopoietic cytokine receptors. Ann. Rev. Immunol. 13: 369-98

9 Bazan Jf (1989) A novel family of growth factor receptors. Biochem. biophys. Res. commun. 164: 788-795

10 Patthy L (1990) Homology of a domain of the growth hormone/prolactin receptor family with type III modules of fibronectin. Cell 61: 13-14

11 Waters MJ (1997) "The Growth Hormone Receptor" Handbook of Physiology

12 vol5 ch14; editor JL Kostyo; Oxford University Press.

13 Frank SJ, Yi W, Zhao Y, Goldsmith JF, Gilliland G, Jiang J, Sakai I, Kraft AS (1995) Regions of the JAK2 tyrosine kinase required for coupling to the growth hormone receptor. J. Biol. Chem. 270: 14776-14785

14 Goujon L, Allevato G, Simonin G, Paquereau L, Le Cam A, Clark J, Nielsen JH, Djiane J, Postel-Vinay M-C, Edery M, Kelly PA (1994) Cytoplasmic sequences of the growth hormone receptor necessary for signal transduction. Proc. Natl. Acad. USA 91: 957-961

15 VanderKuur J, Wang X, Zhang L-Y, Campbell G, Billestrup N, Norstedt G, Carter-Su C (1994) Critical cytoplasmic domains of the growth hormone

receptor for JAK2 activation, MAP kinase activation and tyrosine phosphorylation of cellular proteins. J. Biol. Chem.. 269: 21709-21717

16 Wang YD, Wong K, Wood WI (1995) Intracellular tyrosine residues of the human growth hormone receptor are not required for the signalling of proliferation or JAK-STAT activation. J. Biol. Chem 270: 7021-7024

17 Tanner JW, Chen W, Young RL, Longmore GD, Shaw AS (1995) The conserved box 1 motif of cytokine receptors is required for association with JAK kinases. J. biol. Chem. 270:6523-6530

18 Wells JA, Cunningham BC, Fuh G, Lowman HB, Bass SH, Mulkerrin MG, Ultsch M, Devos AM (1993) The molecular basis for growth hormone - receptor interactions. Rec. Prog. Horm. Res. 48: 253-275

19 De Vos A. M., Ultsch M. and Kossiakoff A. A. (1992) Human growth hormone and extracellular domain of its receptor: Crystal structure of the complex. Science 255:306-312.

20 Cunningham BC, Ultsch M, DeVos AM, Mulkerrin MG, Clauser KR, Wells JA (1991) Dimerization of the extracellular domain of the hGH receptor by a single hormone molecule. Science 254: 821-825

21 Fuh G, Cunningham BC, Fukunaga R, Nagata S, Goeddel DV, Wells JA 1992 Rational design of potent antagonists to the human growth hormone receptor. Science 256: 1677-1680

22 Cunningham BC, Wells JA (1993) Comparison of a structural and functional epitope. J. Mol. Biol. 234: 554-563

23 Chen WY, Wight DC, Wagner TE, Kopchick JJ (1990) Expression of a mutated bovine GH gene suppresses growth of transgenic mice. Proc. Natl. Acad. Sci. USA87: 5061-5065

24 Ross RJ, Esposito N, Shen XY, Von Laue S, Chew SL, Dobson PR, Postel-Vinay MC, Finidori J (1997) A short isoform of the human growth hormone receptor functions as a dominant negative inhibitor of the full length receptor and generates large amounts of binding protein. Mol. Endocrinol. 11: 265-273

25 Waters MJ, Rowlinson SW, Clarkson RW et al., Signal transduction by the growth hormone receptor. Proc. Soc. Exp. Biol. Med. 206: 216-220

26 Kishimoto T, Taga T, Akira S (1994) Cytokine signal transduction. Cell 76: 253-262

27 Livnah O, Stura EA, Johnson DL, Middleton SA, Mulcahy LS, Wrighton NC, Dower WJ, Jolliffe LK, Wilson IA (1996) Functional mimicry of a protein hormone by a peptide agonist. Science 273: 464-471

28 Argetsinger LS, Campbell GS, Yang X, Witthun BA, Silvennoinen O, Ihle JN, Carter-Su C. (1993) Identification of JAK2 as a growth hormone receptor associated kinase. Cell 74:1-20

29 Foster C. M., Shafer J. A., Rozsa F. W., Wang X., Lewis S. D., Renken D. A., Natale J. E., Schwartz J. and Carter-Su C. (1988) Growth hormone promoted tyrosyl phosphorylation of growth hormone receptors in murine 3T3-F442A fibroblasts and adipocytes. Biochemistry 27:326-334.

30 Silva CM, Day RN, Weber MJ, Thorner MO (1993) Human growth hormone receptor is characterized as the 134 kDa tyrosine phosphorylated protein activated by GH treatment in IM-9 cells. Endocrinology 133: 2307-2312

31 Carter-Su C, King AP, Argetsinger LS, Smit LS, Vanderkuur J, Campbell GS (1996) Signalling pathway of GH. Endocrin. J. 43 Suppl:S65-70

32 Smit LS, Meyer DJ, Billestrup N, Norstedt G, Schwartz J, Carter-Su C (1996) The role of the growth hormone receptor and JAK1 and JAK2 kinases in the activation of STATs1,3 and 5 by growth hormone. Mol. Endocrinol. 10: 519-533

33 Johnston JA, Kawamura AM, Kirken RA, Chen YQ, Blake TB, Shibuya K, Ortaldo JR, McVicarDW, O'Shea JJ (1994) Phosphorylation and activation of Janus kinase 3 in response to interleukin 2. Nature 370: 151-153

34 Klingmuller U, Lorenz U, Cantley LC, Neel BG, Lodish HF (1995) Specific recruitment of SH-PTP1 to the erythropoietin receptor causes inactivation of JAK2 and termination of the proliferative signals. Cell 80: 729-738

35 Kim SO, Jiang J, Yi WS, Feng GS, Frank SJ (1997) Interaction of the SH2 containing protein tyrosine phosphatase SHP-1 and SHP-2 with the GH receptor-JAK2 complex. Proc. 79th Ann. Meet. Endocrine Society (Abstract 205)

36 Hackett RH, Wang YD, Sweitzer S, Feldman G, Wood WI, Larner AC (1997) Mapping of a cytoplasmic domain of the human growth hormone receptor that regulates rates of inactivation of JAK2 and STAT proteins. J. Biol. Chem. 272: 11128-11132

37 Yin T, Shen R, Feng GS, Yang YC (1997) Molecular characterization of specific interactions between SHP-2 phosphatase and JAK tyrosine kinases. J. Biol. Chem. 272: 1032-1037

38 Vanderkuur JA, Allevato G, Billestrup N, Norstedt G, Carter-Su C (1995) Growth hormone promoted tyrosyl phosphorylation of shc proteins and shc association with Grb2. J. Biol. Chem. 270: 7587-7593

39 HeTC, Jiang N, Zhuang H, Wojchowski DM (1995) Eryhtropoietin induced recruitment of shc via a receptor phosphotyrsoine independent JAK2 associated pathway. J. Biol. Chem. 270: 11055-11061

40 Giordano V, DeFalco G, Chiari R, Quinto I, Pelicci PG, Bartholomew L, Delmastro P, Gadina M (1997) Shc mediates IL-6 signalling by interacting with gp130 and JAK2 kinase. J. Immunol. 158: 4097-4103

41 Takahashi-Tezuka M, Hibi M, Fujitani Y, Fukuda T, Yamaguchi T, Hirano T (1997) Tec tyrosine kinase links the cytokine receptors to PI-3 kinase probably through JAK. Oncogene 14: 2273-2282

42 Uddin S, Sher DA, Alsayed Y, Pons S, Colamonici OR, Fish EN, White MF, Platanias LC (1997) Interaction of p59 fyn with interferon activated JAK kinases. Biochem. Biophys. Res. Commun. 235: 83-88

43 Sotiropoulos A, Moutoussamy S, Renaudie F, Clauss M, Kayser C, Gouilleux F, Kelly PA, Finidori J (1996) Differential activation of STAT3 and STAT5 by distinct regions of the growth hormone receptor. Mol. Endocrinol. 10: 998-1009

44 Yi W, Kim SO, Jiang J, Park SH, Kraft AS, Waxman DJ, Frank SJ (1996) Growth hormone receptor cytoplasmic domain differentially promotes tyrosine phosphorylation of signal transducers and activators of transcription 5b and 3 by activated JAK2 kinase. Mol. Endocrinol. 10: 1425-1443

45 Wood TJJ, Sliva D, Lobie PE, Gouilleux F, Mui AL, Groner B, Norstedt G, Haldosen LA (1997) Specificity of transcriptiona enhancement via the STAT responsive element in the serine protease inhibitor 2.1 promoter. Mol. Cell. Endocrinol. 130: 69-81

46 Moldrup A, Nielsen JH, Billestrup N (1995) Association of p59 fyn with the intracellular domain of the GH receptor. 77th Endocrine Soc. Meet. (Abstract)

47 Yu CL, Burakoff SJ (1997) Involvement of proteasomes in regulating JAK-STAT pathways upon interleukin-2 stimulation. J. Biol. Chem. 272: 14017-14020

48 Endo TA, Masuhara M, Yokouchi M,Suzuki R, Sakamoto H, Mitsui K, Matsumoto A, Tanimura S, Ohtsubo M, Misawa H, Miyazaki T, Leonor N, Taniguchi T, Fujita T, Kanakura Y, Komiya S, Yoshimura A (1997) A new protein containing an SH2 domain that inhibits JAK kinases. Nature 387: 921-924

49 Wiedermann CJ, Reinisch N, Braunsteiner H 1993 Stimulation of monocyte chemotaxis by human growth hormone and its deactivation by somatostatin. Blood 82: 954-960

50 Goh ELK, Pircher TJ, Wood TJJ, Norstedt G, Graichen R, Lobie PE (1997) Growth hormone induced reorganization of the actin cytoskeleton is not required for STAT5 mediated transcription. Endocrinology 138: 3207-3215

51 Guan JL, Chen HC (1996) Signal transduction in cell matrix interactions. Int. Rev. Cytol. 168:81-121

52 Ilic D, Damsky CH, Yamamoto T (1997) Focal adhesion kinase : at the crossroads of signal transduction. J. Cell. Sci. 110: 401-407

53 Chen HC, Guan JL (1994) Stimulation of phosphatidylinositol 3 kinase association with focal adhesion adhesion kinase by platelet derived growth factor. J. Biol. Chem. 269: 31229-31223

54 Schlaepfer DD, Hunter T (1997) Focal adhesion kinase overexpression enhances ras dependent integrin signalling to ERK2/mitogen activated protein kinase through interactions with and activation of c-src. J. Biol. Chem. 272:13189-13195

55 Ridderstrale M, Tornqvist H (1994) PI-3 kinase inhibitor wortmannin blocks the insulin like effects of growth hormone in isolated rat adipocytes. Biochem. Biophys. Res. Commun. 203: 306-10

56 Winston LA, Hunter T (1995) JAK2, Ras, Raf are reuired for activation of ERK/ mitogen activated protein kinase by growth homrone. J. Biol. Chem. 270: 30837-30840

57 Hanks SK, Polte TR (1997) Signalling through focal adhesion kinase. Bioessays 19: 137-145

58 Treisman R (1996) Regulation of transcription by MAP kinase cascades. Curr. Opin. Cell. Biol. 8: 205-215

59 Campbell GS, Pang L, Miyasaka T, Saltiel AR, Carter-Su C (1992) Stimulation by growth hormone of MAP kinase activity in 3T3-F442A fibroblasts. J. Biol. Chem. 267: 6074-6080

60 Möller C, Hansson A, Enberg B, Lobie PE, Norstedt G (1992) Growth hormone induction of tyrosine phosphorylation and activation of mitogen activated protein kinases in cells transfected with rat GH receptor cDNA. J. Biol. Chem. 267: 23403-23408

61 Winston LA, Bertics PJ (1992) Growth hormone stimulates the tyrosine phosphorylation of 42-kDa and 45 K-Da ERK related proteins. J. Biol. Chem. 267: 4747-4751

62 Tanaka S, Ouchi T, Hanafusa H (1997) Downstream of crk adaptor signalling pathway: activation of jun kinase by v-crk through the guanine nucleotide exchange protein C3G. Proc. Natl. Acad. Sci . USA. 94: 2356-2361

63 Han Y, Leaman DW, Watling D, Rogers NC, Groner B, Kerr IM, Wood WI, Stark GR (1996) Participation of JAK and STAT proteins in growth hormone induced signalling. J. Biol. Chem. 271: 5947-5952

64 Vanderkuur JA, Butch ER, Waters SB, Pessin JE, Guan KL, Carter-Su C (1997) Signalling molecules involved in cuopling growth hormone receptor to mitogen activated protain kinase activation. Endocrinology 138: 4301-4307

65 Kilgour E, Gout I, Anderson NG (1996) Requirement for phosphoinositide 3-OH kinase in growth hormone signalling to the mitogen activated protein kinase and p70 s6k pathways. Biochem. J. 315:517-522

66 Davis RJ (1993) The mitogen activated protein kinase signal transduction pathway. J. Biol. Chem. 268: 14553-14556

67 Pircher TJ, Flores-Morales A, Mui AL, Saltiel AR, Norstedt G, Gustafsson JA, Haldosen LA (1997) Mitogen activated protein kinase kinase inhibition decreases growth hormone stimulated transcription mediated by STAT5. Mol. Cell. Endocrinol., in press.

68 Anderson NG (1993) Similtaneous activation of p90rsk and p70s6k by growth hormone in 3T3-F442A pre-aadipocytes. Biochem. Biophys. Res. Commun. 193: 284-290

69 Tollet P, Hamberg M, Gustafsson JA, Mode A (1995) Growth hormone signalling leading to CYP2C12 gene expression in rat hepatocytes involves phospholipase A2. J. Biol. Chem. 270: 12569-12577

70 Gurland G, Ashcom G, Cochran BH, Schwartz J (1990) Rapid events in growth hormone action. Induction of c-fos and c-jun transcription in 3T3-F442A preadipocytes. Endocrinology 127: 3187-3195

71 Hodge CL, Liao JF, Ho PF, Schwartz J (1997) GH stimulated phosphorylation of ELK-1 or related protein contributes to c-fos transcription through the SRE. Proc. 79th Meet. Endo. Soc. (Abstract p73)

72 Meyer DJ, Campbell GS, Cochran BH, Argetsinger LS, Larner AC, Finbloom DS, Carter-Su C 1994 Growth hormone induces a DNA binding factor related

to interferon stimulated 91 KDa transcription factor. J. Biol. Chem. 269: 4701-4704

73 Campbell GS, Meyer DJ, Raz R, Levy DE, Schwartz J, Carter-Su C (1995) Activation of acute phase response factor (APRF) / STAT3 transcription factor by growth hormone. J. Biol. Chem. 270: 3974-9

74 Sotiropoulos A, Moutoussamy S, Binart N, Kelly PA (1995) The membrane proximal region of the cytoplasmic domain of the growth hormone receptor is involved in the activation of STAT3. FEBS Letters 369: 169-72

75 Wood TJJ, Sliva D, Lobie PE, Pircher TJ, Gouilleux F, Wakao H, Gustafsson J-Å, Groner B, Norstedt G, Haldosen L-A 1995 Mediation of growth hormone-dependent transcriptional activation by mammary gland factor/stat5. J Biol Chem 270:9448-53

76 Waxman DJ, Ram PA, Park SH, Choi HK (1995) Intermittent plasma growth hormone triggers tyrosine phosphorylation and nuclear translocation of a liver expressed, Stat5 related DNA binding protein. Proposed role as a regulator of male specific gene transcription. J. Biol. Chem. 270: 13262-70

77 Bergard PL, Shih HM, Towle HC, Schwarzenberg SJ, Berry SA 1995 Growth hormone induction of hepatic serine protease inhibitor 2.1 transcription is mediated by a STAT5 related factor binding synergistically to two gamma activated sites. J. Biol. Chem. 270: 24903-24910

78 Rivera RM, Miranti CK, Misra RP, Ginty DD,Chen RH, Blenis J, Greenberg ME (1993) A growth factor induced kinase phosphorylates the serum response factor at a site that regulates its DNA binding activity. Mol. Cell. Biol. 13: 6260-6273

79 Meyer DJ, Stephenson EW, Johnson L, Cochran BH, Schwartz J (1993) The serum response element can mediate the induction of c-fos by growth hormone. Proc. Natl. Acad. Sci. USA 90: 6721-6725

80 Horvath CM, Darnell JE (1997) The state of the STATs. Curr. Opin. Cell. biol. 9: 233-239

81 Okazaki K, Sagata N (1995) The mos/map kinase pathway stabilizes c-fos by phosphorylation and augments its transforming activity in NIH 3T# cells. EMBO. J. 14: 5048-5059

82 Davidson MB (1987) Effect of growth hormone on carbohydrate and lipid metabolism. Endocrine Rev. 8: 115-131

83 Eisenhauser KM, Chun SY, Billig H, Hseuh AJ (1995) Growth hormone suppression of apoptosis in preovulatory rat follicles and partial neutralization by insulin like growth factor binding protein. Biol. Reprod. 53: 13-20

84 Argetsinger LS, Hsu GW, Myers MG, Billestrup N, White MF, Carter-Su C (1995) Growth hormone, interferon gamma and leukaemia inhibitory factor promoted tyrosyl phosphorylation of insulin receptor substrate-1. J. Biol. Chem. 270: 14685-14692

85 Argetsinger LS, Norstedt G, Billestrup N, White MF, Carter-Su C (1996) Growth hormone, interferon gamma and leukaemia inhibitory factor utilize insulin receptor substrate 2 in intracellular signalling. J. Biol. Chem. 271: 29415-29421

86 Ridderstrale M, Degerman E, Tornqvist H (1995) Growth hormone stimulates the tyrosine phosphorylation of the insulin receptor substrate-1 and its association with phosphatidylinositol 3-kinase in primary adipocytes. J. Biol. Chem. 279: 3471-3474

87 Souza SC, Frick GP, Yip R, Lobo RB, Tai LR, Goodman HM (1994) Growth hormone activation of insulin receptor substrate-1. J. Biol. Chem. 269: 30085-30088

88 Lavan BE, Fantin VR, Chang ET, Lane WS, Keller SR, Lienhard GE (1997) A novel 160 kDa phosphotyrosine protein in insulin treated embryonic kidney cells is a new member of the insulin receptor substrate family. J. Biol. Chem. 272: 21403-21407

89 Yenush L, White MF (1997) The IRS-signalling system during insulin and cytokine action. BioEssays 19: 491-500

90 Yamauchi K, Holt K, Pessin JE (1993) Phosphatidylinositol 3-kinase functions upstream of ras and raf in mediating insulin stimulation of c-fos transcription. J. biol. Chem. 268: 14597-14600

91 Lam K, Carpenter CL, Ruderman NB, Friel JC, Kelly KL (1994) The phosphatidlyinositol 3 kinase serine phosphorylates IRS-1; stimulation by insulin and inhibition by wortmannin. J. Biol. Chem. 269: 20648-20652

92 Pfeffer LM, Mullersman JE, Pfeffer SR, Murti A, Shi W, Yang CH (1997) STAT3 as an adaptor to couple phosphatidlyinositol 3 kinase to the IFNAR1 chain of the type 1 interferon receptor. Science 276: 1418-1420

93 Beitner-Johnson D, Balkesley VA, Shen-Orr Z, Jiminez M, Stannard B, Wang LM, Pierce J, LeRoith D (1996) The proto oncogene product c-Crk associates with insulin receptor substrate 1 and 4PS. J. Biol. Chem. 271: 9287-9290

94 Smal J, DeMeyts P (1987) Role of protein kinase C in the insulin like effects of growth hormone in rat adipocytes. Biochem. biophys. Res. Communic. 147: 1232-1240

95 Slootweg MC, Degroot RP, Herrmann-Erlee MP, Koornneef I, Kruijer W, Kramer Ym (1991) Growth hormone induces expression of c-jun and jun B oncogenes and employs a protein kinase C signal transduction pathway for the induction of c-fos oncogene expression. J. Mol. Endocrinol. 6: 179-188

96 Tollet P, Legraverand C, Gustafsson JA, Mode A (1991) A role for protein kinases in the growth hormone regulation of cytochrome P4502c12 and insulin like growth factor -1 mRNA expression in primary rat heaptocytes. Mol. Endocrinol. 5: 1351-1358

97 Gaur S, Yamaguchi H, Goodman HM (1996) Growth hormone increases calcium uptake in rat fat cells by a mechanism dependent on protein kinase C. Am. J. Physiol. 270:C1485-C1492

98 Clarkson RW, Chen CM, Harrison S, Wells C, Muscat GE, Waters MJ (1995) Early responses of transactivating factors to growth hormone in pre-adipocytes. Mol Endocrinol. 9: 108-120

99 Dekker LV, Parker PJ (1994) Protein kinase C: a question of specificity. TIBS 19: 73-77

100 Okada S, Kopchick JJ (1997) Growth hormone and insulin stimulate different subsets of protein kinase C isotypes through different mechanisms. Proc. 79th Ann. Meet. Endocrine Soc. (Abstract p576)

101 100.Doglio A, Dani C, Grimaldi P, Ailhaud G (1989) Growth hormone stimulates c-fos gene expression by means of protein kinase C without increasing inositol lipid turnover. Proc. Natl. Acad. Sci. USA 86: 1148-1152

102 Johnson RM, Napier MA, Cronin MJ, King KL (1990) Growth hormone stimulates the formation of sn-1,2-diacylglycerol in rat heaptocytes. Endocrinology 127: 2099-2103

103 Rogers SA, Hammerman MR (1989) Growth hormone activates phospholipase C in proximal tubular basolateral membranes from canine kidney. Proc. Natl. Acad. Sci. USA 86: 6363-6366

104 Roupas P, Herington AC (1995) Signalling mechanisms involved in the production of diacylglycerol by growth hormone. Proc 77th Meet Endocrine Soc. (Abstract p345)

105 Argetsinger LS, Carter-Su C (1996) Mechanism of signalling by growth hormone receptor. Physiol. Rev. 76: 1089-1107

106 Ilondo MM, DeMeyts P, Bouchelouche P (1994) Human growth hormone increases cytosolic free calcium in cultured human IM-9 lymphocytes: a novel mechanism mechanism of growth hormone transmembrane signalling. Biochem. Biophys. Res. communic. 202: 391-397

107 Schwartz Y, Goodman HM (1990) Refractoriness to the insulin like effects of growth hormone depends upon calcium. Endocrinology 126: 170-176

108 Schwartz Y, Yamaguchi H, Goodman HM (1992) Growth hormone increases intracellular free calcium in rat adipocytes: correlations with actions on carbohydrate metabolism. Endocrinology 131: 772-778

109 Billestrup N, Bouchelouche P, Allevato G, Ilondo M, Nielsen JH (1995) Growth hormone receptor C-terminal domains required for growth hormone-induced intracellular free calcium oscillations and gene transcription. Proc. Natl. Acad. Sci. USA 92: 2725-2729

110 Udy GB, Towers RP, Snell RG, Wilkins RJ, Park SH, Ram PA, Waxman DJ, Davey HW (1997) Requirement of STAT5b for sexual dimorphism of body growth rates and liver gene expression. Proc. Natl. Acad. Sci. USA 94: 7239-7244

111 Stahl N, Farruggella T, Boulton TG, Zhong Z, Darnell JE, Yancopoulos GD (1995) Modular tyrosine based motifs in cytokine receptors specify choice of STATs and other substrates. Science 267: 1349-1353

112 Smit LS, Vanderkuur JA, Stimage A, Han Y, Luo G, Yu-Lee LY, Schwartz J, Carter-Su C (1997) Growth hormone induced tyrosyl phosphorylation and DNA binding activity of STAT5a and STAT5b. Endocrinology 138: 3426-3434

113 Lobie PE, Allevato G, Nielsen JH, Norstedt G, Billestrup N (1995) Requirement of tyrosine residues 333 and 338 of the growth hormone receptor for selected GH stimulated function. J. Biol. Chem. 270: 21745-21750

114 Hansen LH, Wang X, Kopchick JJ, Bouchelouche P, Nielsen JH, Galsgaard ED, Billestrup N (1996) Identification of tyrosine residues in the intracellular domain of the growth hormone receptor required for transcriptional signalling and STAT5 activation. J. Biol. Chem. 27: 12669-12673

115 Finbloom DS, Petricoin EF, Hackett RH, David M, Feldman GM, Igarashi K-I, Fibach E, Weber MJ, Thorner MO, Silva CM, Larner AC (1994) Growth hormone and erythropoietin differentially activate DNA-binding proteins by tyrosine phosphorylation. Mol. Cell. Biol. 14: 2113-2118

116 Silva CM, Lu H, Weber MJ, Thorner MO (1994) Differential tyrosine phosphorylation of JAK1, JAK2 and STAT1 by growth hormone and interferon gamma in IM-9 cells. J. Biol. Chem. 269: 27532-27539

117 Luo G, Yu-Lee LY (1997) Differential activities of STAT5 at growth related versus differentiation specific promoters. Proc 79 Meet. Endocr. Soc. (Abstract p571)

118 Stocklin E, Wissler M, Gouilleux F, Groner B (1996) Functional interactions between STAT5 and the glucocorticoid receptor. Nature 383: 726-728

119 Lisanti MP, Scherer PE, Vidugiriene J, Tang ZL, Hermanowski-Vosatka A, Tu YH, Cook RF, Sargiacomo M (1994) Characterization of caveolin rich membrane domains isolated from an endothelial-rich source: Implications for human disease. J Cell Biol 126: 111-126

120 Lobie PE, Mertani H, Morel G, Morales-Bustos O, Norstedt G, Waters MJ (1994) Receptor mediated nuclear translocation of growth hormone. J Biol Chem 269:21330-39

121 Lobie PE, Wood TJJ, Chen CM, Waters MJ, Norstedt G (1994) Nuclear translocation and anchorage of the growth hormone receptor. J. Biol. Chem. 269: 31375-31746

122 Stout LE, Svensson AM, Sorenson RL (1997) Prolactin regulation of islet derived INS-1 cells: characteristics and immunocytochemical analysis of STAT5 translocation. Endocrinology 138: 1592-1603

123 Lobie PE, Ronsin B, Silvennoinen O, Haldosen LA, Norstedt G, Morel G (1996) Constitutive nuclear localization of Janus kinases 1 and 2. Endocrinology 137: 4037-4045

124 Wang Y, Yu-Lee L (1996) Multiple STAT complexes .interact at the interferon regulatory factor-1 interferon gamma activation sequence in prolactin stimulated Nb2 T cells. Mol. Cell. Endocrinol. 121: 19-28

125 Ram PA, Waxman DJ (1997) Interaction of growth hormone activated STATs with SH2 containing phosphotyrosine phosphatase SHP-1 and nuclear JAK2 tyrosine kinase. J. Biol. Chem. 272: 17694-17702

3 GROWTH HORMONE BINDING PROTEINS

Gerhard Baumann

Center for Endocrinology, Metabolism and Molecular Medicine
Department of Medicine
Northwestern University Medical School, Chicago, Illinois, USA

Growth hormone binding proteins (GHBP) were first recognized over a decade ago in plasma of pregnant mouse, rabbit and man. Initial skepticism about the existence of bound forms of circulating polypeptide hormones has given way to over 430 publications dealing with various aspects of the GHBP, as listed in Medline at the time of this writing. Space limitations for this chapter do not permit citation of all; I apologize to those colleagues whose work is not mentioned for this reason. Much has been learned about the nature, source and generation of the GHBP, some of its biological activities have been characterized, and much effort has been expended at defining the regulation of plasma GHBP levels. Interestingly, despite a decade of intensive investigation, the ultimate biological role and physiological significance of the GHBP remains unknown.

GHBPs are present in plasma of many vertebrate species ranging from, at least, reptiles (1) to humans (2, 3). This evolutionary conservation suggests an important biological role for the GHBP. The most detailed information is available for human, rodent, and rabbit GHBP; much of this treatise will therefore focus on these species. GHBPs can be divided into high affinity GHBP and low affinity GHBP; the latter are less well characterized and can be heterogeneous depending on species. In general, I will use the term GHBP to denote the high affinity GHBP.

Nature of the high affinity GHBP

The high affinity GHBP represents the extracellular domain of the growth hormone receptor (GHR) (4, 5). It is a single chain glycoprotein of ~245-273 amino acids and a molecular weight of 50-60,000, half of which is contributed by carbohydrate moieties. The carboxy terminus varies depending on the species and has not been precisely defined, the type and location of glycosylation are unknown, but key features of the GHBP are the GH-binding domain, absence of transmembrane and cytoplasmic domains, and water solubility. Similar soluble receptor ectodomains exist for many members of the cytokine receptor family to which the GHR belongs (5, 6). The GHBP is generated from the GHR either by proteolytic "shedding" (man, rabbit) (4, 7) or as an alternative splice product of GHR/BP pre-mRNA (rat, mouse) (8, 9). In the latter, an extra exon (exon 8A) encodes a unique hydrophilic tail of 17 (rat) or 27 (mouse) residues in place of the transmembrane domain (10,

11). In the classical scenario of GHBP generation, it is either proteolytically released from the membrane-anchored GHR (humans, rabbits) or directly synthesized and secreted as such (rodents). However, recent evidence suggests that the situation may be more complex. On the one hand, the rat GHBP may partly retained in the plasma membrane of adipocytes by an as yet unknown anchor (12); on the other, the human GHBP may be in part derived from a truncated splice variant of the GHR ($GHR_{1\text{-}279}$) (13, 14) This truncated GHR retains the transmembrane domain but contains a very short intracellular portion. For unknown reasons, it has a greater propensity to yield GHBP than the full-length GHR and hence may contribute a disproportionate amount to GHBP production (13, 14). In the rhesus monkey, both proteolysis and alternative splicing have been shown to contribute to GHBP generation (15). In that case, the splice variant results from readthrough into intron 7/8 which lies at the junction between the extracellular and transmembrane domains. In many other species, the mechanism of GHBP generation is not well characterized; proteolytic shedding seems to predominate.

The enzyme involved in shedding is not known, nor is it clear whether proteolysis occurs at the plasma membrane, intracellularly, or both. GHBP can be liberated from GHRs by treatment with sulfhydryl inactivating agents (16, 17); but it is not clear how this phenomenon relates to physiological shedding. Efforts designed to identify responsible protease(s) based on substrate specificity or inhibitor characteristics have been largely unsuccessful (18, 19). Recent evidence has suggested the possible involvement of a metalloprotease in GHBP generation (20).

Function of the GHBP

From a functional standpoint, the *human* GHBP is the best characterized; it can serve as a model for GHBPs in other species. It acts to bind a significant proportion (~45% at steady state, up to 80% transiently) of GH in the circulation (21, 22); its association rate with GH is sufficiently high to permit complex formation *in vivo* after a GH secretory event. For much of the time in-between GH pulses, the GHBP is largely unsaturated. Its affinity ($K_d \sim 2$ nM) and plasma concentration (0.5-2 nM) are in a range that allows it to act as buffer, with a dynamic equilibrium between free and bound GH. Unlike the GH-GHR complex which exhibits almost exclusively 1:2 stoichiometry, the GH-GHBP complex in plasma exists primarily a 1:1 complex – a result of the relatively low plasma concentration of GHBP (23). Complexed GH is cleared much more slowly than free GH (24), a property which creates a circulating GH reservoir and results in prolonged GH bioavailability. At the tissue level, the GHBP modulates GH action by competing with GHRs for ligand, and perhaps also by forming GHR/GHBP dimers at the cell surface. Both events tend to antagonize GH action, and it has been demonstrated that *in vitro* the GHBP inhibits GH binding to GHRs and GH action in a dose-dependent manner (25, 26). While biologically unproductive GHR/GHBP dimers have not been demonstrated to exist in normal physiology, an analogous paradigm occurs in a familial syndrome of short stature where overexpression of a truncated GHR from a mutant heterozygous allele exerts a dominant negative effect on GH action (27). Inhibition of normal GHR signaling by co-expression of the truncated $GHR_{1\text{-}279}$ form *in vitro* has also been demonstrated (14).

Thus, the GHBP can theoretically both enhance GH action (by prolonging bioavailability) and inhibit GH action (by interfering with GHR function). The net effect of these opposing actions *in vivo* has been difficult to predict or determine. Early experiments with limited amounts of natural GHBP demonstrated no effect of GHBP on the growth-promoting activity of GH in a hypophysectomized rat model (28). Carrying out the same experiments with larger doses of recombinant GHBP showed a dose-dependent enhancement of GH action (29). However, a recent study, using GH autofeedback on its own secretion as the endpoint, failed to show an effect of recombinant GHBP on GH action in the rat (30). Correlative studies between plasma GHBP levels and the effect of GH therapy in humans have given similarly conflicting results, with both positive (31), negative (32), and no (33-36) correlations being reported. Finally, correlations between plasma GHBP levels and spontaneous growth rates or stature have not yielded consistent results (37-41). Part of the problem with correlative studies lies in the fact that plasma GHBP levels probably reflect tissue GHR abundance, which has its own influence on GH action quite separate from direct GHBP effects.

Recent studies have suggested that the GHBP can regulate GHR expression (and hence its own production) in a human hepatoma cell line (42). This action is independent of the GH binding function. Although the biological significance of this finding remains to be determined, it is intriguing in the sense that GHBP may exert actions that are not linked to its GH binding properties. Earlier experiments examining this question were designed to identify a receptor for the GHBP but failed to find one (G. Baumann, unpublished). Nevertheless, the fact that GHBPs in animals frequently have relatively low affinity for the homologous (species-specific) GH (43-45) raises the possibility of such a GH-independent role.

Source and Distribution of the GHBP

The GHR gene is expressed ubiquitously (46-49). An organ particularly rich in GHRs is the liver. In species where the GHBP is derived from the GHR by proteolytic cleavage, abundant GHRs can be expected to be associated with abundant GHBP generation if proteolysis is a constitutive process as suggested (17, 18). Hence, the liver is assumed to be an important source of GHBP. However, it should be noted that no direct studies on hepatic GHBP production *in vivo* have been reported. Since the proteolytic mechanism of GHBP generation and its regulation are poorly understood, it is uncertain whether GHR abundance is necessarily reflected in GHBP generation. A preliminary study of GHBP concentrations in the central venous circulation in humans suggests that several visceral organs contribute to circulating GHBP (50). In contrast, peripheral tissues appear to contribute little GHBP, as suggested by the low GHBP levels in lymph derived from lower limb adipose tissue and muscle (51). In rodents, the source of the GHBP is more easily identified because it is derived from its own mRNA, and because its unique tail domain permits specific immunorecognition. In rats and mice, the GHBP is co-expressed ubiquitously with the GHR (52, 53), though GHR and GHBP mRNA are not necessarily regulated in parallel (46-48, 53). GHBP in rat serum is primarily if not exclusively derived from the alternate splicing mechanism (54). It has been suggested that in the rodent, the use of different upstream exons coding for alternative 5'-untranslated regions (GHR1 vs. GHR2 mRNA) partially determines whether GHR or GHBP is produced (55). In rhesus monkeys, where the GHBP can be produced by both the proteolytic and alternative splicing mechanism, sources for the latter appear restricted to heart, stomach and kidney (15). In general, however, very little is known in any species about the actual contribution of different tissues to the extracellular pool of GHBP.

The GHBP is widely distributed in biological fluids, including blood, urine (56), lymph (51), amniotic fluid (57), milk (58-60), follicular fluid (61), semen and ascites (Mercado M, Baumann G, unpublished data). Cerebrospinal fluid contains little or no detectable GHBP (62). GHBP concentrations are generally highest in plasma, although data available for other fluids are limited. Because of the interaction between GHBP, GHRs and GH at the tissue level, it would be of great interest to know the local interstitial GHBP concentration near its site of production/release, but no such information is available at present. Based on studies with the GHBP-GH complex, the vascular tree is permeable to the GHBP (as well as to the complex), with distribution of the complex to about twice the intravascular space (24, 63).

Regulation of GHBP expression and plasma levels

Regulation of expression of the GHBP is partly tissue-specific, and marked differences occur among species. Major and universal physiological regulators of plasma GHBP levels are ontogeny and nutrition. In fetal and neonatal life, GHBP expression and plasma levels are low, with a progressive surge occurring early postnatally, coinciding with upregulation of GHRs and the emergence of GH responsivity (46, 53, 64-69). In humans, GHBP levels continue to rise until puberty, then remain stable until about age 70, after which they decline progressively (70, 71). Nutrition upregulates plasma GHBP levels (72, 73), and body mass index is positively correlated with GHBP levels in numerous studies. Recent evidence suggests that it is primarily visceral fat that is responsible for this association (41, 74). Both leptin and insulin levels are positively correlated with GHBP levels (41, 75); it has been suggested that insulin directly regulates GHBP and GHR expression (76-80).

Plasma GHBP levels exhibit sexual dimorphism, with females having higher levels. In the rat, this difference is very marked (81-83), whereas in mice, rabbits and humans, it is minor (38, 39, 65, 67, 70, 84-87). The gender difference has been ascribed to sex hormones. In general, estrogen tends to promote GHBP expression, whereas androgen diminishes it. However, there are some species differences and other complexities. In the rat, GHBP expression and plasma level is decreased by ovariectomy or anti-estrogen treatment, whereas estrogen treatment enhances GHBP expression (82, 83). Estrogen promotes the expression of a liver-specific GHR/GHBP transcript (GHR1; see above) (83). Interestingly, in the rat, this estrogen effect appears to be dependent on GH, or more precisely the female pattern of GH secretion, as GH-deficient rats do not exhibit it (82). Orchiectomy in rats mildly increases GHBP levels (88). In humans, estrogen treatment by the oral route elevates plasma GHBP, whereas transdermal estrogen does not (89). This is likely a dose phenomenon at the hepatic level. Interestingly, a (male) patient with complete estrogen deficiency due to aromatase deficiency was reported to have a normal GHBP level (90). Similarly, untreated patients with Turner syndrome have normal to slightly elevated GHBP levels (38, 91, 92). Testosterone and, to a lesser extent, dihydrotestosterone treatment lowers GHBP in humans, though GHBP is not increased in adults with hypogonadism (93-95). In normally cycling women, GHBP is inversely correlated with estradiol levels but estrogen therapy elevates GHBP levels (96). Human puberty is associated with either no change (97, 98) or a minimal decline (99) in GHBP levels in either sex. In the male rabbit, both estrogen and testosterone treatment result in lower serum GHBP levels (100). Additional data on sex hormone-related effects on GHBP have been derived from patients treated with gonadotropin releasing hormone (GnRH) analogues for precocious puberty. GnRH-induced gonadal arrest led to an increase of GHBP levels in both sexes in most studies, with one showing no change (93, 101, 102). Treatment of *rats* with GnRH analogue, however, decreases GHBP in females and produces no change in males (88). The possibility must be considered that GnRH has a direct effect on GHBP expression independent of sex steroids. Thus, the gonadal hormone environment has complex and diverse effects on GHBP depending on the species, circumstance, and dose. It appears that the effects of sex hormone therapy are of a pharmacological nature that is difficult to extrapolate to physiology. Whether gender itself determines GHBP expression independently of sex hormone status is unknown.

Pregnancy is another physiological condition that influences GHBP expression. Here again, marked species differences exist. In the mouse, pregnancy induces a massive, progressive upregulation of GHBP expression (primarily in the liver), and consequently maternal plasma GHBP levels rise 30-fold (53, 84, 103). This phenomenon provided the basis for the original description of the GHBP by Peeters and Friesen (104). The rat also upregulates GHBP expression during pregnancy, although to a lesser degree (47, 105). In human pregnancy, there is a slight increase in plasma GHBP during the first trimester, with a progressive fall to non-gravid levels thereafter (96, 106, 107). Maternal and fetal blood or amniotic fluid GHBP levels do not correlate (57). In pigs, GHBP levels are highest at the beginning of the third trimester (108).

Diurnal variation of serum GHBP levels is minimal in humans (109, 110). However, a mild (~20%) seasonal variation, with a nadir in summer, has been recently reported for prepubertal children (40). This was correlated to seasonal changes in weight and IGF-I. It is unknown what factors drive this seasonal variation.

Glucocorticoids have relatively minor effects on GHBP. Again, there appear to be differences among species and between pharmacological and physiological effects. In the rat, adrenalectomy has no effect, but dexamethasone or methylprednisolone treatment lowers GHBP expression (83). In the rabbit, dexamethasone treatment upregulates hepatic GHRs and plasma GHBP (111). Humans with Cushing's disease have normal GHBP levels (112), but treatment with supraphysiological doses (4 mg/day) of dexamethasone decreases plasma GHBP (113).

Thyroid hormone status is positively correlated with GHBP levels in humans (114, 115). GHBP expression studies in rat liver have given conflicting results, perhaps because of sexual dimorphism (116, 117).

Insulin appears to be important for GHBP expression, as suggested by its effect in diabetic rats and humans (76, 77, 118, 119). There is a correlation between insulin dose or endogenous insulin secretion and GHBP levels (78, 79, 119). Whether insulin acts directly on GHBP expression or by normalizing metabolism – or both – remains to be determined. The above mentioned positive correlation between nutrition/body mass index and GHBP is probably mediated at least in part through insulin.

GH itself would be expected to regulate its own receptor/BP system. This is partially the case, but the literature on this subject is quite diverse. Discrepant results can in part be attributed to species differences, pharmacological vs. physiological effects, different modes of administration of GH, etc. However, there remain unexplained discrepancies between studies and laboratories. Initial data in humans and rodents suggested that GHBP expression and serum levels are GH dependent (93, 120, 121), but subsequent studies yielded conflicting evidence ranging from upregulation (36, 122-124) to variable or no effect (31, 32, 34, 38, 91, 125-131) to downregulation (132-134) of GHBP levels by GH. Different results have been seen depending on whether GH was administered by a continuous infusion or by injection (124, 125, 135). In normally growing children, an inverse correlation exists between GH secretion rate and plasma GHBP (136). Patients with severe, long-standing GH deficiency of genetic etiology have normal GHBP levels (137, 138). Similarly, the Ames mouse, a model of genetic GH (as well as prolactin and TSH) deficiency, has normal GHBP levels (139). The majority of studies of adult patients with organic, long-standing, severe GH deficiency have reported normal GHBP levels before GH treatment (32, 126, 127), although both high and low GHBP levels have also been reported. Conversely, patients with acromegaly have low to low normal GHBP levels (70, 128, 134, 140-142), and transgenic mice overexpressing GH have elevated GHBP levels (123). It is presently not possible to reconcile all these observations except to conclude that GHBP expression is as variable in GH deficiency as it is in GH sufficiency. Small scale studies may yield conflicting conclusions about the regulation of GHBP by GH. Taken together, the available evidence suggests that there is no consistent effect of GH on GHBP expression or plasma level.

Relatively little is known about IGF-I as a regulator of GHBP. In one study, treatment of GHBP-positive Laron syndrome (GHR-deficient) patients (see below) with IGF-I led to a decrease in GHBP levels (143). In sheep, liver GHR expression is downregulated by IGF-I treatment (144), although no direct data are available on GHBP. In the dwarf rat, IGF-I treatment does not affect GHBP levels (135).

Altered GHBP Levels in Pathological Conditions

A variety of disease states are associated with altered GHBP expression or plasma levels. Some have already been mentioned above in the section on GHBP regulation. In most cases, the level of GHBP expression parallels the degree of GH sensitivity, lending credence to the view that the GHBP level reflects GHR abundance in tissues.

a) Decreased GHBP levels

The most severe abnormality of GHBP is seen in Laron syndrome which is due to genetic GHR deficiency or dysfunction (145). Since in humans, the GHBP is derived from the GHR, absence of the GHR also results in absence of the GHBP. However, mutations in the GHR gene in Laron syndrome are genetically heterogeneous, ranging from partial gene deletions to point mutations affecting various aspects of GHR expression and function. Depending on how a mutation affects the GH binding domain, GHR synthesis, processing, and membrane display, plasma GHBP activity may be absent, decreased, normal, or even increased. The majority of mutations affect the extracellular domain of the GHR, leading to abnormally low GHBP levels or activity (145). In some patients, the GHBP is present but unable to bind GH (146). Yet in others, GHBP is present but the transmembrane and/or intracellular GHR domains are absent or defunct (27, 147-150). Among animals, the sex-linked dwarf chicken is the only known naturally occurring animal with GHR mutations. They are also genetically heterogeneous, but no GHBP measurements have been reported (151). The other animal model of GHR deficiency is the laboratory-generated GHR/BP knockout mouse (Laron mouse; 85) which was designed to have both GHR and GHBP expression disrupted. The Laron mouse exhibits no detectable GHBP expression at both the mRNA and protein level (85).

Milder mutations in the GHR are the basis for idiopathic short stature in a few patients (152, 153). The reason for the low GHBP expression is not clear in those cases. Patients with idiopathic short stature as a group have decreased GHBP levels, although there is substantial overlap with the normal range (38). Here again, the mechanism responsible for low GHBP expression is not clear; mutations have been demonstrated only in a small minority (152, 153).

Other conditions of genetic short stature include pygmies, both in Africa and in New Guinea (154-156). In that case, the problem appears to be a failure to upregulate GHBP (and by inference GHRs) during childhood (155). Pygmies have partial GH resistance, although *in vitro* studies with cells from African pygmies have also suggested resistance to IGF-I (157). Low GHBP levels were also reported in dwarf pigs (158).

Acquired conditions with decreased GHBP levels include insulin-deficient (both type 1 and type 2) diabetes (76-79, 118, 119), malnutrition and acute fasting (69, 72, 73, 159), liver cirrhosis (65, 160, 161), renal failure (65, 162, 163), hypothyroidism (114, 115), critical illness (164), all of which are associated with GH insensitivity. In these cases, GHBP likely mirrors GHR expression or abundance. Interestingly, in nephrotic syndrome without renal insufficiency, the serum GHBP level is normal (165). This suggests that homeostatic mechanisms maintain normal blood GHBP levels despite renal losses – a possible indication of the importance of circulating GHBP. Acromegaly is another condition where most studies have shown decreased GHBP levels (70, 134, 140-142). It is difficult to assess GH sensitivity in acromegaly, and nothing is known about GHR abundance. However, chronic exposure to high GH concentrations downregulates GHRs *in vitro* (166), and it is reasonable to speculate that this may also occur *in vivo*. Thus, paradoxically, acromegaly may well represent a partially GH-resistant state where the resistance is overcome by high GH levels.

b) Increased GHBP levels

Conditions with elevated GHBP levels are far fewer than those with decreased levels. Genetic causes include some of the GHR mutations where the transmembrane domain is lacking or the intracellular domain is truncated (27, 147-150). The absent transmembrane domain fails to provide a membrane anchor, and truncated GHRs have a propensity to generate large amounts of GHBP (see above). The phenotype of these mutations is GH insensitivity with short stature because of disabled GHR signaling in the absence of an intracellular domain and/or a dominant negative effect of the mutant GHR on normal GHR function. It is not clear to what extent the elevated plasma GHBP level contributes to GH insensitivity. Patients with Turner syndrome tend to have slightly elevated GHBP levels (33, 91), perhaps because of altered body composition (167). Among the acquired conditions with elevated GHBP levels is rodent pregnancy (47, 84, 103-105), early human pregnancy (96, 106, 107), and obesity/overnutrition (41, 72, 74, 75, 168). The latter is associated with well-recognized hyper-responsivity to GH, with normal or enhanced growth and high normal IGF-I levels despite low GH secretion (168, 169).

Clinical significance and utility of GHBP determinations

The measurement of plasma high affinity GHBP theoretically provides a "window" on the overall or at least hepatic GHR status. In general, GHBP determinations have fulfilled that promise in that they correlate with GH responsivity in most circumstances. As a result, insights have been gained into GH action and its derangements in a wide variety of conditions. However, the normal range of plasma GHBP levels is very wide, stretching over at least one order of magnitude (70). In addition, absolute values differ depending on the assay used. The variability of GHBP regulation has already been addressed above. As a consequence of these factors, GHBP measurements are useful as research tool and meaningful when applied to groups, but they are generally less informative when applied to individuals because of overlap between normal and abnormal ranges. The only firmly established clinical use for GHBP to date is in Laron syndrome, where the absence of high affinity GH binding in plasma largely confirms the diagnosis (it should however be noted that 20-25% of Laron syndrome patients are GHBP positive because of the nature of their mutation [170]). The place of GHBP measurements in a wider diagnostic armamentarium will have to await elucidation of the physiological role of the GHBP.

The low affinity GHBP(s)

A low affinity, high capacity GHBP has been recognized in human plasma since the initial description of the high affinity GHBP (2); it was subsequently partially purified and characterized (67, 171). Similar binding components are present in animal plasma (81, 172). It has been proposed that transformed a_2-macroglobulin represents the low affinity GHBP (173). This conclusion is based on the GH-binding properties of pure a_2-macroglobulin; however the positive identification of low affinity GHBP extracted from plasma as a_2-macroglobulin remains to be accomplished. The low affinity GHBP is regulated separately from the high affinity

GHBP; it plays a minor role in terms of the amount of GH bound in the circulation (65).

REFERENCES

1. Sotelo AI, Partata WA, Marques M, Turyn D. (1997) Identification and initial characterization of serum growth hormone binding protein in the turtle Chrysemys dorbigni. Arch Physiol Biochem 105:167-174.
2. Baumann G, Stolar MW, Amburn K, Barsano CP, DeVries BC. (1986) A specific growth hormone-binding protein in human plasma: initial characterization. J Clin Endocrinol Metab 62:134-141.
3. Herington AC, Ymer S, Stevenson J. (1986) Identification and characterization of specific binding proteins for growth hormone in normal human sera. J Clin Invest 7:1817-1823.
4. Leung DW, Spencer SA, Cachianes G, Hammonds RG, Collins C, Henzel WJ, Barnard R, Waters MJ, Wood WI. (1987) Growth hormone receptor and serum binding protein: purification, cloning and expression. Nature 330:537-543.
5. Rose-John S, Heinrich PC. (1994) Soluble receptors for cytokines and growth factors: generation and biological function. Biochem J 300:281-290.
6. Baumann G. (1995) Growth hormone binding to a circulating receptor fragment – the concept of receptor shedding and receptor splicing. Exp Clin Endocrinol 103:2-6.
7. Sotiropoulos A, Goujon L, Simonin G, Kelly P, Postel-Vinay M-C, Finidori J. (1993) Evidence for generation of the growth hormone-binding protein through proteolysis of the growth hormone membrane receptor. Endocrinology 132:1863-1865.
8. Smith WC, Kuniyoshi J, Talamantes F. (1989) Mouse serum growth hormone (GH) binding protein has GH receptor extracellular and substituted transmembrane domains. Mol Endocrinol 3:984-990.
9. Baumbach WR, Horner DL, Logan JS. (1989) The growth hormone-binding protein in rat serum is an alternatively spliced form of the rat growth hormone receptor. Genes Devel 3:1199-1205.
10. Edens A, Southard JN, Talamantes F. (1994) Mouse growth hormone-binding protein and growth hormone receptor transcripts are produced from a single gene by alternative splicing. Endocrinology 135:2802-2805.
11. Zhou Y, He L, Kopchick JJ. (1994) An exon encoding the mouse growth hormone binding protein (mGHBP) carboxy terminus is located between exon 7 and 8 of the mouse growth hormone receptor gene. Receptor 4:223-227.
12. Frick GP, Tai L-R, Goodman HM. (1994) Subcellular distribution of the long and short isoforms of the growth hormone (GH) receptor in rat adipocytes: Both forms participate in specific binding of GH. Endocrinology 134:307-314.
13. Dastot F, Sobrier M-L, Duquesnoy P, Duriez B, Goossens M, Amselem S. (1996) Alternatively spliced forms in the cytoplasmic domain of the human growth hormone (GH) receptor regulate its ability to a generate soluble GH-binding protein. Proc Natl Acad Sci USA 93:10723-10728.

14. Ross RJM, Esposito N, Shen XY, Von Laue S, Chew SL, Dobson PRM, Postel-Vinay M-C, Finidori J. (1997) A short isoform of the human growth hormone receptor functions as a dominant negative inhibitor of the full length receptor and generates large amounts of binding protein. Mol Endocrinol 11:265-273.
15. Martini J-F, Pezet A, Guezennec CY, Edery M, Postel-Vinay M-C, Kelly PA. (1997) Monkey growth hormone (GH) receptor gene expression. Evidence for two mechanisms for the generation of the GH binding protein. J Biol Chem 272:18951-18958.
16. Trivedi B, Daughaday WH. (1988) Release of growth hormone binding protein from IM-9 lymphocytes by endopeptidase is dependent on sulfhydryl group inactivation. Endocrinology 123:2201-2206.
17. Bick T, Amit T, Mansur M, Bar-Am O, Youdim MB, Hochberg Z. (1996) Regulation of cellular rabbit growth hormone (GH) receptor and GH-binding protein generation in vitro. Endocrinology 137:3977-3985.
18. Harrison SM, Barnard R, Ho KY, Rajkovic I, Waters MJ. (1995) Control of growth hormone (GH) binding protein release from human hepatoma cells expressing full-length GH receptor. Endocrinology 136:651-659.
19. Amit T, Bick T, Youdim MB, Hochberg Z. (1996) In search of the cellular site of growth hormone (GH)-binding protein cleavage from the rabbit GH receptor. Endocrinology 137:3986-3991.
20. Alele J, Jiang J, Goldsmith JF, Yang X, Maheshwari HG, Black RA, Baumann G, Frank SJ. (1998) Blockade of growth hormone receptor shedding by a metalloprotease inhibitor. Endocrinology 139:1927-1935.
21. Baumann G, Amburn K, Shaw MA. (1988) The circulating growth hormone (GH)-binding protein complex: a major constituent of plasma GH in man. Endocrinology 122:976-984.
22. Veldhuis JD, Johnson ML, Faunt LM, Mercado M, Baumann G. (1993) Influence of the high affinity growth hormone (GH)-binding protein on plasma profiles of free and bound GH and on the apparent half-life of GH. J Clin Invest 91:629-641.
23. Baumann G, Lowman HB, Mercado M, Wells JA. (1994) The stoichiometry of growth hormone-binding protein complexes in human plasma: comparison with cell surface receptors. J Clin Endocrinol Metab 78:1113-1118.
24. Baumann G, Amburn KD, Buchanan TA. (1987) The effect of circulating growth hormone-binding protein on metabolic clearance, distribution and degradation of human growth hormone. J Clin Endocrinol Metab 64:657-660.
25. Lim L, Spencer SA, McKay P, Waters MJ. (1990) Regulation of growth hormone (GH) bioactivity by a recombinant human GH-binding protein. Endocrinology 127:1287-1291.
26. Mannor DA, Winer LM, Shaw MA, Baumann G. (1991) Plasma growth hormone binding proteins: effect on growth hormone binding to receptors and on growth hormone action. J Clin Endocrinol Metab 73:30-34.
27. Ayling RM, Ross R, Towner P, Von Laue S, Finidori J, Moutoussamy S, Buchanan CR, Clayton PE, Norman MR. (1997) A dominant-negative mutation of the growth hormone receptor causes familial short stature. Nature Genet 16:13-14.

28. Mannor DA, Shaw MA, Winer LM, Baumann G. (1988) Circulating growth hormone-binding protein inhibits growth hormone (GH) binding to GH receptors but not in vivo GH action. Clin Res 36:870A.

29. Clark RG, Mortensen DL, Carlsson LM, Spencer SA, McKay P, Mulkerrin M, Moore J, Cunningham BC. (1996) Recombinant human growth hormone (GH)-binding protein enhances the growth-promoting activity of human GH in the rat. Endocrinology 137:4308-4315.

30. Tzanela M, Wagner C, Tannenbaum GS. (1997) Recombinant human growth hormone-binding protein fails to enhance the in vivo bioactivity of human growth hormone in normal rats. Endocrinology 138:5316-5324.

31. Martha PM, Jr., Reiter EO, Davila N, Shaw MA, Holcombe JH, Baumann G. (1992) Serum growth hormone (GH)-binding protein/receptor: An important determinant of GH responsiveness. J Clin Endocrinol Metab 75:1464-1469.

32. Johannsson G, Bjarnason R, Bramnert M, Carlsson LMS, Degerblad M, Manhem P, Rosen T, Thoren M, Bengtsson B-A. (1996) The individual responsiveness of growth hormone (GH) treatment in GH-deficient adults is dependent on the level of GH-binding protein, body mass index, age and gender. J Clin Endocrinol Metab 81:1575-1581.

33. Attie KM, Julius JR, Stoppani C, Rundle AC. (1997) National cooperative growth study substudy VI: The clinical utility of growth-hormone-binding protein, insulin-like growth factor I, and insulin-like growth factor-binding protein 3 measurements. J Pediatr 131:S56-S57.

34. Boguszewski M, Bjarnarson R, Rosberg C, Carlsson LM, Albertson-Wikland K. (1997) Growth hormone (GH)-binding protein in prepubertal short children born of small for gestational age: effects of growth hormone treatment. Swedish study group for growth hormone treatment. J Clin Endocrinol Metab 82:1014-1019.

35. Mandel S, Moreland E, Nichols V, Hanna C, Lafranchi S. (1995) Changes in insulin-like growth factor-I (IGF-I), IGF-binding protein-3, growth hormone (GH)-binding protein, erythrocyte IGF-I receptors, and growth rate during GH treatment. J Clin Endocrinol Metab 80:190-194.

36. Leger J, Noel M, Czernichow P, Postel-Vinay MC. (1995) Progressive normalization of growth hormone-binding protein and IGF-I levels in treated growth hormone-deficient children. Pediatr Res 37:731-735.

37. Phillip M, Chalew SA, McCarter RJ Jr, Amit T, Kowarski AA, Hochberg Z. (1996) The relationship of growth rate, plasma growth hormone (GH) concentration, and GH-binding protein. Metabolism 45:424-429.

38. Carlsson LM, Attie KM, Compton PG, Vitangcol RV, Merimee TJ. (1994) Reduced concentration of serum growth hormone-binding protein in children with idiopathic short stature. National cooperative growth study. J Clin Endocrinol Metab 78:1325-1330.

39. Baumann G, Shen A, Waters M, Liu K. (1997) Serum growth hormone binding protein/receptor (GHBP/GHR) levels: relationship to adult stature. Progr 79th Meet Endocr Soc, p 156.

40. Gelander L, Bjarnason R, Carlsson LM, Albertson-Wikland K. (1998) Growth hormone-binding protein levels over one year in healthy prepubertal

children: intraindividual variation and correlation with height velocity. Pediatr Res 43:256-261.
41. Kratzsch J, Dehmel B, Pulzer F, Keller E, Englaro P, Blum WF, Wabitsch M. (1997) Increased serum GHBP levels in obese pubertal children and adolescents: relationship to body composition, leptin and indicators of metabolic disturbances. Int J Obes Relat Metab Disord 21:1130-1136.
42. Mullis PE, Wagner JK, Eble A, Nuoffer JM, Postel-Vinay MC. (1997) Regulation of human growth hormone receptor gene transcription by human growth hormone binding protein. Mol Cell Endocrinol 131:89-96.
43. Shaw MA, Baumann G. (1988) Growth hormone-binding proteins in animal plasma: a survey. Progr 70th Meet Endocr Soc, No. 240.
44. Davis SL, Graf M, Morrison CA, Hall TR, Swift PJ. (1992) Identification and partial purification of serum growth hormone binding protein in domestic animal species. J Anim Sci 70:773-780.
45. Amit T, Hochberg Z, Waters MJ, Barkey RJ. (1992) Growth hormone- and prolactin-binding proteins in mammalian serum. Endocrinology 131:1793-1803.
46. Carlsson B, Billig H, Rymo L, Isaksson OG. (1990) Expression of the growth hormone-binding protein messenger RNA in the liver and extrahepatic tissues in the rat: co-expression with the growth hormone receptor. Mol Cell Endocrinol 73:R1-6.
47. Tiong TS, Herington AC. (1991) Tissue distribution, characterization, and regulation of messenger ribonucleic acid for growth hormone receptor and serum binding protein in the rat. Endocrinology 129:1628-1634.
48. Walker JL, Moats-Staats BM, Stiles AD, Underwood LE. (1992) Tissue-specific developmental regulation of the messenger ribonucleic acids encoding the growth hormone receptor and the growth hormone binding protein in rat fetal and postnatal tissues. Pediatr Res 31:335-339.
49. Mercado M, Davila N, McLeod JF, Baumann G. (1994) Distribution of growth hormone receptor messenger ribonucleic acid containing and lacking exon 3 in human tissues. J Clin Endocrinol Metab 78:731-735.
50. Segel S, Nemcek A, Baumann G. (1996) Tissue sources of the growth hormone binding protein in humans. Progr 10th Internatl Congress of Endocrinology, p 254.
51. Maheshwari H, Mercado M, Hollands J, Lillioja S, Castillo C, Baumann G. (1995) Growth hormone binding protein levels in human lymph. J Clin Endocrinol Metab 80:3582-3584.
52. Lobie PE, Garcia-Aragon J, Wang BS, Baumbach WR, Waters MJ. (1992) Cellular localization of the growth hormone binding protein in the rat. Endocrinology 130:3057-3065.
53. Ilkbahar YN, Wu K, Thordarson G, Talamantes F. (1995) Expression and distribution of messenger RNAs for growth hormone receptor and growth hormone-binding protein in mice during pregnancy. Endocrinology 136:386-392.
54. Sadeghi H, Wang BS, Lumanglas AL, Logan JS, Baumbach WR. (1990) Identification of the origin of the growth hormone-binding protein in rat serum. Mol Endocrinol 4:1799-1805.

55. Southard JN, Barrett BA, Bikbulatova L, Ilkbahar Y, Wu K, Talamantes F. (1995) Growth hormone (GH) receptor and GH-binding protein messenger ribonucleic acids with alternative 5'-untranslated regions are differentially expressed in mouse liver and placenta. Endocrinology 136:2913-2921.
56. Hattori N, Shimatsu A, Kato Y, Imura H. (1990) Growth hormone and growth hormone binding protein in human urine. Kidney Int 37:951-954.
57. Harada I, Tsutsumi O, Momoeda M, Horikawa R, Yasunaga T, Tanaka T, Taketani Y. (1997) Comparative concentrations of growth hormone-binding protein in maternal circulation, fetal circulation, and amniotic fluid. Endocr J 44:111-116.
58. Postel-Vinay MC, Belair L, Kayser C, Kelly PA, Djiane J. (1991) Identification of prolactin and growth hormone binding proteins in rabbit milk. Proc Natl Acad Sci U S A 88:6687-6690.
59. Mercado M, Baumann G (1994) A growth hormone/prolactin-binding protein in human milk. J Clin Endocrinol Metab 79:1637-1641.
60. Amit T, Dibner C, Barkey RJ (1997) Characterization of prolactin- and growth hormone-binding proteins in milk and their diversity among species. Mol Cell Endocrinol 130:167-180.
61. Amit T, Dirnfeld M, Barkey RJ, Peleg I, Hacham H, Abramovici H, Blumenfeld Z. (1993) Growth hormone-binding protein (GH-BP) levels in follicular fluid from human preovulatory follicles: correlation with serum GH-BP levels. J Clin Endocrinol Metab 77:33-39.
62. Nixon DA, Jordan DM. (1986) Conversion of CSF monomeric growth hormone to large growth hormone with exposure to serum. Acta Endocrinol 111:289-295.
63. Baumann G, Shaw MA, Buchanan TA (1989) In vivo kinetics of a covalent growth hormone-binding protein complex. Metabolism 38:330-333.
64. Daughaday WH, Trivedi B, Andrews BA. (1987) The ontogeny of serum GH binding protein in man: a possible indicator of hepatic GH receptor development. J Clin Endocrinol Metab 65:1072-1074.
65. Baumann G, Shaw MA, Amburn K. (1989) Regulation of plasma growth hormone-binding proteins in health and disease. Metabolism 38:683-689.
66. Massa G, de Zegher F, Vanderschueren-Lodeweyckx M. (1992) Serum growth hormone-binding proteins in the human fetus and infant. Pediatr Res 32:69-72.
67. Tar A, Hocquette JF, Souberbielle JC, Clot JP, Brauner R, Postel-Vinay MC. (1990) Evaluation of the growth hormone-binding proteins in human plasma using high pressure liquid chromatography. J Clin Endocrinol Metab 71:1202-1207.
68. Holl RW, Snehotta R, Siegler B, Scherbaum W, Heinze E. (1991) Binding protein for human growth hormone: effects of age and weight. Hormone Res 35:190-197.
69. Mulumba N, Massa G, Ketelslegers J-M, Maes M. (1991) Ontogeny and nutritional regulation of the serum growth hormone-binding protein in the rat. Acta Endocrinol 125:409-415.
70. Mercado M, Carlsson L, Vitangcol R, Baumann G. (1993) Growth hormone-binding protein determination in human plasma: a comparison of

immunofunctional and growth hormone-binding assays. J Clin Endocrinol Metab 76:1291-1294.
71. Maheshwari H, Sharma L, Baumann G. (1996) Decline of plasma growth hormone binding protein in old age. J Clin Endocrinol Metab 81:995-997.
72. Hochberg Z, Hertz P, Colin V, Ish-Shalom S, Yeshurun D, Youdim MBH, Amit T. (1992) The distal axis of growth hormone (GH) in nutritional disorders: GH-binding protein, insulin-like growth factor-I (IGF-I), and IGF-I receptors in obesity and anorexia nervosa. Metabolism 41:106-112.
73. Counts DR, Gwirtsman H, Carlsson LM, Lesem M, Cutler GB Jr. (1992) The effect of anorexia nervosa and refeeding on growth hormone-binding protein, the insulin-like growth factors (IGFs), and the IGF-binding proteins. J Clin Endocrinol Metab 75:762-767.
74. Roelen CA, Koppeschaar HP, de Vries WR, Snel YE, Doerga ME, Zelissen PM, Thijssen JH, Blankenstein MA. (1997) Visceral adipose tissue is associated with circulating high affinity growth hormone-binding protein. J Clin Endocrinol Metab 82:760-764.
75. Bjarnason R, Boguszewski M, Dahlgren J, Gelander L, Kristrom B, Rosberg, S, Carlsson B, Albertsson-Wikland K, Carlsson LM. (1997) Leptin levels are strongly correlated with those of GH-binding protein in prepubertal children. Eur J Endocrinol 137:68-73.
76. Mercado M, Molitch ME, Baumann G. (1992) Low plasma growth hormone binding protein in IDDM. Diabetes 41:605-609.
77. Massa G, Verhaeghe J, Vanderschueren-Lodeweyckx M, Bouillon R. (1993) Normalization of decreased plasma concentrations of growth hormone-binding protein by insulin treatment in spontaneously diabetic-BB rats. Horm Metab Res 25:325-326.
78. Hanaire-Broutin H, Sallerin-Caute B, Poncet MF, Tauber M, Bastide R, Chale JJ, Rosenfeld R, Tauber JP. (1996) Effect of intraperitoneal insulin delivery on growth hormone binding protein, insulin-like growth factor (IGF)-I, and IGF-binding protein-3 in IDDM. Diabetologia 39:1498-1504.
79. Kratzsch J, Keliner K, Zilkens T, Schmidt-Gayk H, Selisko T, Scholz GH. (1996) Growth hormone-binding protein related immunoreactivity is regulated by the degree of insulinopenia in diabetes mellitus. Clin Endocrinol 44:673-678.
80. Baxter RC, Turtle JR. (1978) Regulation of hepatic growth hormone receptors by insulin. Biochem Biophys Res Comm 84:350-357.
81. Massa G, Mulumba N, Ketelslegers JM, Maes M. (1990) Initial characterization and sexual dimorphism of serum growth hormone-binding protein in adult rats. Endocrinology 126:1976-1980.
82. Carmignac DF, Gabrielsson BG, Robinson IC. (1993) Growth hormone binding protein in the rat: effects of gonadal steroids. Endocrinology 133:2445-2452.
83. Gabrielsson BG, Carmignac DF, Flavell DM, Robinson IC. (1995) Steroid regulation of growth hormone (GH) receptor and GH-binding protein messenger ribonucleic acids in the rat. Endocrinology 136:209-217.

84. Cramer SD, Barnard R, Engbers C, Ogren L, Talamantes F. (1992) Expression of the growth hormone receptor and growth hormone-binding protein during pregnancy in the mouse. Endocrinology 131:876-882.

85. Zhou Y, Xu BC, Maheshwari HG, He L, Reed M, Lozykowski M, Okada S, Cataldo L, Coschigamo K, Wagner TE, Baumann G, Kopchick JJ. (1997) A mammalian model for Laron syndrome produced by targeted disruption of the growth hormone receptor/binding protein gene (the Laron mouse). Proc Natl Acad Sci USA 94:13215-13220.

86. Barnard R, Waters MJ. (1986) Serum and liver cytosolic growth-hormone-binding proteins are antigenically identical with liver membrane 'receptor' types 1 and 2. Biochem J 237:885-892.

87. Hattori N, Kurahachi H, Ikekubo K, Ishihara T, Moridera K, Hino M, Saiki Y, Imura H. (1991) Effects of sex and age on serum GH binding protein levels in normal adults. Clin Endocrinol 35:295-297.

88. Gevers EF, Wit JM, Robinson IC. (1998) Effects of long-term gonadotrophin-releasing hormone analog treatment on growth, growth hormone (GH) secretion, GH receptors, and GH-binding protein in the rat. Pediatr Res 43:111-120.

89. Weissberger AJ, Ho KK, Lazarus L. (1991) Contrasting effects of oral and transdermal routes of estrogen replacement therapy on 24-hour growth hormone (GH) secretion, insulin-like growth factor I, and GH-binding protein in postmenopausal women. J Clin Endocrinol Metab 72:374-381.

90. Morishima A, Grumbach MM, Simpson ER, Fisher C, Qin K. (1995) Aromatase deficiency in male and female siblings caused by a novel mutation and the physiological role of estrogens. J Clin Endocrinol Metab 80:3689-3698.

91. van Teunenbroek A, de Muinck Keizer-Schrama SM, Stijnen T, Mouton JW, Blum WF, Mercado M, Baumann G, Drop SL. (1993) Effect of growth hormone administration frequency on 24-hour growth hormone profiles and levels of other growth related parameters in girls with Turner's syndrome. Clin Endocrinol 39:77-84.

92. Mauras N, Carlsson LM, Murphy S, Merimee TJ. (1994) Growth hormone-binding protein levels: studies of children with short stature. Metabolism 43:357-359.

93. Postel-Vinay MC, Tar A, Hocquette JF, Clot JP, Fontoura M, Brauner R, Rappaport R. (1991) Human plasma growth hormone (GH)-binding proteins are regulated by GH and testosterone. J Clin Endocrinol Metab 73:197-202.

94. Ip TP, Hoffman DM, O'Sullivan AJ, Leung KC, Ho KK. (1995) Do androgens regulate growth hormone-binding protein in adult man? J Clin Endocrinol Metab 80:1278-1282.

95. Keenan BS, Richards GE, Mercado M, Dallas JS, Eakman GD, Baumann G. (1996) Androgen regulation of growth hormone binding protein. Metabolism 45:1521-1526.

96. Massa G, Igout A, Rombauts L, Frankenne F, Vanderschueren-Lodeweyckx M. (1993) Effect of oestrogen status on serum levels of growth hormone-binding protein and insulin-like growth factor-I in non-pregnant and pregnant women. Clin Endocrinol 39:569-575.

97. Massa G, Bouillon R, Vanderschueren-Lodeweyckx M. (1992) Serum growth hormone-binding protein and insulin-like growth factor-I during puberty. Clin Endocrinol 37:175-180.
98. Martha PM Jr, Rogol AD, Carlsson LM, Gesundheit N, Blizzard RM. (1993) A longitudinal assessment of hormonal and physical alterations during normal puberty in boys. I. Serum growth hormone-binding protein. J Clin Endocrinol Metab 77:452-457.
99. Argente J, Barrios V, Pozo J, Munoz MT, Hervas F, Stene M, Hernandez M J. (1993) Normative data for insulin-like growth factors (IGFs), IGF-binding proteins, and growth hormone-binding protein in a healthy Spanish pediatric population: age- and sex-related changes. J Clin Endocrinol Metab 77:1522-1528.
100. Yu YM, Domene HM, Sztein J, Counts DR, Cassorla F. (1996) Developmental changes and differential regulation by testosterone and estradiol of growth hormone receptor expression in the rabbit. Eur J Endocrinol 135:583-590.
101. Oliveira SB, Donnadieu M, Chaussain JL. (1993) Changes in growth hormone-binding protein in girls with central precocious puberty treated with a depot preparation of luteinizing hormone-releasing hormone analogue. Horm Res 39:42-46.
102. DiMartino-Nardi J, Wu R, Varner R, Wong WL, Saenger P. (1994) The effect of luteinizing hormone-releasing hormone analog for central precocious puberty on growth hormone (GH) and GH-binding protein. J Clin Endocrinol Metab 78:664-668.
103. Smith WC, Talamantes F. (1988) Gestational profile and affinity cross-linking of the mouse serum growth hormone-binding protein. Endocrinology 123:1489-1494.
104. Peeters S, Friesen HG. (1977) A growth hormone binding factor in the serum of pregnant mice. Endocrinology 101, 1164-1183.
105. Amit T, Barkey RJ, Bick T, Hertz P, Youdim MB, Hochberg Z. (1990) Identification of growth hormone binding protein in rat serum. Mol Cell Endocrinol 70:197-202.
106. Blumenfeld Z, Barkey RJ, Youdim MB, Brandes JM, Amit T. (1992) Growth hormone (GH)-binding protein regulation by estrogen, progesterone, and gonadotropins in human: the effect of ovulation induction with menopausal gonadotropins, GH, and gestation. J Clin Endocrinol Metab 75:1242-1249.
107. Barnard R, Chan FY, McIntyre HD. (1997) Growth hormone-binding protein in normal and pathologic gestation: correlations with maternal diabetes and fetal growth. J Clin Endocrinol Metab 82:1879-1884.
108. Mullins TM, Davis SL. (1992) Assessment of factors regulating serum growth hormone binding protein in pigs. J Anim Sci 70:2741-2745.
109. Snow KJ, Shaw MA, Winer LM, Baumann G. (1990) Diurnal pattern of plasma growth hormone-binding protein in man. J Clin Endocrinol Metab 70:417-420.
110. Carlsson LM, Rosberg S, Vitangcol RV, Wong WL, Albertsson-Wikland K. (1993) Analysis of 24-hour plasma profiles of growth hormone (GH)-binding

protein, GH/GH-binding protein-complex, and GH in healthy children. J Clin Endocrinol Metab 77:356-361.
111. Heinrichs C, Yanovski JA, Roth AH, Yu YM, Domene HM, Yano K, Cutler GB Jr, Baron J. (1994) Dexamethasone increases growth hormone receptor messenger ribonucleic acid levels in liver and growth plate. Endocrinology 135:1113-1118.
112. Magiakou MA, Mastorakos G, Gomez MT, Rose SR, Chrousos GP. (1994) Suppressed spontaneous and stimulated growth hormone secretion in patients with Cushing's disease before and after surgical cure. J Clin Endocrinol Metab 78:131-137.
113. Miell JP, Buchanan CR, Norman MR, Maheshwari HG, Blum WF. (1994) The evolution of changes in immunoreactive serum insulin-like growth factors (IGFs), IGF-binding proteins, circulating growth hormone (GH) and GH-binding protein as a result of short-term dexamethasone treatment. J Endocrinol 142:547-554.
114. Amit T, Hertz P, Ish-Shalom S, Lotan R, Luboshitzki R, Youdim MB, Hochberg Z. (1991) Effects of hypo or hyper-thyroidism on growth hormone-binding protein. Clin Endocrinol 35:159-162.
115. Miell JP, Taylor AM, Zini M, Maheshwari HG, Ross RJ, Valcavi R. (1993) Effects of hypothyroidism and hyperthyroidism on insulin-like growth factors (IGFs) and growth hormone- and IGF-binding proteins. J Clin Endocrinol Metab 76:950-955.
116. Shen XY, Rodriguez-Arnao J, Ross RJ. (1995) Regulation of hepatic mRNA levels for the growth hormone receptor in rats with altered thyroid status. Growth Regul 5:199-202.
117. Romero GS, Stephan DA, Sperling MA, Menon RK. (1996) Distinct sexual dimorphism in the effect of hypothyroidism on the expression of the growth hormone receptor and growth hormone-binding protein gene in rat liver. Horm Res 45:273-278.
118. Menon RK, Arslanian S, May B, Cutfield WS, Sperling MA. (1992) Diminished growth hormone-binding protein in children with insulin-dependent diabetes mellitus. J Clin Endocrinol Metab 74:934-938.
119. Clayton KL, Holly JM, Carlsson LM, Jones J, Cheetham TD, Taylor AM, Dunger DB. (1994) Loss of the normal relationships between growth hormone, growth hormone-binding protein and insulin-like growth factor-I in adolescents with insulin-dependent diabetes mellitus. Clin Endocrinol 41:517-524.
120. Hochberg Z, Barkey RJ, Even L, Peleg I, Youdim MB, Amit T. (1991) The effect of human growth hormone therapy on GH binding protein in GH-deficient children. Acta Endocrinol 125:23-27.
121. Sanchez-Jimenez F, Fielder PJ, Martinez RR, Smith WC, Talamantes F. (1990) Hypophysectomy eliminates and growth hormone (GH) maintains the midpregnancy elevation in GH receptor and serum binding protein in the mouse. Endocrinology 126:1270-1275.
122. Zadik Z, Lieberman E, Altman Y, Chen M, Limoni Y, Landau H. (1993) Effect of timing of growth hormone administration on plasma growth-hormone-binding activity, insulin-like growth factor-I and growth in children

with a subnormal spontaneous secretion of growth hormone. Horm Res 39:188-191.
123. Sotelo AI, Dominici FP, Engbers C, Bartke A, Talamantes F, Turyn D. (1995) Growth hormone-binding protein in normal mice and in transgenic mice expressing bovine growth hormone gene. Am J Physiol 268:E745-E751.
124. Carmignac DF, Wells T, Carlsson LMS, Clark RG, Robinson ICAF. (1992) Growth hormone (GH)-binding protein in normal and GH-deficient dwarf rats. J Endocrinol 135:447-457.
125. Tauber M, De Bouet Du Portal H, Sallerin-Caute B, Rochiccioli P, Bastide R. (1993) Differential regulation of serum growth hormone (GH)-binding protein during continuous infusion versus daily injection of recombinant human GH in GH-deficient children. J Clin Endocrinol Metab 76:1135-1139.
126. Ho KK, Jorgensen JO, Valiontis E, Waters MJ, Rajkovic IA, Christiansen JS. (1993) Different modes of growth hormone (GH) administration do not change GH binding protein activity in man. Clin Endocrinol 38:143-148.
127. Davila N, Alcaniz J, Salto L, Estrada J, Barcelo B, Baumann G. (1994) Serum growth hormone-binding protein is unchanged in adult panhypopituitarism. J Clin Endocrinol Metab 79:1347-1350.
128. Rajkovic IA, Valiontis E, Ho KK. (1994) Direct quantitation of growth hormone binding protein in human serum by a ligand immunofunctional assay: comparison with immunoprecipitation and chromatographic methods. J Clin Endocrinol Metab 78:772-777.
129. Pampori NA, Shapiro BH. (1996) Feminization of hepatic cytochrome P450s by nominal levels of growth hormone in the feminine plasma profile. Mol Pharmacol 50:1148-1156.
130. Jammes H, Disenhaus C, Ouriet V, Kayser C, Postel-Vinay MC, Djiane J. (1996) Growth hormone-binding protein in the goat: characterization, evolution under exogenous growth hormone treatment, and correlation with liver growth hormone receptor levels. Domest Anim Endocrinol 13:477-489.
131. Jorgensen JO, Pedersen SB, Borglum J, Frystyk J, Ho KK, Christiansen JS, Orskov H, Blum WF, Richelsen B. (1995) Serum concentrations of insulin-like growth factors (IGFs), IGF binding proteins 1 and 3 and growth hormone binding protein in obese women and the effects of growth hormone administration: a double-blind, placebo-controlled study. Eur J Endocrinol 133:65-70.
132. Roelen CA, Koppeschaar HP, de Vries WR, Zelissen PM, Snel YE, Doerga ME, Thijssen JH, Blankenstein RA. (1996) High-affinity growth hormone binding protein, insulin-like growth factor I and insulin-like growth factor binding protein 3 in adults with growth hormone deficiency. Eur J Endocrinol 135:82-86.
133. Baum HB, Biller BM, Katznelson L, Oppenheim DS, Clemmons DR, Cannistraro KB, Schoenfeld DA, Best SA, Klibanski A. (1996) Assessment of growth hormone (GH) secretion in men with adult-onset GH deficiency compared with that in normal men – a clinical research center study. J Clin Endocrinol Metab 81:84-92.
134. Fisker S, Frystyk J, Skriver L, Vestbo E, Ho KK, Orskov H. (1996) A simple, rapid immunometric assay for determination of functional and growth

hormone-occupied growth hormone-binding protein in human serum. Eur J Clin Invest 26:779-785.

135. Barnard R, Mulcahy J, Garcia-Aragon J, Wyse B, Owens PC, Rowlinson SW, Talamantes F, Baumbach WR, Waters MJ. (1994) Serum growth hormone binding protein and hepatic GH binding sites in the Lewis dwarf rat: effects of IGF-I and GH. Growth Regul 4:147-154.
136. Martha PM Jr, Rogol AD, Blizzard RM, Shaw MA, Baumann G. (1991) Growth hormone-binding protein activity is inversely related to 24-hour growth hormone release in normal boys. J Clin Endocrinol Metab 73:175-181.
137. Perez Jurado LA, Argente J, Barrios V, Pozo J, Munoz MT, Hernandez M, Francke U. (1997) Molecular diagnosis and endocrine evaluation of a patient with a homozygous 7.0 kb deletion of the growth hormone (GH) gene cluster: response to biosynthetic GH therapy. J Pediatr Endocrinol Metab 10:185-190.
138. Baumann G, Maheshwari H. (1997) The dwarfs of Sindh: severe growth hormone (GH) deficiency caused by a mutation in the GH-releasing hormone receptor gene. Acta Paediatr, Suppl 423:33-38.
139. Sotelo AI, Dominici FP, Bartke A, Turyn D (1997) Concentration of free growth hormone-binding protein in the serum of mice is not regulated by growth hormone. J Endocrinol 153:319-325.
140. Amit T, Ish-Shalom S, Glaser B, Youdim MB, Hochberg Z (1992) Growth-hormone-binding protein in patients with acromegaly. Horm Res 37:205-211.
141. Roelen CA, Donker GH, Thijssen JH, Koppeschaar HP, Blankenstein MA. (1992) High affinity growth hormone binding protein in plasma of patients with acromegaly and the effect of octreotide treatment. Clin Endocrinol 37:373-378.
142. Kratzsch J, Blum WF, Ventz M, Selisko T, Birkenmeyer G, Keller E. (1995) Growth hormone-binding protein-related immunoreactivity in the serum of patients with acromegaly is regulated inversely by growth hormone concentration. Eur J Endocrinol 132:306-312.
143. Silbergeld A, Klinger B, Keret R, Eshet R, Almonte AS, Laron Z. (1994) Serum growth hormone-binding protein (GHBP) activity is decreased by administration of insulin-like growth factor I in three Laron syndrome siblings with normal GHBP. Proc Soc Exp Biol Med 206:324-327.
144. Min SH, MacKenzie DDS, Breier BH, McCutcheon SN, Gluckman PD. (1996) Response of young energy-restricted sheep to chronically administered insulin-like growth factor-I (IGF-I): Evidence that IGF-I suppresses the hepatic growth hormone receptor. Endocrinology 137:1129-1137.
145. Rosenfeld RG, Rosenbloom AL, Guevara-Aguirre J. (1994) Growth hormone (GH) insensitivity due to primary GH receptor deficiency. Endocr Rev 15:369-390.
146. Oakes SR, Haynes KM, Batch J, Ennis G, Waters MJ, Daughaday W, Herington AC, Werther GA. (1994) Immunoreactive growth hormone receptor/binding protein is present on fibroblasts and in serum of Laron-type dwarfs. Mol Cell Endocrinol 99:125-132.

147. Rieu M, Le Bouc Y, Villares SM, Postel-Vinay M-C (1993) Familial short stature with very high levels of growth hormone binding protein. J Clin Endocrinol Metab 76:857-860.
148. Woods KA, Fraser NC, Postel-Vinay M-C, Savage MO, Clark AJL. (1996) A homozygous splice site mutation affecting the intracellular domain of the growth hormone (GH) receptor resulting in Laron syndrome with elevated GH-bindig protein. J Clin Endocrinol Metab 81:1686-1690.
149. Silbergeld A, Dastot F, Klinger B, Kanety H, Eshet R, Amselem S, Laron Z. (1997) Intronic mutation in the growth hormone (GH) receptor gene from a girl with Laron syndrome and extremely high serum GH binding protein: extended phenotypic study in a very large pedigree. J Pediatr Endocrinol Metab 10:265-274.
150. Iida K, Takahashi Y, Kaji H, Nose O, Okimura Y, Abe H, Chihara K. (1998) Growth hormone (GH) insensitivity syndrome with high serum GH-binding protein levels caused by a heterozygous splice site mutation of the GH receptor gene producing a lack of intracellular domain. J Clin Endocrinol Metab 83:531-537.
151. Burnside J, Liou SS, Cogburn LA. (1991) Molecular cloning of the chicken growth hormone receptor cpmplementary deoxyribonucleic acid: mutation of the gene in sex-linked dwarf chickens. Endocrinology 128:3183-3192.
152. Goddard AD, Covello R, Luoh SM, Clackson T, Attie KM, Gesundheit N, Rundle AC, Wells JA, Carlsson LM and the Growth Hormone Insensitivity Study Group. (1995) Mutations of the growth hormone receptor in children with idiopathic short stature. N Engl J Med 333:1073-1078.
153. Goddard AD, Dowd P, Chernausek S, Geffner M, Gertner J, Hintz R, Hopwood N, Kaplan S, Plotnick L, Rogol A, Rosenfield R, Saenger P, Mauras N, Hershkopf R, Angulo M, Attie K. (1997) Partial growth hormone insensitivity: the role of growth-hormone receptor mutations in idiopathic short stature. J Pediatr 131:S51-S55.
154. Baumann G, Shaw MA, Merimee TJ. (1989) Low levels of high-affinity growth hormone-binding protein in African pygmies. N Engl J Med 320:1705-1709.
155. Merimee TJ, Baumann G, Daughaday W. (1990) Growth hormone-binding protein: II. Studies in pygmies and normal statured subjects. J Clin Endocrinol Metab 71:1183-1188.
156. Baumann G, Shaw MA, Brumbaugh RC, Schwartz J. (1991) Short stature and decreased serum growth hormone-binding protein in the Mountain Ok people of Papua New Guinea. J Clin Endocrinol Metab 72:1346-1349.
157. Geffner ME, Bersch N, Bailey RC, Golde DW. (1995) Insulin-like growth factor I resistance in immortalized T cell lines from African Efe Pygmies. J Clin Endocrinol Metab 80:3732-3738.
158. Lauterio TJ, Trivedi B, Kapadia M, Daughaday WH. (1988) Reduced ^{125}I-hGH binding by serum of dwarf pigs but not by serum of dwarfed poodles. Comp Biochem Physiol A 91:15-19.
159. Baumann G, Shaw MA, Merimee TJ, Clemmons DR. (1988) Growth hormone-binding protein in human plasma: downregulation by prolonged fasting in lean but not obese subjects. Clin Res 36:477A.

160. Baruch Y, Amit T, Hertz P, Enat R, Youdim MB, Hochberg Z. (1991) Decreased serum growth hormone-binding protein in patients with liver cirrhosis. J Clin Endocrinol Metab 73:777-780.
161. Hattori N, Kurahachi H, Ikekubo K, Ishihara T, Moridera K, Hino M, Saiki Y, Imura H. (1992) Serum growth hormone-binding protein, insulin-like growth factor-I, and growth hormone in patients with liver cirrhosis. Metabolism 41:377-381.
162. Maheshwari HG, Rifkin I, Butler J, Norman M. (1992) Growth hormone binding protein in patients with renal failure. Acta Endocrinol 127:485-488.
163. Tönshoff B, Cronin MJ, Reichert M, Haffner D, Wingen AM, Blum WF, Mehls O, et al. (1997) Reduced concentration of serum growth hormone (GH)-binding protein in children with chronic renal failure: correlation with GH insensitivity. J Clin Endocrinol Metab 82:1007-1013.
164. Ross RJ, Miell JP, Holly JM, Maheshwari H, Norman M, Abdulla AF, Buchanan CR. (1991) Levels of GH binding activity, IGFBP-1, insulin, blood glucose and cortisol in intensive care patients. Clin Endocrinol 35:361-367.
165. Haffner D, Tönshoff B, Blum WF, Vickers M, Siebler T, Cronin MJ, Baxter RC, Mehls O. (1997) Insulin-like growth factors (IGFs) and IGF binding proteins, serum acid-labile subunit and growth hormone binding protein in nephrotic children. Kidney Int 52:802-810.
166. Lesniak MA, Roth J. (1976) Regulation of receptor concentration by homologous hormone: effect of human growth hormone on its receptor in IM-9 lymphocytes. J Biol Chem 251:3720-3729.
167. Gravholt CH, Naeraa RW, Fisker S, Christiansen JS. (1997) Body composition and physical fitness are major determinants of the growth hormone-insulin-like growth factor axis aberrations in adult Turner's syndrome, with important modulations by treatment with 17 beta-estradiol. J Clin Endocrinol Metab 82:2570-2577.
168. Martha PM Jr, Reiter EO, Davila N, Shaw MA, Holcombe JH, Baumann G. (1992) The role of body mass in the response to growth hormone therapy. J Clin Endocrinol Metab 75:1470-1473.
169. Forbes GB. (1977) Nutrition and growth. J Pediatr 91:40-42.
170. Woods KA, Dastot F, Preece MA, Clark AJ, Postel-Vinay MC, Chatelain PG, Ranke MB, Rosenfeld RG, Amselem S, Savage MO. (1997) Phenotype: genotype relationships in growth hormone insensitivity syndrome. J Clin Endocrinol Metab 82:3529-3535.
171. Baumann G, Shaw MA. (1990) A second, lower affinity growth hormone-binding protein in human plasma. J Clin Endocrinol Metab 70:680-686.
172. Ymer SI, Stevenson JL, Herington AC. (1997) Guinea pig serum contains a specific high affinity growth hormone-binding protein with novel ligand specificity. Endocrinology 138:3577-3586.
173. Kratzsch J, Selisko T, Birkenmeier G. (1995) Identification of transformed alpha 2-macroglobulin as a growth hormone-binding protein in human blood. J Clin Endocrinol Metab 80:585-590.

4 REGULATION OF INSULIN-LIKE GROWTH FACTOR-I AND ITS ROLE IN CONTROLLING GROWTH AND METABOLISM

David R. Clemmons

Department of Medicine, The University of North Carolina School of Medicine, Chapel Hill, North Carolina 27599 USA

BACKGROUND

Insulin-like growth factor-I (IGF-I) is a small polypeptide growth factor, also termed Somatomedin-C, that mediates many of the growth promoting actions of human growth hormone (1). IGF-I was discovered based on two of its known biologic actions, its ability to stimulate cartilage sulfation in *in vitro* bioassay systems and to stimulate the incorporation of glucose into fat cells (2,3). Since both of these biologic assay systems were utilized to isolate the factor, two operational names were utilized; the first being Somatomedin; to mediate the growth promoting effects of Somatotropin, and the second insulin-like growth factor, that is, a substance with similar but not identical structure to immunoreactive insulin that had similar biologic activity. Final purification of these peptide activities to homogeneity with subsequent amino acid sequence determination showed that they were identical peptides. Sequence analysis also showed that insulin like growth factor-I IGF-I was 48% identical to human proinsulin, thus confirming a presumed evolutionary relationship between the two peptides (4). It is important to point out, however, that there are distinct amino acid differences that account for major differences in IGF-I and insulin physiology, including their ability to react with distinct transmembrane signaling receptors and their ability to bind to the insulin-like growth factor binding proteins.

With the discovery of a unique receptor for growth hormone and a signal transduction pathway that is clearly distinct from the pathway that is used by IGF-I receptor, it has become obvious that growth hormone stimulates many unique biochemical reactions that are not stimulated by IGF-I (5). Many, but not all of the growth promoting effects of growth hormone can be mimicked by administration of IGF-I *in vivo*; however, since the two separate receptor systems exist and two separate sets of genes are activated by administration of these polypeptides to experimental animals, it is clear at present that IGF-I induction is best viewed as a component of the total somatic response to growth hormone, but not as a sole substitute for growth hormone. The extent to which these other biochemical reactions that are independent of increases in IGF-I are required for normal growth and development is still being definitively determined. Therefore at the present time the best approximation of the role of IGF-I and understanding growth hormone physiology is that of a molecule that can stimulate many biochemical reactions that are required for normal growth and development but that clearly does not replicate all the actions of growth hormone in stimulating these processes.

Gene and Peptide Structure

The insulin-like growth factor family is composed of three peptides, proinsulin, IGF-I and IGF-II. All are encoded by single copy genes (6). The gene for prepro IGF-I is localized on the long arm of chromosome 12, whereas IGF-II is localized on the short arm of chromosome 11. The IGF-II gene is localized near that of human proinsulin. The IGF-I gene is very large, and it is composed of 6 exons that are differentially utilized and separated by 5 introns in the human IGF-I gene. The human IGF-II gene is also large and complex and composed of 9 exons. Both genes have multiple promoters and polyadenylation signals. The encoded prepro peptides of the gene products have been divided into subdomains termed A-E (7). IGF-I and IGF-II are synthesized with leader sequences of 25 and 24 amino acids, respectively. The B region of IGF-I and IGF-II corresponds to the B domain of human proinsulin as do the C and A domains. There is a high degree of conservation in the A domain between IGF-I and IGF-II, with 23 of 27 aminoacids conserved. Similarly, 14 of these 23 amino acids are conserved in human proinsulin. In contrast, the C peptide domains are highly divergent between the three peptides. In the A domain, there is even greater conservation, with 18 of 21 amnioacids being identical between IGF-I and IGF-II and 13 of 21 being identical in human proinsulin. Unlike proinsulin, IGF-I and II have D peptide domain extensions in the mature secreted peptide that has been isolated from blood. The D extension for IGF-I contains 8 amino acids, and IGF-II contains 6. Two longer distinct translation products of IGF-I have been identified, and this extension is termed the E domain. The A variant of the E domain consists of 35 additional amino acids beyond the D domain, and the B variant contains 77 additional amino acids (8). These forms arise by alternative splicing of primary IGF-I mRNA transcripts. In IGF-II there is a single E peptide which contains 89 amino acids. The mechanism by which this occurs and the biologic function of these E peptide extensions have not been elucidated. Three disulfide linkages are present in each peptide. This data has been utilized to predict folding patterns based on known tertiary structural models of human proinsulin (9). Based on the structural models it has been predicted that critical amino acids in the B domain are necessary for optimum receptor binding. This has been confirmed by *in vitro* mutageneis studies showing that alteration of tyrosines at position 24 and 31 in human IGF-I results in major reduction in loss of receptor reactivity (10). Specifically, substitution with leucine in position 24 results in 100 fold reduction in affinity for the type I IGF receptor. Substitution for position 31 also results in significant loss of binding to the type I receptor, but not the Type II receptor.

Control of Cellular Secretion

Unlike classic endocrine hormones such as insulin, the insulin-like growth factors are synthesized and secreted by multiple connective tissue cell types, principally fibroblast-like cells, although *in situ* hybridization studies in animals have shown that cells of epithelial origin can also synthesize IGF-I gene transcripts (11). Although an extensive amount of literature has accumulated regarding the factors that control the synthesis of IGF-I in target cells in culture, much less experimental data exists regarding factors that control the peripheral synthesis of IGF-I *in vivo.*

Analysis for messenger IGF-I mRNA for multiple tissues indicates that in adult animals the greatest expression occurs in the liver (12). Likewise administration of growth hormone to hypophysectomized animals has clearly shown that IGF-I expression in the liver is under the direct control of growth hormone, in that its mRNA abundance can be increased several fold in these animals after single injection of growth hormone (13). Direct evidence indicates that this is due to increased gene transcription (14). In addition, growth hormone has been shown to regulate gene transcription in the brain. Other organs wherein hormonal regulation has been demonstrated include the testes and ovaries, where gonadotrophins can increase IGF-I gene expression and the uterus, ovary and bone where estrogen stimulates gene expression. Glucocorticoids, in contrast, have been shown to decrease an IGF-I expression in bone (15). Insulin has been shown to increase liver expression. Fasting has also been shown to be a potent inhibitor of IGF-I gene expression in the liver, and prolonged fasting in rats can make them refractory to stimulation of IGF-I gene transcription by growth hormone.

Several cell types, such as osteoblasts, have been studied in culture and shown to respond to polypeptide growth factors, such as, fibroblast growth factor or platelet derived growth factor, with increased IGF-I mRNA expression. Similarly, other connective cell types, such as skeletal myoblasts and fibroblasts, also respond to peptide growth factors with increases in IGF-I mRNA expression and synthesis. One important consequence of these responses that has been demonstrated in *in vivo* studies is the change in IGF-I expression that occurs in response to injury. Following multiple different types of injury, whether surgical wounding or thermal injury, for example, satellite cells in muscle that ultimately differentiate into myocytes or prechondrocytes in cartilage cells express IGF-I immediately following the injury and for several days during their differentiation into more mature cell types (16). This wave of IGF-I expression by these precursor cell types appears to be an important component to response to injury that may be necessary for normal wound repair. This expression pattern is clearly independent of growth hormone, since it can occur and proceed normally in hypophsectomized animals (17).

Following IGF-I transcription, the IGF-I transcripts are translated into the prepro peptides and often secreted by cells and culture. These propeptides may contain an E peptide extension (18). The exact point at which these peptides are further processed to mature IGF-I prior to its appearance in the vasculature is unclear. Likewise it has not been definitively determined whether hepatocytes secrete gene products with the peptide extensions. Since the liver is the most important source of IGF-I in blood, it is possible that processing is not required prior to entrance into the circulation, and higher molecular weight forms of IGF-I and IGF-II have been found to occur in serum. These most often occur during pathophysiologic conditions such as, the production of IGF-II precursors by retroperitoneal sarcomas. In summary, the regulation of IGF-I gene expression by cell types and tissues is a tightly coordinated event which can lead to major increases in local concentration of this peptide.

Clearly this process appears to be important in response to injury and tissue regeneration as well as in normal fetal development. The extent to which this autocrine/paracrine control IGF-I secretion by peripheral tissues is required for normal growth and development during childhood has not been definitively determined.

Measurement of IGF in serum: Interactions with IGF binding proteins

Measurement of IGF-I in serum has always been technically challenging due to the presence of the IGF binding proteins. A complete discussion of IGF binding proteins is taken up in Chapter 5 and therefore will not be discussed herein. However, two important issues regarding IGF binding proteins (IGFBPs) are necessary for understanding the problems that occur when interpreting changes in IGF-I concentrations in serum. First is the technical problem of IGF binding protein interference in the radioimmunoassay. Unlike insulin, the IGF's have four amino acids in their B chain that determine their capacity to bind to the IGFBPs(19). Because of this high affinity, binding can be difficult to remove the IGFBPs sufficiently to validly measure IGF-I. The methodologies that have been used include acid gel filtration chromatology, acid ethanol extraction and hydrophobic separation by HPLC. Each of these methods has strengths and weakness. The most important determinant however, is to validate the assay by showing all the IGF binding protein has been removed or that the minimal fraction that remains is no longer interfering in the assay. Unless this type of validation is performed, no assay should be considered accurate. The interference by binding proteins is a particularly difficult problem when patient's with growth hormone deficiency are being assessed. This is because the remaining binding protein in their serum is often unsaturated and therefore has greater potential to interfere. Two newer methods for removing binding protein interference also show promise. The first is to use highly IGF-I specific antiserum and add additional IGF-II to the serum sample to saturate the remaining amount of IGF binding protein. If the primary antisera has sufficient specificity for IGF-I as compared to IGF-II, this is a valid technique. A second method has been to develop two site IRMA assays in which two antibodies are directed against IGF-I such that competitive ionated IGF-I displacement is not required in the assay and therefore binding protein interference is not an issue. For this type of method it is important to determine that the sites that react with the IGF-I antisera are not altered when IGF-I binds to binding protein.

A second major consideration in interpreting changes in serum IGF-I concentrations is the fact that the binding proteins are often either inversely or directly regulated by factors that also regulate IGF-I. For example, IGFBP-3 increases in parallel with IGF-I after growth hormone administration to humans (20). Therefore much of the change in total IGF-I that occurs after growth hormone administration may be due to changes in IGFBP-3. In contrast, in poorly controlled diabetes, IGFBP-1 concentrations rise 6-8 fold, whereas IGF-I concentrations fall; therefore, changes in total IGF-I may be related to changes in specific forms of IGFBPs (21).

Regulation of IGF-I in blood

IGF-I concentrations in the serum of normal children and adults vary with age. Therefore, IGF-I levels are relatively low at birth (mean plasma concentrations in the range of 50 ng/ml) then increase progressively to puberty, peaking at concentrations of approximately 535 ng/ml (22). They then fall rapidly between puberty and the third decade of life but still continue to decline slowly up to age 70 years. This decline is substantial, since values in subjects in the third decade may be twice the value of subjects in the 8th decade. These changes are directly paralleled by changes in IGFBP-3 and no doubt reflect changes in total serum binding capacity, which is the major determinate of IGF-I half life. Female sex has a small determining factor on IGF-I values since values may be 10-15% higher in females during puberty. Children with hypopituitarism tend to have definitive decreases in serum IGF-I. Cross sectional studies have shown that less than 10% of children with severe growth hormone deficiency have IGF-I values that are within the normal range, and generally these children have values that are near the lower limit of normal (23). Because IGF-I values change so much during childhood, they can only be interpreted correctly with good normative data for developmental age. Administration of growth hormone to growth hormone deficient subjects results in a substantial rise in IGF-I occurring approximately 4 to 6 hours following injection (24). These values peak at 24 hours and then begin to attenuate over the next 24 hour period and decline back to baseline if no further growth hormone is administered. Because growth hormone also increases IGFBP-3 and ALS, part of the change in IGF-I may be due to altered IGF-I clearance, since this large tertiary complex is cleared much more slowly from serum than free IGF-I, e.g. 16 hours vs. less than 5 minutes (25). In contrast, in adults with severe GH deficiency, a higher percentage of subjects will have normal IGF-I values. This is thought to be due to the fact that control of IGF-I is multihormonal, and therefore changes in other hormonal profiles may partially compensate for GH deficiency. Again, although mean values are usually substantially lower than normal age-matched subjects, there is significant overlap between growth hormone deficient and nongrowth hormone deficient individuals (26). A further complicating problem is the fact that in normal subjects growth hormone secretion declines with age; therefore, in the elderly it may be extremely difficult to differentiate between normal and abnormal IGF-I values. Because multiple conditions other than GH deficiency result in low IGF-I concentrations in adults and children, its measurement has less value in screening for growth hormone deficiency, that is, children with short stature who are not growth hormone deficient may have low IGF-I values (27). The variables that contribute to low IGF-I values in these children include other hormonal deficiencies, such as low thyroxine, poor nutritional status or the effect of chronic disease states, all of which can act to suppress IGF-I concentrations. This is less of a problem in adults wherein the most common difficulty is in interpreting a normal IGF-I value in a patient with definite evidence of impaired GH secretion.

Acromegaly

In states of GH excess, IGF-I values are almost invariably increased and large cross sectional studies of acromegalic patients have shown almost uniform increases in

IGF-I. The mean value is approximately 7-8 times higher than the mean for age-adjusted normals (28). In acromegaly IGF-I values are not only elevated but correlate with disease activity. In a large study of 57 acromegalic subjects, there was an excellent correlation between heel pad thickness or fasting serum glucose and basal serum IGF-I concentrations (29). There are very few other physiologic conditions that can result in higher IGF-I. These include the normal pubertal growth spurt and the last trimester pregnancy wherein there is some overlap between these values and acromegaly. Following ablation of growth hormone secreting tumors, IGF-I values decrease dramatically and can enter the normal range if the disease is cured. Because it is a very stable measurement, IGF-I represents a rapid and accurate means for assessing degree of improvement in acromegaly.

Other hormonal variables

Plasma IGF-I concentrations are very low in thyroid hormone deficiency and rise after thyroxine replacement (30). Values are not suppressed in patients with Turner's Syndrome, and physiologic replacement of estrogen results in little change in serum IGF-I; however pharmacologic administration of large doses of estrogen can reduce plasma IGF-I concentrations (31). Prolactin has a very weak stimulatory effect on IGF-I, and patients must be severely growth hormone deficient in order to detect the effect of prolactin. Human placental lactogen clearly increases IGF-I and the increasing IGF-I that occurs during pregnancy is caused by a placental growth hormone variant (32). This change is clearly distinct from that caused by pituitary-derived growth hormone, since a change can be seen in pregnant acromegalics.

Nutritional Status

Nutrient intake is a dominant variable controlling plasma IGF-I concentrations. Specifically, normal protein and caloric intake must be maintained in order to maintain adequate serum IGF-I concentrations (33, 34). Fasting for periods as short as three days results in substantial reduction in total serum IGF-I. Most importantly, a three day fast will result in refractoriness to growth hormone, and administration of growth hormone to subjects that have been fasted for these periods of time results in very little change in serum IGF-I compared to control, well fed subjects (35). Ten days of fasting leads to declines in serum IGF-I that are comparable to those seen in severe growth hormone deficiency, and subjects must be re-fed normal diet for at least six days in order to restore IGF-I secretion towards normal (36). These changes in IGF-I correlate with the catabolic response to starvation and suggest that maintenance of a normal IGF-I concentration is necessary to maintain the normal anabolic state. Hepatic failure, chronic illness and renal failure are associated with low serum IGF-I concentrations (37). In the case of hepatic failure, it is not clear whether this is due to the catabolism that is associated with a chronic disease state or whether it is due to loss of IGF-I secretion due to hepatocyte destruction. However other chronic illnesses that result in catabolism, such as inflammatory bowel disease, are also associated with low serum IGF-I concentrations (38). In patients with renal failure, IGF-I concentrations are reduced; however the serum binding capacity is increased, and therefore assay measurements must remove all binding proteins before the levels can be reliably assessed.

Assessment of IGF-I actions in vitro

A major problem in determining IGF-I actions *in vitro* has been to account for the effects of interference by binding proteins. Even in isolated cell types in cell culture, IGFBPs are almost invariably present, and therefore the addition of IGF-I to target cell systems relies on the assumption that adequate IGF-I is available to bind the receptors. In cases where there are large concentrations of binding proteins interfering or where binding proteins are being proteolytically cleaved, these assumptions may not always be accurate. In spite of these limitations, much useful data has been generated studying the cellular responses to IGF-I addition.

Almost all the cellular effects of IGF-I are mediated by binding to the IGF-I receptor. IGF-I receptors are relatively ubiquitous and are present in almost all cell types. This probably accounts for the ability of this peptide to stimulate generalized growth. The receptor contains a 706 amino acid extracellular α subunit which contains the binding domain and a 626 amino acid transmembrane β subunit which contains the signaling apparatus (40). The two α and two β subunits are linked by disulfide bonds to form a heterotetramer. Ligand binding specificity is conferred solely by the extracellular domain. The cytoplasmic domains contain intrinsic tyrosine kinase activity. This is similar to the domain structure of the insulin receptor. Both the IGF-I receptor and insulin receptor contain tyrosine kinase domains that are approximately 84% homologous (41). The affinity for the IGF-I receptor for IGF-I is approximately 0.2-1 nanomolar. Insulin cross reacts with 1000 fold lower affinity. IGF-II has approximately six fold lower affinity. Insulin receptors have approximatel 100 fold lower affinity for IGF-I (42). Several monoclonal antibodies that react against the extracellular domain of the IGF-I receptor inhibit IGF-I binding, and the most useful of these (αIR3) has been used to probe the effects of IGF-I on several cell types. Activation of the receptor results in direct activation of the tyrosine activity with autophosphorylation of specific tyrosine residues in response to ligand binding. This is a transreaction wherein the tyrosine kinase activity of one β subunit transphosphorylates residues on the opposite β subunit. Mutations that eliminate the ability of the receptor to autophosphorylate abrogate biologic activity (43). Activation of tyrosine phosphorylation results in binding of the docking protein, insulin receptor substrate-1 (IRS-1) to the activated receptor (44). IRS-1 contains 21 potential tyrosine phosphorylation sites that occur in the sequence YMXM, which is a src homology SH-2 recognition motif. Phosphorylation of IRS-1 in turn leads to activation of two separate signaling cascades. One occurs by binding of the regulatory subunit P85 of phosphatydylinositol-3 kinase (PI3 kinase) which then leads to downstream activation of protein tyrosine kinase B (45). A second pathway is activated by binding of the growth factor receptor leads to formation of a complex between IRS-1, Grb-2 and son of sevenless (SOS). This complex then activates the GTP exchange factor Ras, which leads to further activation of MEK and MAP kinases (46). This pathway is thought to be critical for growth stimulation.

In addition to the IGF-I receptor, a second receptor exists. The cation dependent mannose-6 phosphate receptor. This receptor binds IGF-II with 50 fold higher affinity than IGF-I, and it does not bind to insulin (47). This is a membrane associated receptor that cycles from the plasma membrane on the cell surface to the trans golgi network. In addition to lysosomal enzymes, other mannose-6 containing glycoproteins bind to this receptor, including renin, proliferin, thyroglobulin and transforming growth factor beta. These receptors bind IGF-II with a high affinity in the range of 0.17-0.7 nanomolar. The binding of IGF-II causes its internalization and degradation. This receptor is not known to have classic transmembrane signaling since no known definitive evidence of growth promoting signal transduction has been identified for this receptor. It has been assumed that it may have an important role in IGF clearance.

In Vitro Biologic Actions

An extensive list of cell types which IGF-I has been shown to stimulate proliferation has been reviewed by Lowe (48). The most extensively studied specific cell types would include the following: chondrocytes, osteoblasts, fibroblasts myoblasts and granuloma cells. The addition of IGF-I in concentrations that occur *in vivo* clearly stimulates ^{3}H-thymidine incorporation into the DNA in cultured chondrocytes from multiple species. Likewise, it promotes clonal growth, suggesting that it is a potent mitogen (49). Finally, as originally demonstrated by Salmon and Daughaday, it stimulates multiple anabolic effects in cartilage, including the incorporation of sulfate in the glycosaminoglycans side chains, an important growth regulatory parameter for cartilage *in vivo*. IGF-I is equally potent in stimulating DNA and RNA synthesis in bone cells. It is also a potent stimulant of type I collagen synthesis by bone, which leads to deposition of collagen in bone matrix. Rat fetal calvarial cells appear to be very sensitive to IGF-I, and this has been the major model system for analysis (50). IGF-I is a potent stimulator of homatopoietic stem cells, and it stimulates erythroid burst forming units in bone marrow cultures. Likewise it will stimulate colony formation of granulocytic precursor cells. In this system IGF-I also functions to enhance the effects of granulocyte colony stimulating factor (51). IGF-I stimulates a variety of both male and female gonadal cell types.

The most extensively studied cell type is the ovarian granulosa cell. IGF functions to enhance the effect of EGF on DNA synthesis (52). More importantly, IGF-I is a very potent enhancer of steroidogenesis by these cells, and in multiple species, it markedly potentiates the effects of FSH on progesterone production (53). IGF-I can also stimulate steroidogenic effects of ACTH and angiotensin II on adrenal cells as well as the effect of LH on androgen production by leydig cells of the testes. IGF-I is also a potent stimulant of DNA synthesis in thyroid follicular cells. It can potentiate the effects of TSH (54).

IGF-I is a potent stimulant of differentiation. Specifically, in skeletal myoblasts cell lines, both IGF-I and II have been shown to markedly potentiate differentiation and synthesis of differentiation specific genes (55). Several neural cell types have been studied for IGF sensitivity. IGF-I is a potent mitogen for glial cells derived from rat brain. It has also been shown to inhibit apoptosis in multiple types of neurons, and in poorly differentiated neuronal cells it can directly stimulate cell division. Permanent cell lines, such as PC12 cells or SHSY5Y neuroblastoma cells, have been shown to be markedly sensitive to IGF-I (55). Cultured skin fibroblasts are very sensitive to IGF-I. This effect appears to be definitively mediated through the type I receptors since it can be blocked with "IR3. IGF-I syngerizes with several other growth factors, such as PDGF, to further stimulate fibroblast DNA synthesis (56). Finally, vascular smooth muscle cells are also extremely sensitive to IGF-I. IGF-I synergizes with both FGF and PDGF to further stimulate DNA synthesis. Endothelial cells are IGF-I responsive (57). IGF-I stimulates not only DNA synthesis but also amino acid uptake, glucose uptake and proteoglycan synthesis by pulmonary artery and capillary endothelial cells. Hepatocytes derived from young animals are IGF-I sensitive and can be made to grow in response to concentrations as low at 50 ng/ml of IGF-I (58). In summary, IGF stimulates the growth of a variety of cells in culture when concentrations which are known to occur in vivo are used. These effects are modulated greatly by changes in receptor number and by the presence or absence of IGF binding proteins.

In vivo biologic actions

In general IGF-I is thought of as a potent comitogen, meaning that it enhances the effects of other growth factors. However, as pointed out in several of these examples cited above, IGF-I has many mitogenic effects in its own right. Recently there has been a tremendous surge of interest in the ability of IGF-I to enhance differentiated functions, and more importantly, to inhibit cell death. Obviously all of these effects can lead to increased trophic effects in target tissue. The somatomedin model of growth hormone action predicted that the growth promoting effects of GH would be mediated by IGF-I. The recent ability to delete IGF genes and to administer IGF-I to animals and humans has made it possible to directly test this hypothesis. However, one factor complicating the interpretation of changes that occur following administration of IGF-I to animals is the presence of IGF binding proteins. Since over 99% of the circulating IGF-I is bound to carrier proteins, systemic administration of IGF-I relies on adequate IGF-I being delivered outside the vasculature to IGF-I responsive tissues. In addition, the binding protein profiles in the pericellular environment will determine whether adequate IGF-I is delivered to receptors to reasonably reflect the tissue exposure that occurs during normal growth and development. No doubt perturbations in binding protein levels occur, and they may alter the response to in vivo administration.

IGF-I infusions to normal rats causes hypoglycemia primarily by stimulating peripheral glucose uptake and suppressing hepatic glucose production (59). IGF-I has been reported to be potently anabolic, and stimulates protein synthesis and enhances nitrogen balance in animals or humans that have been made catabolic by administration of corticosteroids (60), dietary restriction (61) or the induction of acute renal failure (62). IGF-I under these circumstances has positive effects on wound healing, recovery of renal function and total body protein accretion. When changes in growth or body weight are measured in hypophysectomized rats, growth hormone is consistently more potent than IGF-I (63). However if IGF-I is administered in sufficient concentrations, it causes substantial increases in body weight and in long bone growth in both normal and hypophysectomized rats (64). IGF-I causes an increase in glomerular filtration rate and renal blood flow (25). It also has a direct trophic effect on gut epithelium if delivered directly into the lumen. Diabetic rats respond to IGF-I with increased peripheral blood glucose uptake and enhanced glycogen synthesis (65). Insulin resistant animals respond with an improvement in insulin sensitivity. Growth rate can be restored toward normal with administration of IGF-I to severely insulin deficient rats.

Effects of overexpression of IGF-I in transgenic animals

In contrast to the studies that have directly compared IGF-I and growth hormone administration to normal animals, overexpression of IGF-I generally leads to impressive changes in somatic growth and body weight. In a mouse model developed by Palmiter and co-workers, growth hormone is deleted by insertion of diphtheria toxin into growth hormone producing cells in fetal life. This results in severe, isolated growth hormone deficiency. Overexpression of IGF-I transgene in this model system in multiple organs leads to restoration of IGF-I concentrations that approach normal (66). In spite of the fact that these animals have IGF binding protein abnormalities in their serum, for example extremely high concentrations of IGFBP-2, they undergo proportionate, balanced growth. If IGF-I concentrations are expressed adequately to maintain normal serum concentrations, these animals grow to a normal size that is comparable to control, non-growth hormone deficient litter mates. Although these animals do not exhibit impressive statural overgrowth, as compared to animals that are transgenic for growth hormone, they do grow normally in spite of the fact that they have no detectable growth hormone. Growth is slightly disproportionate in that the increase in brain growth is greater than in other organs. The livers are also smaller in these animals, and this defect is not completely reversed by overexpression of IGF-I. In summary, the results of these experiments suggest that most of the growth promoting effects of growth hormone are mediated by IGF-I, but there may be additional direct GH actions in the liver. IGF-I may have additional actions in the brain that are independent of growth hormone.

IGF Gene Deletion Experiments

The most illuminating experiments have been those in which the IGF-I or IGF-I receptor genes have been deleted (67, 68). Experiments by Efstradiatis and colleagues using mice with null mutations created by homologous recombination for the genes coding IGF-I, IGF-II and the IGF receptor have been extremely revealing. In these studies important effects of these deletions were detected on fetal and post natal growth as well as in tissue histomorphometry. The phenotype of neonatal animals lacking IGF-I is similar to the those lacking IGF-II at birth; that is, they are approximately 60% of normal size. There is also some placental hypoplasia. Postnatally, there are major differences. The animals that are null for IGF-I have progressive postnatal growth retardation and never experience catchup growth. In contrast, the IGF-II deleted animals are born small (e.g. 50% of normal size) but grow at a normal rate postnatally. Survival is also normal in the mice lacking IGF-II; however, the mice lacking IGF-I have a very high neonatal death rate which can approach 90% prior to adulthood. Deletion of the IGF-I and IGF-II genes results in extremely small mice (30% of normal weight at birth). The animals are unable to take their first breath and die in respiratory failure immediately after delivery. The phenotype of animals lacking IGF-I receptor is similar, except that they are 45% of normal weight at birth but also die immediately after birth, again due to diaphragmatic hypoplasia and inability to take a normal inspiration. Deletion of the IGF-II receptor also results in in utero death, but deletion of IGF-II plus the IGF-II receptor enables the animals to survive, suggesting that one of the main functions of this receptor may be IGF-II clearance. These studies show that IGF-I is required for normal intrauterine and postnatal growth.

Effects of IGF-I administered to humans

Several studies have delivered IGF-I to normal human volunteers. Intravenous administration of 100 mcg/kg caused rapid symptomatic hypoglycemia and it is equally potent with 0.15 unit/kg insulin (69). However IGF-I is less potent than insulin in reducing free fatty acid levels. A continuous infusion of 24 mcg/kg/hr of normal volunteers reduced C-peptide concentrations by 50% in spite of maintenance of euglycemia with intravenous glucose, suggesting that IGF-I has a direct and suppressive effect on insulin release. The study also showed that IGF-I increased glucose uptake and lowered fatty acids and triglycerides as well as decreasing protein breakdown (70). IGF-I also suppresses glucagon, and the counterregulatory response to hypoglycemia. When IGF-I and insulin are compared IGF-I is a more potent suppressor of both the growth hormone and glucagon responses to hypoglycemia. Administration of IGF-I at 12 mcg/kg/hr to catabolic subjects resulted in improved negative nitrogen balance. The degree of improvement was comparable to the administration of 0.05 mg/kg/day of growth hormone (61). Simultaneous administration of growth hormone plus IGF-I, however, resulted in a 2.4 fold greater effect on nitrogen retention as compared to the administration of either growth hormone or IGF-I alone, suggesting that their combine effects are additive (71). Infusion of IGF-I alone results in suppression of IGFBP-3 due to suppression of acid labile subunit (ALS), but if growth hormone is administered, ALS concentrations remain elevated, as does IGFBP-3, suggesting that maintenance of a high level of the IGF-I-IGFBP-3-ALS complex is necessary for the optimum anabolic response. The administration of IGF-I to humans also results in increased glomerular filtration rate and renal plasma flow. Chronic subcutaneous administration of IGF-I results not only in increased anabolic response in patients with growth hormone receptor mutations and growth rate is accelerated during the first year of treatment to the same extent as GH deficient children receiving growth hormone. However, during the second year of treatment there is some attenuation of growth rate that is more than the usual attenuation seen with growth hormone, suggesting that chronic IGF-I therapy, while growth stimulatory, is not equipotent with growth hormone. However final adult heights will have to be assessed in these children in order to determine if IGF-I is unable to equal the effect of growth hormone.

In diabetes, IGF-I is effective in lowering serum glucose and enhancing insulin sensitivity. Patients with type II diabetes had as much as a 2.4 fold improvement in insulin sensitivity as calculated by Bergman modeling (72). Likewise, patients with extreme insulin resistance syndromes respond to IGF-I with improvement in serum glucose and lowering of hemoglobin A1C. Several chronic trials of long term administration of IGF-I in doses of 40 mcg/kg twice daily subcutaneously to patients with type II diabetes show that in large cohorts, for example 40 subjects per group, IGF-I is very effective in lowering hemoglobin A_1C, which was lowered by approximately 1.8%. Likewise, fructosamine is lowered and mean daily blood glucoses improves, suggesting that long term maintenance of improved glucose homeostasis is possible. IGF-I has also been administered chronically to children with type I diabetes. In this case the insulin requirement is markedly lower, particular that of regular insulin, suggesting that IGF-I is also enhancing insulin sensitivity in this population and reducing the need for exogenous insulin administration. The effects of IGF-I in enhancing diabetic complications, such as nephropathy or retinopathy, are unknown. The incidence of progression of retinopathy in these studies was no greater than that observed with insulin in which patients with accelerated retinopathy improved control acutely, and therefore the long term risk of IGF-I in accelerating these complications has not been definitively determined.

SUMMARY

The results of recent studies have reaffirmed the central role of IGF-I growth regulation and maintenance of normal intermediary metabolism. IGF-I functions as both for a mitogen and an anabolic hormone. The exact proportion of normal growth and development that is atributtable to autocrine/paracinee produced IGF-I, as compared to IGF-I that is transported through the vasculature to distant sties of action, remains unknown. An important area of further studies will be to determine how IGF-I functions with other growth hormone inducible genes to facilitate growth promoting an anabolic actions of GH

REFERENCES

1. Daughaday WH, Rotwein P. Insulin-like growth factors I and II. Peptide, messenger ribonucleic acid and gene structures, serum and tissue concentrations. Endocrine Rev 1989; 10: 68-91
2. Salmon Jr WD, Daughaday WH. A hormonally controlled serum factor which stimulates sulfate incorporation by cartilage in vitro. J Lab Clin Med 1957; 49: 825-36
3. Froesch ER, Muller WA, Burgi H. Nonsuppressible insulin-like activity of human serum. II. Biological properties of plasma extracts with nonsuppressible insulin-like activity. Biochim Biophys Acta 1966; 121: 360-74
4. Rinderknecht E, Humbel RE. The amino acid sequence of human insulin-like growth factor I and its structural homology with proinsulin. J Biol Chem 1978; 253: 2769-76

5. Argetsinger LS, Campbell GS, Yang X, Witthuhn BA, Silvennoinen O, Ihle JN, Carter-Su C. Identification of Jak-2 as a growth hormone recptor associated tyrosine kinase. Cell 1993; 74: 237-44

6. van den Brande JL. "Structure of the human insulin-like growth factors: relationship to function," In: "*The insulin-like growth factors. Structure and biological functions,*" Schofield PN; ed. Oxford Oxford University Press, 1992. p. 12-44.

7. LeRoith D, Baserga R, Helman L, Roberts CT. Insulin like growth factors and cancer. Ann Intern Med 1995; 122: 54-9

8. Shimatsu A, Rotwein P. Sequence of two rat insulin-like growth factor-I mRNA's differing in the 5' untranslated region. Nucl Acid Res 1986; 15: 7196

9. Blundell TL, Bedarkar S, Rinderknecht E, Humbel RE. Insulin-like growth factor: a model for tertiary structure accounting for immunoreactivity and receptor binding. Proc Natl Acad Sci USA 1978; 75: 180-4

10. Cascieri MA, Chicchi GC, Applebaum J, Haynes NS, Given BG, Bayne ML. Mutants of human insulin like growth factor I with reduced affinity for the type I insulin like growth factor receptor. Biochemistry 1988; 27: 3229-33

11. Han VKM, D'Ercole AJ, Lund PK. Cellular location of somatomedin (insulin-like growth factor) messenger RNA in the human fetus. Science 1987; 236: 193-7

12. Lund PK, Moats-Staats BM, Hynes MA, Simmons JG, Jansen M, D'Ercole AJ, Van Wyk JJ. Somatomedin-C/Insulin-like growth factor-I and insulin-like growth factor-II mRNAs in rat fetal and adult tissues. J Biol Chem 1986; 261: 14539

13. Roberts CT, Brown AO, Graham DE, Sealy S, Barry S, Gabbay KH, Rechler MM. Growth hormone regulates the abundance of insulin-like growth factor-I RNA in adult rat liver. J Biol Chem 1986; 261: 10025-8

14. Mathews LS, Norstedt G, Palmiter RD. Regulation of insulin-like growth factor I gene expression by growth hormone. Proc Natl Acad Sci USA 1986; 83: 9343

15. McCarthy TL, Centrella M, Canalis E. Cortisol inhibits the synthesis of insulin-like growth factor-I in skeletal cells. Endocrinology 1990; 126: 1569

16. Jennische E, Skottner A, Hansson HA. Satellite cells express the trophic factor IGF-I in regenerating skeletal muscle. Acta Physiol Scand 1987; 129: 9-15

17. Edwall D, Schalling M, Jennische E, Norstedt G. Induction of insulin-like growth factor I messenger ribonucleic acid during regeneration of rat skeletal muscle. Endocrinology 1989; 124: 820-5

18. Rotwein P. Two insulin-like growth factor I messenger RNAs are expressed in human liver. Proc Natl Acad Sci USA 1986; 83: 77-81

19. Cascieri MA, Bayne ML. "Biological activity of IGF-I analogs with reduced affinity for soluble IGF binding proteins," In: "*Insulin-like growth factor binding proteins. Excerpta Medica International Congress Series*

881," Drop SLS, Hintz RL, eds. Amsterdam Elsevier Science Publishers B.V. 1989. p. 207-15.

20. Baxter RC, Martin JL. Binding proteins for insulin-like growth factors: Structure, Regulation and Function. Prog Growth Fact Res 1989; 1: 49-68
21. Westwood M, Gibson JM, Williams AC, Clayton PE, Hamberg O, Flyvbjerg A, White A. Hormonal regulalation of circulating insulin-like growth factor binding protein-1 phosphorylation status. J Clin Endocrinol Metab 1995; 80:(12) 3520-7
22. Argente J, Barrios V, Pozo J, Munoz MT, Hervas F, Stene M, Hernandez M. Normative Data for Insulin-Like Growth Factors (IGFs), IGF Binding Proteins, and Growth Hormone-Binding Protein in a Healthy Spanish Pediatric Population: Age- and Sex-Related Changes. J Clin Endocrinol Metab 1993; 77:(6) 1522-8
23. Nunez SB, Municchi G, Barnes KM, Rose SR. Insulin-Like Growth Factor I (IGF-I) and IGF Binding Protein-3 Concentratios Compared to Stimulated and Night Growth Hormone in the Evaluation of Short Children - A Clinical Research Center Study. J Clin Endocrinol Metab 1996; 81:(5) 1927-32
24. Copeland KC, Underwood LE, Van Wyk JJ. Induction of immunoreactive somatomedin-C in human serum by growth hormone: dose response relationships and effect on chromatographic profiles. J Clin Endocrinol Metab 1980; 50: 690-7
25. Guler H-P, Zapf J, Schmid C, Froesch ER. Insulin-like growth factors I and II in healthy man. Estimations of half-lives and production rates. Acta Endocrinol 1989; 121: 753-8
26. Baum HBA, Biller BMK, Katznelson L, Oppenheim DS, Clemmons DR, Cannistraro KB, Schoenfeld DA, Best SA, Klibanski A. Assessment of Growth Hormone (GH) Secretion in Men with Adult-Onset GH Deficiency Compared with that in Normal Men - A Clinical Research Center Study. J Clin Endocrinol Metab 1996; 81:(1) 84-92
27. Ranke MB, Blum WF, Bierich JR. Clinical relevance of serum measurements of insulin-like growth factors and somatomedin binding proteins. Acta Pediatr Scand 1988; Suppl. 347: 114-26
28. Melmed S. Acromegaly. N Engl J Med 1990; 322: 966-77
29. Clemmons DR, Underwood LE, Ridgway EC. Evaluation of acromegaly by radioimmunoassay of somatomedin-C. N Engl J Med 1979; 301: 1138-42
30. Chernausek SD, Underwood LE, Utiger RD, Van Wyk JJ. Growth hormone secretion and plasma somatomedin-C in primary hypothyroidism. Clin Endocrinol 1983; 19: 337-44
31. Clemmons DR, Underwood LE, Ridgway EG. Estradiol treatment of acromegaly: reduction of immunoreactive somatomedin-C and improvement of metabolic status. Am J Med 1980; 69: 571-5
32. Furlanetto RW, Underwood LE, Van Wyk JJ, D'Ercole AJ. Estimation of somatomedin-C levels in normals and patients with pituitary disease by radioimmunoassay. J Clin Invest 1977; 60: 648-57
33. Isley WL, Underwood LE, Clemmons DR. Dietary components that regulate serum somatomedin-C in humans. J Clin Invest 1983; 71: 175-82

34. Strauss DS, Takemoto CD. Effect of fasting on insulin-like growth factor-I (IGF-I) and growth hormone receptor mRNA levels and IGF-I gene transcription in rat liver. Mol Endocrinol 1990; 4: 91-100

35. Merimee TJ, Zapf J, Froesch ER. Insulin-like growth factors in fed and fasted states. J Clin Endocrinol Metab 1982; 55: 999-1002

36. Clemmons DR, Klibanski A, Underwood LE, McArthur JW, Ridgway EC, Beitens IZ, Van Wyk JJ. Reduction of plasma immunoreactive somatomedin-C during fasting in humans. J Clin Endocrinol Metab 1981; 53: 1247-50

37. Ziegler TR, Young LS, Ferrari-Balivera E, Denling RH, Wilmore DW. Use of human growth hormone combined with nutritional support in a critical care unit. JPEN J Parent Enteral Nutr 1990; 14: 574-81

38. Thomas AG, Holly JM, Taylor F, Miller V. Insulin like growth factor-I, insulin like growth factor binding protein-1, and insulin in childhood Crohn's disease. Gut 1993; 34: 944-7

39. Goldberg AC, Trivedi B, Pelmez JA. Uremia reduces serum insulin-like growth factor-I, increases insulin-like growth factor-II and modifies their serum binding protein. J Clin Endocrinol Metab 1982; 55: 1040-5

40. Lammers R, Gray A, Schlessinger J, Ulrich A. Differential signalling potential of insulin- and IGF-I receptor cytoplasmic domains. EMBO J 1989; 8: 1369

41. Ullrich A, Gray A, Tam AW, Yang-Feng T, Tsubokawa M, Collins C, Henzel WJ, LeBon T, Kathuria S, Chen E, et al. Insulin-like growth factor-I receptor primary structure: Comparison with insulin receptor suggests structural determinants that define functional specificity. EMBO J 1986; 5: 2503-12

42. Werner H, Woloschak M, Stannard B, et al."The insulin-like growth factor I receptor: molecular biology, heterogeneity, and regulation," In: "*Insulin Like Growth Factors: Molecular and Cellular Aspects,*" LeRoith D; ed. Boca Raton,FL CRC Press, 1991. p. 17-46.

43. Kato H, Faria TN, Stannard B, Roberts CT, LeRoith D. Role of tyrosine activity in signal transduction by the insulin-like growth factor-I (IGF-I) receptor. Characterization of kinase-deficient IGF-I receptors and the action of an IGF-I-mimetic antibody (alpha IR-3). J Biol Chem 1993; 268:(4) 2655-61

44. Sun XJ, Rothenberg P, Kahn CR, Backer JM, Araki E, Wilden PA, Cahill DA, Goldstein BJ, White MF. Structure of the insulin receptor subsrate IRS-1 defines a unique signal transduction protein. Nature 1991; 351: 73-7

45. Myers MG, Sun XJ, Cheatham B, Jachna BR, Glasheen EM, Bacher JM, White MF. IRS-I is a common element in insulin and insulin-like growth factor I signalling to the phosphatidylinositol 5'-kinase. Endocrinology 1993; 132: 1421-30

46. Lu K, Campisi J. Ras proteins are essentialand selective for the action of insulin-like growth factor I late in the G_1 phase of the cell cycle in Balb c murine fibroblasts. Proc Natl Acad Sci USA 1992; 89: 3889-93

47. Tally M, Enberg G, Li CH, Hall K. The specificity of the human IGF-2 receptor. Biochem Biophys Res Commun 1987; 147: 1206

48. Lowe WL. "Biologic actions of the insulin-like growth factors," In: "*Insulin-like Growth Factors: Molecular and Cellular Aspects,*" LeRoith D; ed. Boca Raton CRC Press, 1991. p. 49-85.
49. Vetter U, Zapf J, Heit W, Helbing G, Heinze E, Froesch ER, Teller WM. Human fetal and adult chondrocytes. Effect of insulin-like growth factors I and II, insulin, and growth hormone on clonal growth. J Clin Invest 1986; 77: 1903-8
50. Canalis E, Gabbitas B. Skeletal growth factors regulate the synthesis of insulin-like growth factor binding protein-5 in bone cell cultures. J Biol Chem 1995; 270: 10771-6
51. Merchav S, Tatarsky I, Hochberg Z. Enhancement of human granulopoiesis *in vitro* by biosynthetic insulin-like growth factor I/somatomedin-C and human growth hormone. J Clin Invest 1988; 81: 791
52. May JV, Frost JP, Schomberg DW. Differential effects of epidermal growth factor, somatomedin-C/insulin-like growth factor I, and transforming growth factor-$ on porcine granulosa cell deoxyribonucleic acid synthesis and cell proliferation. Endocrinology 1988; 123: 168
53. Adashi EY, Resnick CE, Svoboda ME, Van Wyk JJ. Somatomedin-C synergizes with follicle-stimulating hormone in the acquisition of progestin biosynthetic capacity by cultured rat granulosa cells. Endocrinology 1985; 116: 2135-42
54. Tramontano D, Moses AC, Venezani BM, Ingbar SH. Adenosine 3'5' monophosphate mediates both the mitogenic effect of TSH and its ability to amplify the responses to insulin-like growth factor I in FRTL-5 cells. Endocrinology 1988; 122: 127-32
55. Recio-Pinto E, Ishii DN. Insulin and insulinlike growth factor receptors regulating neurite formation in cultured human neuroblastoma cells. J Neurosci Res 1988; 19: 312
56. Clemmons DR, Van Wyk JJ. Evidence for a functional role of endogenously produced somatomedin-like peptides in the regulation of DNA synthesis in cultured human fibroblasts and porcine smooth muscle cells. J Clin Invest 1985; 75: 1914-8
57. Bar RS, Harrison LC, Baxter RC, Boes M, Drake BL, Booth BA, Cox A. Regulation of IGF binding proteins by vascular endothelial cells. Biochem Biophys Res Commun 1987; 148: 734-9
58. Strain AJ, Hill DJ, Swenne I, Milner RDG. The regulation of DNA synthesis in human fetal hepatocytes by placental lactogen, growth hormone and insulin-like growth factor I/somatomedin-C. J Cell Physiol 1987; 132: 33-40
59. Jacob RJ, Barrett E, Plewe G, Fagin KD, Sherwin RJ. Acute effects of insulin-like growth factor-I on glucose and amino acid metabolism in the awake fasted rat. J Clin Invest 1989; 83: 1717-23
60. Mauras N, Horber FF, Haymond MW. Low dose recominant human insulin like growth factor-I fails to affect protein anabolism but alters islet cell secretion in humans. J Clin Endocrinol Metab 1992; 1192-7

61. Clemmons DR, Smith-Banks A, Celniker AC, Underwood LE. Reversal of diet-induced catabolism by infusion of recombinant insulin-like growth factor-I (IGF-I) in humans. J Clin Endocrinol Metab 1992; 75: 234-8

62. Hirschberg R, Kopple JD. Evidence that insulin-like growth factor I increases renal plasma flow and glomerular filtration rate in fasted rats. J Clin Invest 1989; 83: 326-30

63. Schoenle E, Zapf J, Hauri C, Steiner T, Froesch ER. Comparison of in vivo effects of insulin-like growth factors I and II and of growth hormone in hypophysectomized rats. Acta Endocrinol 1985; 108: 167-74

64. Guler H-P, Zapf J, Scheiwiller E, Froesch ER. Recombinant human insulin-like growth factor I stimulates growth and has distinct effects on organ size in hypophysectomized rats. Proc Natl Acad Sci USA 1988; 85: 4889-93

65. Zapf J, Hauri C, Waldvogel M, Froesch ER. Acute metabolic effects and half-lives of intravenously administered insulinlike growth factors I and II in normal and hypophysectomized rats. J Clin Invest 1986; 77: 1768-75

66. Matthews LS, Hammer RE, Beheringer R, D'Ercole AJ, Bell GI, Brinster RL, Palmiter RD. Growth enhancement of transgenic mice expressing human insulin-like growth factor-I. Endocrinology 1988; 123: 2827-33

67. Liu JP, Baker J, Perkins AS, Robertson EJ, Efstratiadis A. Mice carrying small mutations of the genes encoding insulin like growth factor I (IGF-I) and type 1 IGF receptor (IGF/r). Cell 1993; 75: 73-82

68. Baker J, Liu JP, Robertson EJ, Efstratiadis A. Role of insulin like growth factors in embryonic and postnatal growth. Cell 1993; 75: 83-94

69. Guler H-P, Zapf J, Froesch ER. Short-term metabolic effects of recombinant human insulin-like growth factor-I in healthy adults. N Engl J Med 1987; 317: 137-40

70. Jacob RJ, Sherwin RS, Greenawalt K, Shulman GI. Simultaneous insulin-like growth factor I and insulin resistance in obese Zucker rats. Diabetes 1993; 41: 691-7

71. Kupfer SR, Underwood LE, Baxter RC, Clemmons DR. Enhancement of the anabolic effects of growth hormone and insulin like growth factor-I by the use of both agents simultaneously. J Clin Invest 1993; 91: 391-7

72. Moses AC, Young SCJ, Morrow LA, O'Brien M, Clemmons DR. Recombinant human insulin-like growth factor I increases insulin sensitivity and improves glycemic control in Type II diabetes. Diabetes 1996; 45: 95-100

5 THE INSULIN-LIKE GROWTH FACTOR (IGF) BINDING PROTEINS (IGFBPS)

Jeff MP Holly
Janet K Fernihough

University of Bristol Department of Medicine
Division of Surgery, Bristol Royal Infirmary, BS2 8HW, UK

ABSTRACT

Twenty years after the initial discovery of IGF (Salmon and Daughaday 1957) it was first realised that in the circulation, a huge proportion of IGF was bound to a "carrier protein" increasing its apparent molecular weight from 7.5kDa to between 50 and 150kDa (Zapf et al 1975). It wasn't until 1989 that this carrier protein was analysed in detail and shown to comprise the predominant serum binding protein, now termed IGFBP-3 and an ~85kDa, acid labile subunit (ALS) (Baxter and Martin 1989). To date there are now six specific IGF binding proteins named sequentially from IGFBP-1 to -6 according to their discovery in time and temporal publication of their sequences (See Shimasaki and Ling 1991 for review after IGFBPs 1-6 had been identified, Kim et al 1997 for more recent discussion on biological roles of all IGFBPs). In addition to this, there are other proteins with structural and possible functional homology to these proteins, which will be discussed later. Typically, the affinity of the intact binding proteins for IGF is higher than that of IGF for its receptor, resulting in a high concentration of systemic or locally produced IGF remaining effectively inactive. Therefore, the principle function of the binding proteins would appear to be to simply restrict the availability of IGF to its receptor, although this is achieved through a variety of mechanisms. In addition, tissue specificity can be grossly achieved through differences in expression of the various IGF binding proteins. Other means include not only changes in the absolute amounts of binding protein, but also in their post-translational modifications. The precise means of regulation at any one time can vary depending on physiological or pathological conditions. The following section will outline the individual binding proteins and our understanding of the role that they play in the modulation of IGF effects, as well as novel IGF independent actions. This review will concentrate on IGFBPs 1-6 as these are most fully characterised and have the highest affinity for IGF. Their role in various disease states will also be discussed.

STRUCTURE OF IGFBPS

IGFBPs 1-6 all share structural characteristics, and are highly conserved across species. Amino acid sequencing for the binding proteins in man shows several common features. Each binding protein can be roughly divided into three domains: highly homologous N and C domains, and a less well conserved central domain. A total of eighteen cysteine residues are conserved in four of the binding proteins, IGFBP-4 has an additional two cysteines, and IGFBP-6 possesses only sixteen (Drop et al 1992). Absolute percentage homologies between murine and human sequences for the IGFBPs 1-6 are 73, 89, 81, 93, 97 and 73, (Schuller et al 1994) suggesting evolutionary conservation as well as structural homology between binding proteins. Owing to the extensive homology in the N and the C termini, both these regions may contribute to the common ability of all the binding proteins to bind IGF. This has been demonstrated for some of the binding proteins by site-directed mutagenesis and binding affinity studies using fragments of IGFBPs. These techniques have shown that the N terminus is able to bind IGF in IGFBP-3 and IGFBP-1, and that the C-terminus is important in IGFBP-2 and IGFBP-1 (Brinkman et al 1991 and reviewed by Drop et al 1992). Although these studies are by no means comprehensive, it supports the idea that the tight folding of IGFBPs (as evidenced by multiple cysteine residues forming di-sulphide bridges) enables both the N and the C termini to associate with IGF.

Although the cysteine residues in each binding protein are conserved, forming di-sulphide bridges to stabilise the tertiary structure, there are several structural features individual to each binding protein. Both IGFBP-1 and IGFBP-2 possess an RGD (Arginine-Glycine-Aspartate) sequence which, for IGFBP-1, confers binding to the fibronectin receptor, integrin a5b1 (Jones et al 1993a). The RGD in IGFBP-2 appears to be non-functional in terms of integrin binding presumably owing to its conformational environment.

Serine phosphorylation IGFBP-1, IGFBP-3 and IGFBP-5 (Jones and Clemmons 1995), although the functional relevance of this has only been studied in any depth for IGFBP-1. Work by Jones et al (PNAS 1991) has shown that IGFBP-1 secreted by hepatocytes and decidua cells is predominantly in a phosphorylated form. However, IGFBP-1 purified from human fetal serum or amniotic fluid contains more of the non-phosphorylated form. The apparent 4-6 fold higher affinity for IGF by the phosphorylated form suggests that this may result in an IGF inhibitory action. Phosphorylation of IGFBP-3 appears to have no effect on the affinity for IGF (reviewed by Coverley and Baxter 1997) although it appears to increase the formation of the complex between IGF-IGFBP-3 and the ALS. It is also interesting to note that IGF, through a receptor mediated pathway is able to increase the phosphorylation of IGFBP-3 (Coverley and Baxter 1995). Whether this alters the ability of IGFBP-3 to associate with cell surfaces, or to enhance or inhibit the actions of IGF has not been shown.

In addition to phosphorylation, glycosylation sites have been identified in several of the binding proteins, although its physiological relevance is unclear (Jones and Clemmons 1995). IGFBP-3 contains three available sites for N-linked glycosylation, resulting in an observable doublet by SDS-PAGE separation, owing to differential glycosylation of these three sites *in vivo*. No difference in affinity for IGF has been shown for glycosylated or non-glycosylated IGFBP-3, (Sommer et al 1993) although the association of IGFBP-3 with the ALS in the circulation has a different kinetic profile for each form of the protein. Recent reports have demonstrated the ability of IGFBP-3 to associate with cellular proteins, and the presence of a "receptor" for IGFBP-3 (discussed later). It may be that the carbohydrate moiety of IGFBP-3 may be involved in these interactions, but no evidence has yet been documented for this.

Perhaps the best evidence for a function of the glycosylation of IGFBP-3 is the susceptibility to enzyme degradation (Firth and Baxter 1995).

N-linked glycosylation is also present on IGFBP-4 giving rise to 28kDa, and 24kDa forms. Again, no clear function for this modification has been discussed. Although IGFBP-6 also possesses a potential N-linked glycosylation site, this protein only demonstrates O-linked glycosylation (Bach et al 1993), a feature also of IGFBP-5. Enzymatic deglycosylation of IGFBP-6 has been shown not to alter the affinity of IGFBP-6 for IGF, and, therefore, evidence for a physiological role for O-linked glycosylation is lacking. IGFBP-2 appears to have neither glycosylation of phosphorylation sites.

IGFBP-1

This binding protein has a molecular weight of approximately 30kDa as determined by non-reducing SDS-PAGE, and shows approximately equal preference for IGF-I and IGF-II (Jones and Clemmons 1995). IGFBP-1 is acutely regulated by nutritional status, insulin and glucocorticoids, with insulin being by far the most dominant regulator. In animal models, this regulation has been shown to be mediated via the insulin receptor, and caused by varying the level of hepatic IGFBP-1 gene transcription (Ooi et al 1992). Many independent studies have shown that plasma IGFBP-1 levels are inversely related to levels of insulin (Lee et al 1993a, Conover et al 1990). No independent regulation by glucose or growth hormone has been demonstrated suggesting that the normal dynamic regulation *in vivo* is dependent virtually entirely on circulating levels of insulin (Holly et al 1988). This has resulted in a model being proposed whereby the low level of insulin during fasting maintains a high IGFBP-1 level to ensure that the free IGF in serum not bound to IGFBP-3 is prevented from exerting a hypoglycaemic effect. Although the fasting levels of IGFBP-1 remain inversely correlated to insulin levels in insulin-dependent diabetes mellitus (IDDM) and obesity, studies have shown irregular IGFBP-1 levels in these two conditions. In IDDM, the usual relationships between IGFBP-1 and age and IGF-I are lost, resulting in high post-pubertal IGFBP-1 levels (Wang et al 1989). Conversely in obesity, low circulating levels of IGFBP-1 were observed, possibly leading to a blunting of the response of IGFBP-1 to hypoglycaemia (Conover et al 1992a). This may enable the free IGF in circulation (that is not reduced) to have insulin-like effects, aggravating the pathogenesis of obesity.
The ability of glucocorticoid to stimulate IGFBP-1 levels has been demonstrated in the rat, both in cell culture and whole animal experiments, although the increases seen are by no means as dramatic as that achieved by fasting/low insulin. A comprehensive review of the regulation and function of IGFBP-1 is available by Lee et al (1993b).

In cell cultures, the action of IGFBP-1 has been shown to be both inhibitory (Ritvos et al 1988) and stimulatory (Elgin et al 1987) to the actions of IGF. Potentiation of IGF action appears to require cell surface binding (Busby et al 1988), although independent actions of IGFBP-1 have also been identified once bound to the integrin receptor (Jones et al 1993a). This may be due, in part, to the differential phosphorylation of IGFBP-1 which affects the affinity for cell surface binding sites. IGFBP-1 is novel amongst the IGFBPs in that its expression is limited to a very few tissues in the body. In addition to the liver, IGFBP-1 is only expressed in the endometrium, decidua and ovary. IGFBP-1 predominates in amniotic fluid and foetal plasma (Nonoshita et al 1994) reflecting its role in a variety of reproductive functions (Giudice et al 1994, Adashi 1994).

IGFBP-2

Although this protein shows some similarities with IGFBP-1 its actions have not been as well characterised. It is able to show both inhibitory and slightly stimulatory effects on IGF-II and IGF-I actions respectively, and demonstrates a preference for IGF-II binding (Jones and Clemmons 1995, Kelley et al 1996). IGFBP-2 predominates in cerebrospinal fluid (Lamson et al 1989), and is expressed in a variety of tissues (Giudice et al 1994). Although not as acutely as IGFBP-1, metabolic factors also affect IGFBP-2 expression (Thissen et al 1994).

IGFBP-3 and the acid labile subunit

The amino acid sequence of IGFBP-3 demonstrates a molecular weight of 28.7kDa (Wood et al 1988). The observed doublet on SDS gels at around 40-43kDa is owing to the presence in vivo of differentially N-glycosylated forms, as mentioned previously (Martin and Baxter 1986).

IGFBP-3, complexed with the acid labile subunit, constitutes the major carrier of serum IGF, with no preference for IGF-I or IGF-II. Acid labile subunit synthesis has been identified in the rat liver and kidney, but not in brain, heart, lung, muscle, spleen, testis or ovary (Dai and Baxter 1994). Outside the circulation, this protein has also been identified by immunoblotting in synovial fluid (Schneiderman et al 1995b). This is unlikely to be derived from the vascularised bone, as the acid labile subunit is absent in culture medium from bone cells (Kanzaki et al 1995). The 150kDa complex in the circulation has been shown to have a half-life of between 12-15 hours compared to only 10-12 minutes for IGF alone (Guler et al 1989). Owing to its size, it has been suggested that this complex is unable to passively diffuse across the epithelial cell layer (Binoux and Hossenlopp 1988). The high levels of essentially inert IGF in this form, therefore, function as a pool of bioavailable IGF in the absence of a tissue store.

The ternary complex in serum was further characterised by Baxter and Martin in 1989 when it was shown that the ALS can only bind IGFBP-3 when IGFBP-3 is already bound to IGF-I or IGF-II. Factors affecting the formation of this ternary complex include low pH, high salt concentration and heparin which all lead to the dissociation of this complex (Clemmons et al 1983, Baxter 1990). The former report indicates that this dissociation additionally involves the action of proteases. Since this time, the affinity of IGF and the acid-labile subunit for IGFBP-3 have been further characterised, and the binding between ALS and IGFBP-3 has been shown to be relatively weak with an association constant for ALS binding to IGF-I-IGFBP-3 of 2.5×10^{8} $lmol^{-1}$ (Holman and Baxter 1996). The low affinity of the acid labile subunit for IGFBP-3 and its relatively low abundance outside the circulation supports the original hypothesis that it functions to enable transport of IGF to the tissues from the circulation, by dissociation of IGFBP-3 from the ALS at the endothelial surface (Binoux and Hossenlopp 1988). The precise mechanisms by which IGFBP-3 achieves its varied effects are still not clear.

Growth hormone regulation of IGFBP-3 and the ALS

IGFBP-3 has classically been considered to be dependent on growth hormone (Moses et al 1976). Many papers in the early eighties described IGFBP-3 as the growth hormone dependent IGFBP (e.g. Baxter and Martin 1986), and indeed, a radioimmunoassay for IGFBP-3 was developed as a determinant of growth hormone deficiency (Blum et al 1990). Clinical studies using these radioimmunoassays have demonstrated that disorders of growth hormone metabolism e.g. acromegaly and growth hormone deficiency are associated with alterations in the circulating levels of IGFBP-3 (Baxter and Martin 1986, Ranke et al 1989). Additionally, the circulating levels of ALS and tissue mRNA for the ALS are also thought to be regulated by growth hormone (Dai and Baxter 1994, Ooi et al 1997).

Although the classical site for IGFBP-3 regulation is thought to be the liver, work in rats by Lemmey et al (1997) shows that the skin may be as important, in the regulation of circulating levels of IGFBP-3, since it demonstrates higher mRNA levels for IGFBP-3 than the liver. Skin, muscle and liver all responded to growth hormone treatment with a rapid increase in IGFBP-3 mRNA that was sustained for 24 hours.

It is also worthy of note that Chin et al (1994) have demonstrated that although the spatial localisation of IGFBP-3 production was in the Kupffer cells (and not the hepatocytes) of the liver, these cells do not express mRNA for the growth hormone receptor. It is hard, therefore, to construct a model whereby growth hormone has a direct effect on the cells producing IGFBP-3. However, an alternative model has been provided for the regulation of IGFBP-3 by insulin in liver cell cultures. Scharf et al (1996) have demonstrated that insulin cannot directly regulate the production of IGFBP-3 in isolated Kupffer cells, but that once co-cultured with hepatocytes separated by a porous membrane IGFBP-3 is responsive to insulin regulation. A similar model for the growth hormone regulation of IGFBP-3 via a presently unidentified diffusable factor may well operate.

It is becoming apparent that the paradigm of liver cells as the main target for growth hormone production and regulation of IGFBP-3 is questionable, and that the true picture may be more complex, and involve other tissues.

Inhibitory actions of IGFBP-3

The inhibitory actions of IGFBP-3 have been characterised in many cell systems and they mirror those of the other binding proteins, in that a molar excess is generally required (reviewed by Jones and Clemmons 1995). This extracellular pool of IGFBP-3 prevents IGF-I from binding to its receptor, owing to the higher affinity of IGF-I for IGFBP-3.

Cell binding of IGFBP-3

In 1993 Oh et al demonstrated that IGFBP-3 was able to inhibit the growth of Hs578T human breast cancer cells (Oh et al 1993a). Owing to the observed dose response of IGFBP-3 action (60% inhibition at 20nM), and the requirement for divalent cations, this was thought to occur via a unique IGFBP-3 binding site or receptor on the cell membrane. The inhibition of growth specifically occurred without the involvement of IGF binding to its receptor which was shown to have no effect on this cell line. However, binding of IGF to IGFBP-3 was able to block the inhibition of growth seen with IGFBP-3 alone. A mechanism was proposed whereby IGF-I binding to cell associated IGFBP-3 disrupted the binding of IGFBP-3 to its receptor, abolishing the inhibition.

This complemented work by Martin et al (1992) where addition of IGF-I was shown to increase the level of IGFBP-3 in the culture medium of fibroblasts by releasing cell associated IGFBP-3. Oh et al (1993b) subsequently demonstrated the presence of specific binding sites for IGFBP-3 in Hs578T cells by affinity cross-linking of IGFBP-3 to cell membranes, and immunoprecipitation of cell lysates with an anti-IGFBP-3 antibody. The same dose dependent binding, and competition for binding by non-radiolabelled IGFBP-3 as shown in their previous paper, was observed. These specific sites were identified as having molecular weights of 20, 26 and 50kDa.

Using Western ligand blotting analysis, Hodgkinson et al (1995) have demonstrated that rat muscle and a range of tissues in sheep expressed proteins that were able to bind IGFBP-3 with molecular weights of 60, 52 and 40kDa. These binding proteins for IGFBP-3 were tissue and IGFBP-3 specific. Heparin was able to prevent binding of IGFBP-3 to these proteins, by interacting with IGFBP-3 directly, rather than the cell proteins. These were not described as receptors, merely binding sites, owing to their relative ease of solubility in aqueous conditions.

Cell surface association of IGFBP-3 to proteoglycans has been described by Martin et al in 1992 and Smith et al in 1994. This could be blocked by heparin, IGF-I and sodium chlorate, an inhibitor of cell surface proteoglycan sulphation. Recently, Fowlkes et al in 1996 have identified glycosaminoglycan binding domains in IGFBP-3 that may be responsible for these associations. Whether IGFBP-3 binds to specific cell membrane receptors, binding sites or cell associated proteoglycans, or a combination of these, this is clearly another means by which the IGF system is regulated locally. The physiological relevance of this has not yet been fully elucidated, but it may be involved in the potentiation of IGF action.

Potentiation of IGF activity

In 1989, Blum et al described the potentiation of IGF-I stimulated DNA synthesis in human fibroblasts with the simultaneous addition of approximately equimolar IGFBP-3 and IGF-I. IGFBP-3 alone had no effect and did not bind to the cells under study. The potentiation of IGF-I was thought to proceed by IGFBP-3 providing a reservoir of local IGF-I that could be supplied to the cells at a low level, over an extended period of time. This was confirmed by the addition of small amounts of IGF-I hourly to the culture medium.

The mechanism by which IGFBP-3 can increase the stimulation by IGF-I has been studied in bovine dermal fibroblasts using both glycosylated and non-glycosylated IGFBP-3 (Conover et al 1992b and 1996). The first study described the ability of IGFBP-3 to bind to the cell surface, and after several hours, to enhance the amino acid uptake into cells stimulated by 2nM IGF-I. This effect required a pre-incubation of cells with IGFBP-3 of at least 24 hours, after which time IGFBP-3 was processed to lower molecular weight forms. A 72 hour pre-incubation demonstrated complete modification of all cell associated IGFBP-3 with an accompanied 10 fold reduction in the affinity of cell associated IGFBP-3 for IGF-I. There was no difference in the use of either glycosylated or non-glycosylated IGFBP-3, indicating that the attachment to cells did not involve the carbohydrate moiety of IGFBP-3. The effect was specific for IGF alone, and no effect was seen using EGF, bFGF, or PDGF. The proteolytic modification of IGFBP-3 was proposed to be the mechanism by which the modified IGFBP-3 then potentiates the effect of IGF-I.

Interestingly, the potentiation of IGF-I action did not require the binding of IGF-I to the binding protein, but to its receptor. This was indicated using IGF analogues that are unable to bind to IGFBPs but retain their affinity and function regarding IGF receptor binding. In this model, IGFBP-3 pre-incubation was able to enhance the effect of these analogues equally as well as native IGF-I. The potentiating effect was shown to be directly proportional to the affinity of the IGF analogues for the receptor, and was unrelated to their affinity for IGFBP-3. The action of IGFBP-3, however, did not alter the number or affinity of IGF receptors on these cells.

These intriguing results where further explored when factors responsible for processing IGFBP-3 at the cell surface were investigated (Conover et al 1996). The effects of inhibiting endocytosis, proteolysis, pH gradients and heparin binding were studied. The weak bases chloroquine and ammonium chloride were the only molecules initially studied that showed any effect on inhibiting IGFBP-3 potentiation of IGF-I action. This suggested that the local pH was important in eliciting this action of IGFBP-3. This was confirmed by the ability of bafilomycin, an inhibitor of H+ extrusion from cells, to completely block the enhancing effect of IGFBP-3. The effects of these factors on the processing of IGFBP-3 at the cell demonstrated that agents that were able to block the action of IGFBP-3 prevented the proteolysis of IGFBP-3 to its lower molecular weight forms. The lack of any action of protease inhibitors, was suggested to be related to difficulties in presenting enough inhibitor to the cell surface without causing toxicity.

A model was finally proposed that involved the initial association of IGFBP-3 to the cell surface, possibly through specific receptors. This IGFBP-3 would then be proteolytically modified in an acidic pericellular environment to a lower molecular weight form, which has a higher affinity for the proposed IGFBP-3 receptor. Signalling through this receptor then interacts with the intracellular signalling pathway of the IGF receptor to enhance the response of IGF. This does not require internalisation of IGFBP-3 itself.

The recent acknowledgement of the wide variety of effects that IGFBP-3 can have on the action of IGF-I has opened up exciting areas of research. It is clear that an understanding of the actions of IGF demand some perception of the binding protein environment, the properties of the cell in question and its ability to process IGFBPs.

IGFBP-4

This was discovered in human bone cell conditioned medium and named as inhibitory IGF binding protein (Mohan et al 1989) as it inhibited the action of IGF in all conditions tested (reviewed in Jones and Clemmons 1995). This is exemplified by its role in inhibiting cartilage synthesis in vitro (Schiltz et al 1993). It shows no preference for IGF-I over IGF-II, although binding of either one enhances proteolysis of IGFBP-4 by a specific protease (Fowlkes and Freemark 1992, Kanzaki et al 1994). The fragmented form of the protein shows negligible binding to either IGF.

IGFBP-5

This IGFBP is the predominant IGFBP in bone, and has a particular ability to bind to bone via hydroxyapatite (Mohan et al 1994). Unbound, it shows inhibitory effects in molar excess, similar to the other binding proteins (Kiefer et al 1992). IGFBP-5 has the highest number of glycosaminoglycan binding motifs within its sequence, compared to the other binding proteins (Hodgkinson et al 1994) facilitating its high affinity associations with the extracellular matrix. Once bound to the extracellular matrix, IGFBP-5 is able to potentiate the effects of IGF, probably a consequence of the lowered affinity for IGF-I once IGFBP-5 has bound to the matrix (Jones et al 1993b). The binding of IGFBP-5 to proteoglycans and IGF also has important consequences on the extent of IGFBP-5 proteolysis which will be discussed in the next section. In rat chondrocytes, the level of IGFBP-5 is regulated by IGF, involving both IGF receptor dependent and independent mechanisms (Matsumoto et al 1996b).

IGFBP-5 and -4 appear to have opposing effects on regulating the availability of IGF-I to its receptor on bone cells (Mohan et al 1995). The opposing actions of IGFBP-4 and -5, therefore, act in tandem to regulate this pool of IGF. In bone there is a high level of stored IGFs, and during osteoblast and osteoclast turnover of bone, this IGF may be released to regulate the subsequent actions of these two cell types (Hayden et al 1995). There is also evidence to show that IGFBP-4 and -5 expression in the skeleton is site dependent (Malpe et al 1997), and that there is an age related decrease in the cortical bone content of IGFBP-5 (Nicolas et al 1995). This suggests that not only do IGFBP-4 and -5 regulate normal bone cells, but that they may also play a role in the pathogenic states of arthritis and osteoporosis.

IGFBP-6

Contrasting the other binding proteins, IGFBP-6 has a markedly lower affinity for IGF-I than IGF-II (Roghani et al 1991). Although it contains a heparin binding motif (Hodgkinson et al 1994) IGFBP-6 does not specifically bind endothelial cell monolayers, or possess a high affinity for heparin (Bar et al 1994). Little has been published on the physiological role of IGFBP-6. In serum, its concentration, like IGFBP-5 is very low, and it is unlikely to play a role in the circulation (Jones and Clemmons 1995). Individual reports have demonstrated potential roles for this protein in cerebrospinal fluid, and in female reproductive function (Baxter and Saunders 1992, Rohan et al 1993). However, the full tissue distribution and function of this binding protein is still unknown.

OTHER BINDING PROTEIN-LIKE PROTEINS

Our understanding of the role of IGF binding proteins has perhaps been limited by the very nomenclature we are using. Until recently, it was believed that the ability to bind IGF defined a unique class of proteins comprising six members. However, a wider family of proteins, that includes the original six IGF binding proteins has now been described. The evolutionary relationships between all these proteins are displayed in figure 1. Although high affinity IGF binding has only been ascribed to IGFBPs 1-6, and limited affinity to Mac 25 (Oh 1996) and connective tissue growth factor (CTGF) (Kim et al 1997), these ten proteins show both sequential and structural homology. Unsurprisingly perhaps, owing to its extensive homology with IGF, insulin is also able to bind to IGFBPs though with a much lower affinity. These findings indicate that there are six proteins with high affinity and direct regulatory roles concerning IGF (IGFBPs 1-6) and other similar proteins that may have IGF independent or, possibly, insulin related roles (IGFBPs 7-10).

IGFBP-7 has recently been described in a report (Oh et al 1996), which reassigns the previously named mac25 to this family of proteins. Mac25 is a protein originally isolated from senescent human mammary epithelial cells (Swisshelm et al 1994). The homology of mac25 to the IGFBPs is 40-45% overall, with 20-25% identity (Oh et al 1996). Eleven of the twelve conserved cysteine residues at the amino-terminus of the IGFBPs are also present in mac25, along with the common IGFBP motif GCGCCXXC. Its affinity for IGF-I is low, similar to that for cleaved IGFBP-3, and as such, it physiological role is still unclear.

Figure 1. Phylogenetic tree demonstrating the evolutionary relationships between IGFBPs

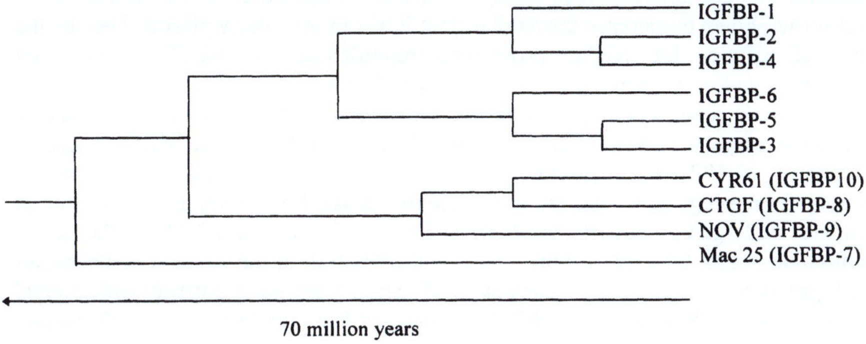

IGFBP PROTEASES

It is impossible to discuss the role of IGF binding proteins without considering the role of the binding protein proteases. Although intact binding proteins typically have a higher affinity for IGF than the IGF receptor, the fragmented forms of the binding proteins do not. This provides the primary mechanism whereby regulated proteolysis of the binding proteins releases IGF to act at the receptor level. Although a few of the papers reviewed above have highlighted potential functions for proteolysed IGFBP-3 in particular, the protease(s) responsible has(have) not yet been definitively identified.

Initial reports identifying a protease for insulin-like growth factor binding proteins came from studies of serum from pregnant women. Despite a report suggesting an increase in the level of IGFBP-3 in the third trimester of pregnancy (Baxter and Martin 1986), this protein could not be identified on Western ligand blots (Hossenlopp et al 1990). Two concurrent papers then described a protease for various IGFBPs in "pregnancy serum" that lowered their affinity for IGF to such an extent that they could no longer be visualised on Western ligand blots (Hossenlopp et al 1990, Giudice et al 1990). The factor responsible for reducing the affinity of IGFBP-3 for IGF was shown to be temperature, pH and time dependent, suggesting the discovery of a novel enzyme(s). This was further characterised as a cation dependent metallo/serine protease, based on the ability to inhibit it with EDTA, aprotinin and partially with PMSF. Addition of normal serum to pregnancy serum resulted in a decrease of intact IGFBP-3 normally detected in serum by Western ligand blotting. This implied that the enzyme in pregnancy serum was not inhibited by the circulating inhibitors that prevent proteolysis of IGFBP-3 in normal physiology. IGFBP-1 remained unaffected by proteases in pregnancy, indeed the levels of this protein increased throughout gestation (Giudice et al 1990).

These papers all used Western ligand blotting to demonstrate the presence or absence of IGFBP-3 in pregnancy serum. This technique relies on disruption of all protein-protein interactions in the sample to be loaded, by boiling of the sample in SDS buffer. In 1991 Davies et al demonstrated, by using associative gel chromatography, that in vivo proteolytically modified IGFBP-3 remains in the 150kDa complex, but with a lower affinity for IGF-I (Davies et al 1991a). This was confirmed by Lassarre and Binoux in 1994 when the altered affinity of IGFBP-3 for IGF-I was shown to increase the level of free IGF-I in pregnancy plasma. Despite the reduced affinity for IGF-I, proteolytic modification of IGFBP-3 does not substantially alter the affinity of IGFBP-3 for IGF-II.

Since this time, serum proteases for IGFBP-3 have been identified in patients following surgery (Cwyfan-Hughes et al 1992) severely ill patients post-surgery (Davies et al 1991b) and in cancer patients (Frost et al 1993). In contrast to the results from pregnancy studies, the decrease in IGFBP-3 seen in the serum of patients undergoing surgery resulted in a redistribution of IGF-I to the lower molecular weight binding proteins, with little formation of the 150kDa complex. As with pregnancy serum, the co-incubation of serum from these patients with normal serum caused a decrease in IGFBP-3 intensity on Western ligand blots. This again appeared to be owing to a cation dependent metallo/serine protease but in contrast to serum obtained during pregnancy, only IGFBP-3 was affected by a proteolysis in these catabolic patients.

Research into characterising the IGFBP-3 protease rapidly spread to looking outside the circulation. In the cell conditioned medium of various cancer and normal cell lines, protease activity for IGFBP-3 was readily detected, although its characteristics were not the same as the protease(s) derived from the circulation (Frost et al 1993). Addition of normal serum to tissue derived enzymes did not result in IGFBP-3 proteolysis in the serum, but conversely inhibited the enzyme(s) in cell conditioned medium. The cell derived enzyme(s) were able to degrade both IGFBP-1 and IGFBP-3, whereas IGFBP-1 is unaffected by proteases during gestation. In addition, the pattern of inhibition by EDTA was not the same as that for the circulation derived protease(s). Addition of known proteases to samples of radiolabelled IGFBP-3 did not generate the same pattern of fragments as seen in the cell conditioned medium (Frost et al 1993), suggesting that plasmin, kallikrein, cathepsin D, and t-PA are not responsible for the tissue proteolysis of IGFBP-3. Plasmin was able to generate the same pattern of fragments as those seen in pregnancy serum, but the action of this protease could not be inhibited by EDTA.

Other studies at the tissue level have identified IGFBP-3 protease activity in seminal plasma (Cohen et al 1994), interstitial fluid (Xu et al 1996) and synovial fluid (Fernihough et al 1996, Matsumoto et al 1996a, Tavera et al 1996). In seminal plasma, prostate specific antigen was shown to degrade IGFBP-3, with inhibition possible using zinc, aprotinin, and potently with PMSF. No inhibition was seen using EDTA or trypsin inhibitor. The enzyme in synovial fluid could be inhibited by EDTA (Fernihough et al 1996) and was described as a metallo/serine protease based on its inhibition profile by Matsumoto et al 1996a. Cathepsin D has also been proposed as an IGFBP-3 protease (Conover and De Leon 1994), although its pH optimum of less than 5.5 make this a highly specialised IGFBP-3 protease with possibly limited physiological relevance, since at this pH IGFs would dissociate from the IGFBPs without the need for proteolysis. It is feasible that the proteolytically cleaved forms of IGFBP-3 may have IGF independent effects different from the parent molecule, that are activated by specific proteolysis.

In human dermal fibroblasts, and rat pregnancy serum, matrix metalloproteases have been proposed as the enzymes responsible for cleavage of IGFBP-3 (Fowlkes et al 1994a and b). Several proteases were identified by this group, spanning molecular weights of 52-72kDa, that could be inhibited by EDTA and TIMP -1. Immunoblotting identified proMMP-1, -2 and -3 at the same molecular weights as those for IGFBP-3 proteases. Removal or inhibition of these enzymes blocked the proteolysis of IGFBP-3 in both rat pregnancy and dermal fibroblast conditioned medium.

Proteases for other binding proteins

Proteases for the other IGF binding proteins have not been studied in as much depth as those for IGFBP-3. An IGF dependent IGFBP-4 protease has been demonstrated in bone cells (Kanzaki et al 1994), chondrocytes (Matsumoto et al 1996b), and dermal fibroblasts (Fowlkes and Freemark 1992). Conversely, proteolysis of IGFBP-5 identified in the former two systems was reduced by the inclusion of IGF-II. The proteolysis of IGFBP-5 has also shown to be inhibited by glycosaminoglycans (Arai et al 1994). Interestingly, a protease for IGFBP-4 was observed in patients undergoing heart surgery (Cwyfan-Hughes et al 1992) 60 minutes after bone damage resulting from surgery. The IGFBP-4 protease in osteoblasts and fibroblasts can be inhibited by the addition of IGFBP-3 (Fowlkes et al 1995, Donnelly and Holly 1996), although the osteoblasts cells do not synthesise IGFBP-3 themselves. This inhibition is reversible by addition of IGF-I or IGF-II. The mechanism by which IGFBP-3 can inhibit IGFBP-4 proteolysis has been investigated further, and shown to be associated with the heparin binding domain in the C-terminus of IGFBP-3. IGFBP-5 and IGFBP-6 both have this heparin binding domain and are also able to inhibit IGFBP-4 proteolysis with the potency related to their respective affinities for heparin (Fowlkes et al 1997). The physiological consequences of these observations are not yet known.

The ability of plasmin, trypsin and kallikrein to degrade the IGFBP-IGF-I/IGF-II complexes purified from decidua or bone cell conditioned medium confirmed previous studies, mentioned above, that none of these enzymes is the one observed in pregnancy serum (Campbell et al 1992). IGFBP-2 proteolysis in serum has been identified as being induced in starvation of new-born animals or young children (McCusker et al 1991, Pucilowska et al 1993). There has been no protease identified for IGFBP-6 to date.

The putative identity of an IGFBP protease has been published by Zumbrunn and Trueb (1996) as a serine protease with domains homologous to an IGFBP, a bacterial protease and a protease inhibitor. As yet, this study is based on the sequence information alone, and the protein has yet to be expressed and demonstrate IGFBP protease activity.

The conclusions drawn from all these studies suggests that the protease(s) for IGFBP-3 in the serum and the cell conditioned medium share the characteristics of cation dependent endoproteases, but that the specificity of the protease(s) in the tissue and circulation, and in differing physiological conditions varies. In a healthy individual the protease is present at a very low level in the circulation, "masked" by the high levels of natural endogenous inhibitors. The marked proteolysis of circulating IGFBP-3 observed in predominantly catabolic conditions favours the release of IGF to the lower molecular weight binding proteins. This in turn facilitates the transport of IGF out of the circulation to the tissues. Here, local regulation by more general proteases enables IGF to act on its receptor. In pregnancy, the increase in free IGF is able to support the continual high rate of metabolic activity demanded of the mother. A more detailed description of the function and relevance of these proteases, and their role in normal physiology is discussed in several papers (Holly et al 1993, Lamson et al 1993, Cwyfan-Hughes et al 1995). To date, there is still no accurate description of the precise site of action of the protease in IGFBP-3, and no sources of pure IGFBP-3 protease for complete characterisation.

CONCLUSIONS

The IGF system as a whole is clearly an intricate and precisely regulated network of proteins about which we are still learning. Post-translational modifications are clearly apparent on several of the IGF binding proteins but, again, we are still limited in our understanding of which have physiological relevance. The classical picture of an inert "binding protein" is being questioned more as the field of research increases, and IGF independent effects continue to emerge.

REFERENCES

ADASHI EY. Regulation of intrafollicular IGFBPs, possible relevance to ovarian follicular selection. In The Insulin-like Growth Factors and their Regulatory Proteins 1994 Baxter RC, Gluckman PD, Rosenfeld RG (Eds) Elsevier Amsterdam 341-350

ARAI T, Arai A, Busby WH Jr, Clemmons DR. Glycosaminoglycans inhibit degradation of insulin-like growth factor-binding protein-5. Endocrinology 1994 **135** *6* 2358-2363

BACH LA, Thotakura NR, Rechler MM. Human insulin-like growth factor binding protein 6 is O-glycosylated. Growth Regulation 1993 **3** 59-62

BAR RS, Boes M, Booth BA, Dake BL, Moser DR, Erondu NE. Vascular endothelium, IGF's and IGF binding proteins In The Insulin-like Growth Factors and their Regulatory Proteins 1994 Baxter RC, Gluckman PD, Rosenfeld RG (Eds) Elsevier Amsterdam 237-244

BAXTER RC, Martin JL. Radioimmunoassay of growth hormone-dependent insulin like growth factor binding protein in human plasma. Journal of Clinical Investigation 1986 **78** *6* 1504-1512

BAXTER RC, Martin JL. Structure of the Mr 140 000 growth hormone-dependent insulin-like growth factor binding protein complex: Determination by reconstitution and affinity labelling. Proceedings of the National Academy of Sciences USA 1989 **86** 6898-6902

BAXTER RC. Glcosaminoglycans inhibit formation of the 140kDa insulin-like growth factor-binding protein complex. Biochemistry Journal 1990 **271** 773-777

BAXTER RC, Saunders H. Radioimmunoassay of insulin-like growth factor-binding protein-6 in human serum and other body fluids. Journal of Endocrinology 1992 **134** 133-139

BINOUX M, Hossenlopp P. Insulin-like growth factor (IGF) and IGF-binding proteins: Comparison of human serum and lymph. Journal of Clinical Endocrinology and Metabolism 1988 **67** *3* 509-514

BLUM WF, Jenne EW, Reppin F, Kietzman K, Ranke MB, Bierich JR. Insulin-like growth factor I (IGF-I)-binding protein complex is a better mitogen than free IGF-I. Endocrinology 1989 **125** 766-772

BLUM WF, Ranke MB, Kietzmann K, Gauggel E, Zeisel HJ, Bierich JR. A specific radioimmunoassay for the growth hormone (GH)-dependent somatomedin-binding protein: Its use for diagnosis of GH deficiency. Journal of Clinical Endocrinology and Metabolism 1990 **70** *5* 1292-1298

BRINKMAN A, Kortleve DJ, Zwarthoff EC, Drop SLS. Mutations in the C-terminal part of insulin-like growth factor (IGF)- binding protein-1 result in dimer

formation and loss of IGF binding capacity. Molecular Endocrinology 1991 **5** *7* 987-994), 1991.
BUSBY WH, Klapper DG, Clemmons DR. Purification of a 31000 dalton insulin-like growth factor binding protein from human amniotic fluid. Journal of Biological Chemistry 1988 **263** 14203-14210
CAMPBELL PG, Novak JF, Yanosick TB, McMaster JH. Involvement of the plasmin system in dissociation of the insulin-like growth factor-binding protein complex. Endocrinology 1992 **130** *3* 1401-1412
CHIN E, Zhou J, Dai J, Baxter RC, Bondy CA. Cellular localization and regulation of gene expression for components of the insulin-like growth factor ternary binding protein complex. Endocrinology 1994 **134** *6* 2498-2504
CLEMMONS DR, Underwood LE, Chatelain PG, van Wyk JJ. Liberation of immunoreactive somatomedin-C from its binding proteins by proteolytic enzymes and heparin. Journal of Clinical Endocrinology and Metabolism 1983 **56** *2* 384-389
COHEN P, Peehl DM, Craves HCB, Rosenfeld RG. Biological effects of prostate specific antigen as an insulin-like growth factor binding protein-3 protease. Journal of Endocrinology 1994 **142** *3* 407-415
CONOVER CA, Butler PC, Wang M, Rizza RA, Lee PDK. Lack of growth hormone effect on insulin-associated suppression of insulin-like growth factor binding protein 1 in humans. Diabetes 1990 **39** 1251-1256
CONOVER CA, Lee PDK, Kanaley JA, Clarkson JT, Jensen MD. Insulin regulation of insulin-like growth factor binding protein-1 in obese and nonobese humans. Journal of Clinical Endocrinology and Metabolism 1992a **74** 1355-1360
CONOVER CA. Potentiation of insulin-like growth factor (IGF) action by IGF-binding protein-3: Studies of underlying mechanism. Endocrinology 1992b **130** *6* 3191-3199
CONOVER CA, De Leon DD. Acid activated insulin-like growth factor-binding protein-3 proteolysis in normal and transformed cells. Role of cathepsin D. Journal of Biological Chemistry 1994 **269** *10* 7076-7080
CONOVER CA, Clarkson JT, Bale LK. Factors regulating insulin-like growth factor-binding protein-3 binding, processing, and potentiation of insulin-like growth factor action. Endocrinology 1996 **137** 2286-2292
COVERLEY JA, Baxter RC. Regulation of insulin-like growth factor (IGF) binding protein-3 phosphorylation by IGF-I. Endocrinology 1995 **136** 12 5778
COVERLEY JA, Baxter RC. Phosphorylation of insulin-like growth factor binding proteins. Molecular and Cellular Endocrinology 1997 **128** 1-5
CWYFAN-HUGHES SC, Cotterill AM, Molloy AR, Cassell TB, Braude N, Hinds CJ, Wass JAH, Holly JMP. The induction of specific proteases for insulin-like growth factor-binding proteins following major heart surgery. Journal of Endocrinology 1992 **135** *1* 135-145
CWYFAN-HUGHES SC, Xu S, Fernihough JK, Hampton A, Mason HD, Franks S, van der Stappen J, Donnelly M, Holly JMP. Tissue IGFBP-3 proteolysis: contrasting pathophysiology to that in the circulation. Progress in Growth Factor Research 1995 **6** *2-4* 293-299
DAI J, Baxter RC. Regulation in vivo of the acid labile subunit of the rat serum insulin-like growth factor binding protein complex. Endocrinology 1994 **135** *6* 2335-2341
DAVIES SC, Holly JMP, Coulson VJ, Cotterill AM, Abdulla AF, Whittaker PG, Chard T, Wass JAH. The presence of cation-dependent proteases for insulin-like

growth factor binding proteases does not alter the size distribution of insulin-like growth factors in pregnancy. Clinical Endocrinology 1991a **34** *6* 501-506

DAVIES SC, Wass JAH, Ross RJM, Cotterill AM, Buchanan CR, Coulson VJ, Holly JMP. The induction of a specific protease for insulin-like growth factor binding protein-3 in the circulation during severe illness. Journal of Endocrinology 1991b **130** *3* 469-473

DONNELLY MJ, Holly JMP. Does IGFBP-3 regulate IGFBP-4 proteolysis in cultured human dermal fibroblasts. Journal of Endocrinology 1996 **149** R1-R7

DROP SLS, Schuller AGP, Lindenbergh-Kortleve DJ, Groffen C, Brinkman A, Zwarthoff EC. Structural aspects of the IGFBP family. Growth Regulation 1992 **2** 69-79

ELGIN RG, Busby WH, Clemmons DR. An insulin-like growth factor (IGF) binding protein enhances the biologic response to IGF-I. Proceedings of the National Academy of Sciences USA 1987 **84** 3254-3258

FERNIHOUGH JK, Billingham MEJ, Cwyfan-Hughes SC, Holly JMP. Local disruption of the IGF system in the arthritic joint. Arthritis and Rheumatism 1996 **39** *9* 1556-1565

FIRTH SM, Baxter RC. The role of glycosylation in the action of IGFBP-3. Progress in Growth Factor Research 1995 **6** *2-4* 223-229

FOWLKES J, Freemark M. Evidence for a novel insulin-like growth factor (IGF)-dependent protease regulating IGF-binding protein-4 in dermal fibroblasts. Endocrinology 1992 **131** 2071-2076

FOWLKES JL, Enghild JF, Suzuki K, Nagase H. Matrix metalloproteinases degrade insulin-like growth factor binding protein-3 in dermal fibroblast cultures. Journal of Biological Chemistry 1994a **269** 25742-25746

FOWLKES JL, Suzuki K, Nagase H, Thrailkill KM. Proteolysis of insulin-like growth factor-binding protein-3 during rat pregnancy: A role for matrix metalloproteinases. Endocrinology 1994b **135** *6* 2810-2813

FOWLKES JL, Serra DM, Rosenberg CK, Thrailkill KM. Insulin-like growth factor (IGF)-binding protein-3 (IGFBP-3) functions as an IGF-reversible inhibitor of IGFBP-4 proteolysis. Journal of Biological Chemistry 1995 **270** *46* 27481-27488

FOWLKES JL, Serra DM. Characterisation of glycosaminoglycan-binding domains present in insulin-like growth factor-binding protein-3. Journal of Biological Chemistry 1996 **271** *25* 14676-14679

FOWLKES JL, Thrailkill KM, George-Nascimento C, Rosenberg CK, Serra DM. Heparin-binding, highly basic regions within the thyroglobulin type-1 repeat of insulin-like growth factor (IGF)-binding proteins (IGFBPs) -3, -5, and -6 inhibit IGFBP-4 degradation. Endocrinology 1997 **138** *6* 2280-2285

FROST VJ, Macaulay VM, Wass JAH, Holly JMP. Proteolytic modification of insulin-like growth factor binding proteins: Comparison of conditioned media from human cell lines, circulating proteases and characterised enzymes. Journal of Endocrinology 1993 **138** *3* 545-554

GIUDICE LC, Farrell EM, Pham H, Lamson G, Rosenfeld RG. Insulin-like growth factor binding proteins in maternal serum throughout gestatin and in the puerperium: effects of a pregnancy-associated serum protease activity. Journal of Clinical Endocrinology and Metabolism 1990 **71** 806-816

GIUDICE LC, Irwin JC, Dsupin BA, de las Fuentes L, Jin IH, Vu TH, Hoffman AR. Insulin-like growth factors (IGFs), IGF-binding proteins (IGFBPs), and IGFBP protease in human uterine endometrium, their potential relevance to endometrial cyclic function and maternal-embryonic interactions In The Insulin-like Growth

Factors and their Regulatory Proteins 1994 Baxter RC, Gluckman PD, Rosenfeld RG (Eds) Elsevier Amsterdam 351-361

GULER HP, Zapf J, Schmid C, Froesch ER. Insulin-like growth factors I and II in healthy man. Estimations of half-lives and production rates. Acta Endocrinologica 1989 **121** *6* 753-758

HAYDEN JM, Mohan S, Baylink DJ. The insulin-like growth factor system and the coupling of formation to resorption. Bone 1995 **17** *2 suppl* 93S-98S

HODGKINSON S, Fowke P, Al Somai N, McQuoid M. Proteins in tissue extracts which bind insulin-like growth factor binding protein-3 (IGFBP-3) Rapid Communication, Journal of Endocrinology 1995 **145** R1-R6

HODGKINSON SC, Napier JR, Spencer GSG, Bass JJ. Glycosaminoglycan binding characteristics of the insulin-like growth factor-binding proteins. Journal of Molecular Endocrinology 1994 **13** *1* 105-112

HOLLY JMP, Biddlecombe RA, Dunger DB, Edge JA, Amiel SA, Howell R, Chard T, Rees LH, Wass JH. Circadian variation of GH-independent IGF-binding protein in diabetes mellitus and its relationship to insulin. A new role for insulin? Clinical Endocrinology 1988 **29** 383-387

HOLLY JMP, Claffey DCP, Cwyfan-Hughes SC, Frost VJ, Yateman ME. Proteases acting on IGFBPs: their occurrence and physiological significance. Growth Regulation 1993 **3** *1* 88-91

HOLMAN SR, Baxter RC. Insulin-like growth factor binding protein-3: factors affecting binary and ternary complex formation. Growth Regulation 1996 **6** 42-47

HOSSENLOPP P, Segovia B, Lassarre C, Roghani M, Bredon M, Binoux M. Evidence of enzymatic degradation of insulin-like growth factor binding proteins in the 150K complex during pregnancy. Journal of Clinical Endocrinology and Metabolism 1990 **71** 797-805

JONES JI, D'Ercole AJ, Camacho-Hubner C, Clemmons DR. Phosphorylation of insulin-like growth factor binding protein in cell culture and *in vivo*, effects on affinity for IGF-I Proceedings of the National Academy of Sciences USA 1991 **88** 7481-7485

JONES JI, Gockerman A, Busby WH, Wright G, Clemmons DR. Insulin-like growth factor binding protein 1 stimulates cell migration and binds to the a5b1 integrin by means of its Arg-Gly-Asp sequence. Proceedings of the National Academy of Sciences 1993a **90** 10553-10557

JONES JI, Gockerman A, Busby WHJ, Camacho-Hubner C, Clemmons DR. Extracellular matrix contains insulin-like growth factor binding protein-5: potentiation of the effects of IGF-I. Journal of Cell Biology 1993b **121** 679-687

JONES JI, Clemmons DR. Insulin-like growth factors and their binding proteins: Biological actions. Endocrine reviews 1995 **16** *1* 3-34 (18)

KANZAKI S, Hilliker S, Baylink DJ, Mohan S. Evidence that human bone cells in culture produce insulin-like growth factor-binding protein-4 and -5 proteases. Endocrinology 1994 **134** *1* 383-392

KANZAKI S, Baxter RC, Knutsen R, Balink DJ, Mohan S. Evidence that human bone cells in culture secrete insulin-like growth factor (IGF)-II and IGF binding protein-3 but not acid-labile subunit both under basal and regulated conditions. Journal of Bone and Mineral Research 1995 **10** *6* 854-858

KELLEY KM, Oh Y, Gargosky SE, Gucev Z, Matsumoto T, Hwa V, Ng L, Simpson DM, Rosenfeld RG. Insulin-like growth factor-binding proteins (IGFBPs) and their regulatory dynamics. International Journal of Biochemistry and Cell Biology 1996 **28** *6* 619-637

KIEFER MC, Schmid C, Waldvogel M, Schlapfer I, Futo E, Masiarz FR, Green K, Barr PJ, Zapf J. Characterisation of recombinant human insulin-like growth factor binding proteins 4, 5 and 6 produced in yeast. Journal of Biological Chemistry 1992 **267** 12692-12699

KIM H-S, Rosenfeld RG, Oh Y. Biological roles of insulin-like growth factor binding proteins (IGFBPs) Experimental and Molecular Medicine 1997 **29** *2* 85-96

KIM H-S, Nagaila SR, Oh Y, Wilson EM, Roberts CT Jr, Rosenfeld RG. Synthesis and characterisation of connective tissue growth factor (CTGF) as insulin-like growth factor binding protein-8. Proceedings of the 79th Annual Meeting of the Endocrine Society 1997 **352** Abs P2 269

LAMSON G, Giudice LC, Cohen P, Liu F, Gargosky S, Muller HL, Oh Y, Wilson KF, Hintz RL, Rosenfeld RG. Proteolysis of IGFBP-3 may be a common regulatory mechanism of IGF action in vivo. Growth Regulation 1993 **3** 91-95

LAMSON G, Pham H, Oh Y, Ocrant I, Schwander J, Rosenfeld RG. Expression of the BRL-3A insulin-like growth factor binding protein (rBP-30) in the rat central nervous system. Endocrinology 1989 **123** 1100-1102

LASSARRE C, Binoux M. Insulin-like growth factor binding protein-3 is functionally altered in pregnancy plasma. Endocrinology 1994 **134** *3* 1254-1262

LEE PDK, Jensen MD, Divertie GD, Heiling VJ, Katz HH, Conover CA. Insulin-like growth factor binding protein-1 response to insulin during suppression of endogenous insulin secretion. Metabolism 1993a **42** 409-414

LEE PDK, Conover CA, Powell DR. Regulation and function of insulin-like growth factor binding protein-1. Proceedings of the Society for Experimental Biology and Medicine 1993b **204** 4-29

LEMMEY AB, Glassford J, Flick-Smith HC, Holly JMP, Pell JM. Differential regulation of tissue insulin-like growth factor binding protein (IGFBP)-3, IGF-I and IGF type 1 receptor mRNA levels, and serum IGF-I and IGFBP concentrations by growth hormone and IGF-I. Journal of Endocrinology 1997 **154** 319-328

MALPE R, Baylink DJ, Linkhart TA, Wergedal JE, Mohan S. Insulin-like growth factor (IGF)-I, -II, IGF binding proteins (IGFBP)- 3, -4, and -5 levels in the conditioned media of normal human bone cells are skeletal site-dependent. Journal of Bone and Mineral Research 1997 **12** *3* 423-430

MARTIN JL, Baxter RC. Insulin-like growth factor binding protein from human plasma: purification and characterisation. Journal of Biological Chemistry 1986 **261** 8754-8760

MARTIN JL, Ballesteros M, Baxter RC. Insulin-like growth factor-I (IGF-I) and transforming growth factor-b release IGF-binding protein-3 from human fibroblasts by different mechanisms. Endocrinology 1992 **131** *4* 1703-1710

MATSUMOTO T, Gargosky SE, Iwasaki K, Rosenfeld RG. Identification and characterisation of insulin-like growth factors (IGFs), IGF binding proteins (IGFBPs) and IGFBP proteases in human synovial fluid. Journal of Clinical Endocrinology and Metabolism 1996a **81** 150-155 (45)

MATSUMOTO T, Gargosky SE, Oh Y, Rosenfeld RG. Transcriptional and post-translational regulation of insulin-like growth factor-binding protein-5 in rat articular chondrocytes. Journal of Endocrinology 1996b **148** 355-369

McCUSKER RH, Cohick WS, Busby WH, Clemmons DR. Evaluation of the development and nutritional changes in porcine insulin-like growth factor binding protein-1 and 2 serum level by immunoassay. Endocrinology 1991 **129** 2631-2638

MOHAN S, Bautista CM, Wergedal CM, Baylink DJ. Isolation of an inhibitory insulin-like growth factor (IGF) binding protein from bone cell-conditioned medium: A potential local regulator of IGF action. Proceedings of the National Academy of Sciences USA 1989 **86** 8338-8342

MOHAN S, Strong DD, Linkhart TA, Baylink DJ. Regulation and actions of insulin-like growth factor binding protein (IGFBP)-4 and IGFBP-5 in bone, physiological and clinical implications In The Insulin-like Growth Factors and their Regulatory Proteins 1994 Baxter RC, Gluckman PD, Rosenfeld RG (Eds) Elsevier Amsterdam 205-215

MOHAN S, Nakao Y, Honda Y, Landale E, Leser U, Dony C, Lang K, Baylink DJ. Studies on the mechanisms by which insulin-like growth factor (IGF) binding protein-4 (IGFBP-4) and IGFBP-5 modulate IGF actions in bone cells. Journal of Biological Chemistry 1995 **270** *35* 20424-20431

MOSES AC, Nessley SP, Cohen KL, Rechler MM Specific binding of a somatomedin-like polypeptide in rat serum depends on growth hormone. Nature 1976 **263** *5573* 137-140

NICOLAS V, Mohan S, Honda Y, Prewett A, Finkelman RD, Baylink DJ, Farley JR. An age-related decrease in the concentration of insulin-like growth factor binding protein-5 in human cortical bone. Calcified Tissue International1995 **57** *3* 206-212

NONOSHITA LD, Wathen NC, Dsupin BA, Chard T, Giudice LC. Insulin-like growth factors (IGFs), IGF-binding proteins (IGFBPs), and proteolyzed IGFBP-3 in embryonic cavities in early human pregnancy, their potential relevance to maternal-embryonic and fetal interactions. Journal of Clinical Endocrinology and Metabolism 1994 **79** 1249-1255

OH Y, Muller HL, Lamson G, Rosenfeld RG. Insulin-like growth factor (IGF)-independent action of IGF-binding protein-3 in Hs578T human breast cancer cells. Journal of Biological Chemistry 1993a **268** *20* 14964-14971

OH Y, Muller HL, Pham H, Rosenfeld RG. Demonstration of receptors for insulin-like growth factor binding protein-3 on Hs578T human breast cancer cells. Journal of Biological Chemistry 1993b **268** *35* 26045-26048

OH Y, Nagalla SR, Yamanaka Y, Kim H-S, Wilson E, Rosenfeld RG. Synthesis and characterisation of insulin-like growth factor-binding protein (IGFBP)-7. Journal of Biological Chemistry 1996 **271** *48* 30322-30325

OOI GT, Tseng LY-H, Tran MQ, Rechler MM. Insulin rapidly decreases insulin-like growth factor binding protein-1 gene transcription in streptozotocin-diabetic rats. Molecular Endocrinology 1992 **6** 2219-2228

OOI GT, Cohen FJ, Tseng LY-H, Rechler MM, Boisclair YR. Growth hormone stimulates transcription of the gene encoding the acid- labile subunit (ALS) of the circulating insulin-like growth factor-binding protein complex and ALS promoter activity in rat liver. Molecular Endocrinology 1997 **11** *7* 997-1007

PUCILOWSKA JB, Davenport ML, Kabir I, Clemmons DR, Thissen JP, Butler T, Underwood LE. The effect of dietary protein supplementation on insulin-like growth factors (IGF's) and IGF binding proteins in children with shigellosis. Journal of Clinical Endocrinology and Metabolism 1993 **77** 1516-1521

RECHLER MM. Editorial: Growth inhibition by insulin-like growth factor (IGF) binding protein-3 - What's IGF got to do with it? Endocrinology 1997 **138** *7* 2645-2647

RITVOS O, Ranta T, Jalkanen J, Suikkari A-M, Voutilainen R, Bohn H, Rutanen EM. Insulin-like growth factor binding protein from human decidua inhibits the binding and biological action of IGF-I in cultured choricarcinoma cells. Endocrinology 1988 **122** 2150-2157

ROGHANI M, Lassarrre C, Zapf J, Povoa G, Binoux M. Two insulin-like growth factor (IGF)-binding proteins are responsible for the selective affinity for IGF-II of cerebrospinal fluid binding proteins. Journal of Clinical Endocrinology and Metabolism 1991 **73** 658-666

ROHAN RM, Ricciarelli E, Kiefer MC, Resnick CE, Adashi EY. Rat ovarian insulin-like growth factor-binding protein-6, a hormonally regulated theca-interstitial-selective species with limited antigonadatropic activity. Endocrinology 1993 **132** 2507-2512

SALMON WD, Daughaday WH. A hormonally controlled serum factor which stimulates sulfate incorporation by cartilage in vitro. Journal of Laboratory and Clinical Medicine 1957 **49** 825-836

SCHARF J-G, Ramadori G, Braulke T, Hartmann H. Synthesis of insulinlike growth factor binding proteins and of the acid- labile subunit in primary cultures of rat hepatocytes, of Kupffer cells, and in cocultures: Regulation by insulin, insulin-like growth factor, and growth hormone. Hepatology 1996 **23** *4* 818-827

SCHILTZ PM, Mohan S, Baylink DJ. Insulin-like growth factor binding protein-4 inhibits both basal and IGF-mediated chick pelvic cartilage growth in vitro. Journal of Bone and Mineral Research 1993 **8** *4* 391-396

SCHNEIDERMAN R, Rosenberg N, Hiss J, Lee P, Liu F, Hintz RL, Maroudas A. Concentration and size distribution of insulin-like growth factor-I in human normal and osteoarthritic synovial fluid and cartilage. Archives of Biochemistry and Biophysics 1995 **324** 1 173-188

SCHULLER AGP, Groffen C, Van Neck JW, Zwarthoff EC, Drop SLS. CDNA cloning and mRNA expression of the six mouse insulin-like growth factor binding proteins. Molecular & Cellular Endocrinology 1994 **104** *1* 57-66

SHIMASAKI S, Ling N. Identification and molecular characterisation of insulin-like growth factor binding proteins (IGFBP-1, -2, -3, -4, -5 and -6). Progress in Growth Factor Research 1991 **3** 243-266

SMITH EP, Lu L, Chernausek SD, Klein DJ. Insulin-like growth factor-binding protein-3 (IGFBP-3) concentration in rat Sertoli cell-conditioned medium is regulated by a pathway involving association of IGFBP-3 with cell surface proteoglycans Endocrinology 1994 **135** *1* 359-364

SOMMER A, Spratt SK, Tatsuno GP, Tressel T, Lee R, Maack CA. Properties of glycosylated and non-glycosylated human recombinant IGF binding protein-3 (IGFBP-3). Growth Regulation 1993 **3** *1* 46-49

SWISSHELM K, Ryan K, Tsuchiya K, Sager R. Enhanced expression of an insulin growth factor-like binding protein (mac 25) in senescent human mammary epithelial cells and induced expression with retinoic acid. Proceedings of the National Academy of Sciences USA 1995 **92** 4472-4476

TAVERA C, Arribat T, Reboul P, Doré S, Brazeau P, Pelletier J-P, Martel-Pelletier J. IGF and IGF-binding protein system in the synovial fluid of osteoarthritic and rheumatoid arthritic patients. Osteoarthritis and Cartilage 1996 4 263-274

THISSEN J-P, Ketelslegers J-M, Underwood LE. Nutritional regulation of the insulin-like growth factors. Endocrine Reviews 1994 **15** 80-101

WANG M, Wilson DM, Lee PDK. Serum levels of the 25kDa insulin-like growth factor binding protein in children with insulin-dependent diabetes mellitus. In

Drop SLS, Hintz RL, Eds Insulin-like growth factor binding proteins 1989. Amsterdam: Excerpta Medica 39-45

WOOD WI, Cachianes G, Henzel WJ, Winslow GA, Spencer SA, Hellmiss R, Martin JL, Baxter RC. Cloning and expression of the growth hormone dependent insulin-like growth factor binding protein. Molecular Endocrinology 1988 **2** *12* 1176-1185

XU S, Cwyfan-Hughes SC, van der Stappen JWJ, Sansom J, Burton JL, Donnelly M, Holly JMP. Altered insulin-like growth factor-II (IGF-II) level and insulin-like growth factor binding protein-3 (IGFBP-3) protease activity in interstitial fluid taken from the skin lesions of Psoriasis. Journal of Investigative Dermatology 1996 **106** *1* 109-112

ZAPF J, Waldvogel M, Froesch ER. Binding of non-suppressible insulin-like activity to human serum. Evidence for a carrier protein. Archives of Biochemistry and Biophysics 1975 **168** 638-645

ZUMBRUNN J, Trueb B. Primary structure of a putative serine protease specific for IGF-binding proteins. FEBS Letters 1996 **398** 187-192

6 THE ETIOLOGY OF GROWTH HORMONE DEFICIENCY IN THE HUMAN ADULT

E.R. Christ
P.V. Carroll
P.H. Sönksen

Division of Medicine
St. Thomas' Hospital, London SE1 7EH, UK

INTRODUCTION

Although growth hormone (GH) deficiency in human adults is now recognised as a specific clinical syndrome (1-2) and the effects of GH replacement have been extensively studied (1-2), little systematic information exists regarding the etiology of GH-deficiency acquired in adulthood. This may be explained by the fact that until recently most physicians considered GH to have little function in adulthood and it is only in the last decade that this topic has received widespread attention.

The definition of GH-deficiency further complicates the categorisation of etiologies of GH-deficiency. Studies to date have focused on patients with absent or profoundly reduced GH secretion on provocative testing (1-2). Adults with partial GH-deficiency were not included in clinical trials and thus the clinical details of patients with partial GH-deficiency are not generally available. In addition the conditions responsible for GH-deficiency differ between childhood and adulthood.

The etiologies of severe GH-deficiency in human adults are detailed in this chapter. The occurrence of GH-deficiency in the context of both partial and total hypopituitarism is given particular attention. In addition, the effects of treatment (surgery and radiotherapy) of pituitary and peri-pituitary lesions on GH secretion are summarised. Finally, interesting, rare causes of hypopituitarism and GH-deficiency are discussed.

PATHOPHYSIOLOGY

GH is synthesised and secreted by somatotrophes located in the anterior pituitary. These processes are regulated by GH releasing hormone (GHRH) and somatostatin, both of which are generated in the hypothalamus (3). The hypothalamic factors in turn are modulated by the central nervous system, other endocrine factors (thyroxine and steroid hormones) and circulating metabolites (glucose and free fatty acids) (3). GH induces both systemic (hepatic) and autocrine/paracrine (most tissues) synthesis and secretion of insulin-like growth factor-I (IGF-I) (4). IGF-I mediates many of the actions of GH, and through negative feedback, inhibits GH secretion at both the pituitary and hypothalamic levels (5). GH also has direct effects on protein and fat metabolism via activation of a specific receptor located on the cell surface (6).

GH-deficiency may, therefore, result from pathological processes involving the hypothalamus, pituitary gland or both of these structures. Additionally, if GH-deficiency is considered as reduced biological activity of GH, conditions associated with resistance or insensitivity to the action of GH, could also be included in the context this chapter. Laron type dwarfism, which results from a defect of the signalling pathway of the GH receptor, is the most recognised of these conditions (7).

Understanding the nature and site of defects in the GH pathway is important, since different conditions, require different treatments. Patients with growth failure resulting from abnormalities of the GH receptor or its intracellular signalling pathway have been successfully treated with rhIGF-I (8). Those with pituitary abnormalities resulting in GH-deficiency can be treated with GH supplementation. Current research indicates that GHRH analogues may soon be available, and it is possible that these agents will have a therapeutic role in the treatment of patients with lesions affecting the hypothalamus but otherwise normal pituitary function.

THE ETIOLOGIES OF ADULT GH-DEFICIENCY

The availability of recombinant GH in the late 1980s, providing large amounts of GH, without the risks associated with human derived products prompted intense investigation of the effects of GH replacement in GH-deficient adults, and simultaneously, criteria for the diagnosis of GH-deficiency were established (9-10). Since the etiology of GH-deficiency in adults is related to the etiology of hypopituitarism in general, we have focused on the following issues:

1. Are the conditions responsible for pituitary based GH-deficiency identical to those that result in hypopituitarism?
2. Is there an association between deficiencies of other pituitary hormones and GH-deficiency?
3. Does GH-deficiency of childhood-onset become GH-deficiency in adult life?
4. Is it rather the treatment (i.e. surgery or radiotherapy) of the pathologic condition in the pituitary-hypothalamic area or the initial pathology itself (i.e. tumours) which result in GH-deficiency?
5. Can genetic conditions result in GH-deficiency?

1. The etiologies of hypopituitarism and GH-deficiency in adulthood

The causes of hypopituitarism from epidemiological studies (11-13) reported in the last seven years are summarised in Table 1. These studies compared mortality, in particular cerebro- and cardiovascular mortality, in patients with hypopituitarism and healthy control subjects. Patients with Cushing's syndrome and acromegaly are known to be at increased risk for cerebrovascular disease, and were excluded from these studies. In addition Bulow et al. (11) only included patients with pituitary tumours requiring neurosurgery. The occurrence of craniopharyngioma in patients with hypopituitarism was similar in all three studies. In the study of Rosen et al. (11), 53 (15%) patients were recorded to have a unknown cause of hypopituitarism.

Twenty of these (6%) had only hypogonadism. Since serum prolactin measurement has been available only since the 1970s, it is possible that the hypogonadism in these patients may have resulted from prolactin-secreting tumour. This effect was likely to be less relevant in the third study (12), given that this study began in 1967.

In summary, based on these three epidemiological studies it appears that adult hypopituitarism results from pituitary tumours (and their treatment) in 65 - 75 % of affected individuals, by cranyopharyngioma in 12-13%, and that idiopathic etiologies may account for about 10% of hypopituitarism with a small number of patients suffering from other rare conditions.

Table 1

The etiologies of hypopituitarism in human adults

Author (year)	Rosen (11) (1990)	Bates (12) (1996)	Bulow (13) (1997)
Number of patients	333	172	344
Study period	1956 - 1987	1967 - 1994	1946 - 1988
Country	City of Gotheborg and West of Sweden	UK, Stoke-on-Trent	South of Sweden
Pituitary tumours	223 (67%)	131 (76 %)	302 (88%)
Craniopharyngioma	34 (10%)	22 (13 %)	42 (12%)
Idiopathic	53 (15%)	14 (8 %)	
Others	23 (7%)	5 (3%)	

The etiologies of GH-deficiency in two study populations (patients at St. Thomas' Hospital from 1987 - 1997 and subjects from a European multicentre trial (14) are summarised in Table 2. The conditions responsible for GH-deficiency in these two trials shows a very similar distribution. The percentages in each category in Tables 1 & 2 (comparison between etiologies of hypopituitarism and etiologies of adult GH-deficiency) are virtually identical with the exception of patients labelled with "other" diagnosis. In Rosen's study (11), pituitary trauma and empty sella syndrome were not mentioned, however these diagnoses accounted for 20 patients in St. Thomas's follow-up and for 9 patients in the multicentre trial (14). Similarly, inflammatory diseases, including lymphocytic hypophysitis, histiocytosis and post-tubercular inflammation were not enumerated as a cause for hypopituitarism whilst these conditions accounted for 4 and 2 patients in St. Thomas's Hospital and the multicentre trial, respectively. The majority of these patients did not present with isolated GH-deficiency. It is therefore unlikely that these conditions result in isolated GH-deficiency. It is rather a feature of the progress in detecting these rare conditions. In summary, those pathologies that result in partial or complete hypopituitarism are also those that cause GH-deficiency.

Table 2
The etiologies of GH-deficiency in human adults

Study population	**Multicentre trial (14) (1997)**	**St. Thomas' Hospital (1987-1997)**
Number of patients	98	238
Country (ies)	European countries	UK
Functional adenoma	30 (31 %)	72 (30 %)
Non-functional adenoma	25 (26 %)	69 (29 %)
Craniopharyngioma	19 (19%)	32 (13%)
Idiopathic	7 (7%)	15 (6%)
Others	18 (17%)	50 (21%)

2. The relationship between deficiencies of other pituitary hormones and GH-deficiency

To address this question we retrospectively analysed the peak plasma GH obtained during an insulin tolerance test (ITT, with a nadir plasma glucose ≤ 2.2 mmol/l) of 88 patients investigated in St. Thomas' Hospital (1995). Those patients with no requirement for other pituitary hormone replacement therapy had the highest GH response whereas those with multiple (3-4) hormone deficiencies had the lowest GH peak. Patients with 1-2 pituitary hormone deficiencies showed an intermediate GH response. Further examination demonstrates that those with 3-4 hormone deficiencies are severely GH-deficient (GH peak < 3mU/l), whereas patients requiring 1-2 hormones are partially deficient (GH peak 3-10 mU/l). There is of course a small number of patients with severe isolated GH-deficiency.

These results are consistent with a recent report by Toogood et al. (15). This study investigated GH status in association with gonadotrophin, ACTH and TSH secretion in 96 male patients with pituitary disease. The patients were divided in four groups according to the number of anterior pituitary deficiencies. Patients with isolated GH-deficiency had the highest mean GH peak to an insulin provocation test (10.0 mU/l) whereas patients with multiple (≥3) pituitary hormone deficiencies had the lowest mean GH response (1.8 mU/l). Patients requiring replacement of 1-2 other pituitary hormones had a peak GH between these extremes (2.0 - 4.0 mU/l). These data suggest that GH-deficiency is usually an accompaniment of multiple pituitary hormone (≥3) deficiency, and that GH secretion is diminished in patients with partial hypopituitarism.

3. Does GH-deficiency presenting in childhood persist as GH-deficiency throughout adult life?

A recent study suggests that GH-deficiency results in different effects when occurring in childhood than the syndrome recognised in GH-deficiency acquired in adulthood (14). These data suggests that the GH-deficiency of childhood results in both developmental and metabolic alterations, whereas the severe GH-deficiency of adulthood is a clear clinical/biochemical entity (14). A small number of studies have assessed the GH status in young adults, who had previously received GH treatment in childhood (16-17). These have demonstrated a normalisation of peak GH response in 20- 25% of the patients using the same test procedures as initially performed. Nicolson et al. (18) reassessed 88 adults who had received GH therapy following the diagnosis of childhood-onset GH-deficiency. A peak GH response of less than 9 mU/L (defined as severe GH-deficiency) was found in 65% at initial assessment, and in 60 % at reassessment. Patients who had suffered organic disease in childhood, and/or radiotherapy had an increased incidence of severe GH-deficiency (from 48.8% to 55.8%). In contrast, the incidence of severe GH-deficiency in the idiopathic group (mostly isolated GH-deficiency) decreased from 78.1% at initial assessment to 53.1% at reassessment.

These studies suggest that all patients with GH-deficiency of childhood-onset should undergo reassessment in young adult life as recently proposed in the guidelines of the Growth Hormone Research Society. Nicolson et al. (18) suggest that between 40% and 60% of these patients may merit consideration for GH replacement therapy in adult life. Thus the majority of patients with childhood-onset GH-deficiency continue to experience GH-deficiency throughout adulthood. Patients with a diagnosis of isolated GH-deficiency in childhood are the least likely to have demonstrable GH-deficiency in adult life, however whether this is due to an error in initial diagnosis or recovery of the somatotrope is not known. Evidence suggests that 'priming' pre-pubertal children with sex steroids prior to performance of provocative testing results in less error in labelling healthy subjects as GH-deficient (19).

4. Is the GH-deficiency associated with pituitary or peri-pituitary tumours caused directly by the tumour or is the result of treatment (surgery or radiotherapy)?

The therapeutic modalities of the adult patients with GH-deficiency detailed in Table 2 are summarised in Table 3. Greater than 80 % of the patients had surgery or radiotherapy and about one third of the patients underwent both of these therapies. These numbers are similar to findings from another study who reported that 82% of patients with adulthood-onset GH-deficiency had undergone surgical intervention for the treatment of their pituitary/peri-pituitary tumour (14).

Table 3
The treatments of pituitary/peri-pituitary tumours in patients with adult GH-deficiency at St. Thomas' Hospital (1987-1997)

	n	%
Number of patients	238	100
Surgery	171	72
Radiotherapy and surgery	72	30
Only Radiotherapy	24	10
No surgery or radiotherapy	43	18

a) Surgical Intervention

Theoretically GH-deficiency may result directly from a pituitary/peri-pituitary lesion or from its treatment. It is questionable whether it is of clinical importance to accurately determine the onset of GH-deficiency in patients in whom operative intervention or radiotherapy are planned. However, a recent study has shown that preoperative administration of GH to patients with Cushing's disease can counteract the catabolic action of excess endogenous glucocorticoids and may therefore have a therapeutic role, especially in patients with Cushing's disease who are GH-deficient prior to surgery (20).

Littley (21) examined anterior pituitary deficiencies in 165 adult patients with tumours of the pituitary gland or anatomically adjacent structures. Forty-five percent of these patients were diagnosed as GH-deficient, prior to surgical intervention. However, of those in whom surgery was performed, 85% of the patients were demonstrated to GH-deficiency, following the operation. In contrast no significant increase in deficiencies of other pituitary deficiencies were demonstrated in this study population after surgery. Thus, it appears that surgical interventions play an important role in the development of GH-deficiency.

b) Radiotherapy

Tan was the first to report the occurrence of GH-deficiency 11 years after successful radiotherapy of a nasopharyngeal carcinoma (22). This and other reports suggest that pituitary hormone deficiencies develop insidiously after many years. With respect to GH-deficiency two questions remain to be addressed:

1. The site of radiation induced damage (hypothalamus or pituitary gland?)
2. What is the relationship between GH-deficiency and deficiencies of other pituitary hormones following radiotherapy?

Site of radiation damage

Post-mortem examination of the pituitary gland and the hypothalamus in patients who have previously undergone radiotherapy indicates that the hypothalamus is more likely to show histological damage than the pituitary, which appears to be more resistant to the effects of radiation (23-24). These findings are somewhat in contrast to the findings from clinical studies, in which administration of the releasing hormones, thyrotropin-releasing hormone (TRH), growth hormone-releasing hormone (GH-RH) and corticotropin-releasing hormone (CRH), indicated that radiotherapy results in significant functional pituitary damage (24-25). Thus the hypothalamus undergoes greater histological change following radiotherapy, but the hypopituitarism arise from both impaired hypothalamic and pituitary function.

Occurrence of GH-deficiency and deficiencies of other pituitary hormones following radiotherapy of the hypothalamic-pituitary area

Radiotherapy is used to treat a wide variety of pathologies, including pituitary/peri-pituitary tumours, nasopharyngeal carcinoma, primary brain tumours, and haematological malignancies (whole-body irradiation) (26). This therapy has also been used for rarer indications such as proliferative diabetic retinopathy prior to laser treatment, metastatic cancer and pituitary tumours strictly confined to the pituitary fossa, usually in the form of localised radioactive implants (90 Yttrium and 198Gold) (26)

Pituitary dysfunction after external radiotherapy for pituitary tumours was prospectively investigated in 251 patients for up to 12 years (21). Follow-up data of this study showed that 5 years after treatment all patients were GH-deficient. Life-table analysis indicate that GH secreting cell mass is the most sensitive to the deleterious effects of radiation and TSH the most resistant. Deficiencies of the gonadotrophins and ACTH were found between these extremes.

In a prospective study in 31 patients with nasopharyngeal carcinoma Lam et al (27) demonstrated that one year following radiotherapy, there was a significant decrease in the GH response to hypoglycaemia. Two years after radiotherapy there was a progressive increase in endocrine dysfunction in the 21 patients who remained alive.

Littley (28) assessed endocrine function in seven adult patients who have survived 5-13 years after radiotherapy for brain tumours (glioma and astrocytoma). Only 1 patient retained a normal GH response to provocative testing 8 years following treatment. Five of the patients were gonadotrophin deficient, three had ACTH-deficiency, and 2 were TSH-deficient.

Littley (28) retrospectively examined pituitary function in 10 patients between 2 and 5 years after whole body irradiation in preparation for bone marrow transplantation. No pituitary hormone deficiencies were detected in these patients. This may be due to the fact that a much lower dose was administered to these patients (12 Gy) compared to the dose which is used to treat brain tumours and pituitary tumours (20-45 Gy). Prospective analysis of the effects of radiotherapy on pituitary function with different dose regimes suggest that patients do not usually develop hypopituitarism below a threshold dose of 20 Gy (28).

After interstitial radiotherapy of the pituitary gland the relative sensitivity of the different hormonal axes are consistent with the findings after external radiotherapy, i.e. GH-deficiency occurs first, followed by gonadotrophins, ACTH and TSH (29).

5. Genetic conditions and GH-deficiency

The gene encoding hGH spans 1.6 kilobases and resides on the 5'-end of a 50 kB gene cluster on chromosome 17 (30). The incidence of GH-deficiency arising from genetic defects is estimated to be between 1/4000 and 1/10000 births (31). These defects result in a spectrum ranging from total/severe GH-deficiency, to partial lack of GH synthesis and secretion. Four Mendelian types have been described based on the mode of inheritance and degree of GH-deficiency: Isolated GH-deficiency (IGHD) type IA is an autosomal recessive condition with absent endogenous GH-secretion (32). IGHD type IB is an autosomal recessive condition with diminished circulating GH; IGHD type II is autosomal dominant with reduced GH, and type III, is X-linked and results in partial GH-deficiency. The abnormalities arise from heterogeneous gene deletions, nonsense, frame and splice mutations or occasionally dominant negative mutations in one allele of the GH gene (32).

In addition to these defects, abnormal expression of the Pit-1/GHF-1 binding factor results in abnormal GH gene activity. Pit-1 is a pituitary-specific POU-domain binding-factor, which binds and trans-activates the promoters of both GH and prolactin genes (33). Pit-1 is also a regulator of the PRL and TSH-beta genes by TRH and cAMP. Point mutations in the encoding of the Pit-1 gene have been described in several families (33) and result in the phenotypic expression of GH and PRL deficiency with reduced circulating levels of thyroid hormones.

RARE ETIOLOGIES OF HYPOPITUITARISM AND GH-DEFICIENCY

The availability of cerebral magnetic resonance imaging has considerably enlarged the spectrum of possible etiologies for hypopituitarism. A decrease in the number of "idiopathic" forms of hypopituitarism can therefore be expected in the future. Many case reports of unusual cases of pan-hypopituitarism have been recently published of which a selection is summarised below.

1. Vascular disorders

The classical cause of post-partum hypopituitarism is Sheehan's syndrome, in which obstetric complications result in global hypotension, with subsequent infarction of the pituitary gland. This diagnosis accounted for 9 of the 238 patients with hypopituitarism in St. Thomas' Hospital. In a retrospective study of 40 patients with Sheehan's syndrome (34) only 6.6% had a normal peak GH response to hypoglycaemia. The other pituitary axes were more resistant to functional damage, consistent with the findings after radiotherapy. Pituitary apoplexy is a rare condition characterised by a sudden-onset of headache, symptoms and signs of hormone deficiencies and a typical image on MRI (35). Pituitary apoplexy was the underlying diagnosis in two patients at St.Thomas's Hospital, both of which were GH-deficient. Haemorrhage within the pituitary gland can result in damage and pan-hypopituitarism. Such haemorrhage may occur as a result of pituitary tumour, a rupture of an intrasellar aneurysm (36) or a Hanta-Virus infection (37).

2. Malignancies

Clinical pan-hypopituitarism caused by cancer with parasellar metastasis and/or hypothalamic invasion are rare conditions. The primary cancer manifestation is usually lung-carcinoma or breast-carcinoma (38-39).

3. Hypopituitarism after tuberculous meningitis in childhood.

Forty-nine patients who had tuberculous menigitis in childhood were reassessed in adulthood (40). Hypopituitarism was present in 10 patients (20%). Seven patients were GH-deficient, four of whom had also gonadotrophin insufficiency. Only three suffered from ACTH-deficiency.

4. Lymphocytic hypophysitis

This diagnosis should be considered in postpartum women with general malaise and persistent amenorrhoa, as well as in women who develop visual impairment in the last trimester of pregnancy without a past medical history of pituitary disease (41). Four of 238 patients at St. Thomas's Hospital had lymphocytic hypophysitis in the last ten years. All of them were GH-deficient.

Other rare conditions leading to GH-deficiency include trauma, extrapituitary tumours (dysgerminoma, glioma, ependynoma, pinealoma, menigioma), histiocytosis X (typical in children), developmental disorders (empty sella, ectopic gland and rare congenital syndromes, e.g. The "CHARGE" syndrome)

A summary of pathological conditions responsible for GH-deficiency is listed in Table 4. The respective frequency of each diagnosis is indicated, based on the patients attending St. Thomas's Hospital.

SUMMARY

Epidemiological data suggest that any cause and any degree of hypopituitarism can be associated with GH-deficiency in adults, the most common being pituitary/peri-pituitary tumours and their treatments. Studies have demonstrated that the greater the number of pituitary hormone deficiencies, the greater the likelihood of GH-deficiency. Reassessment of patients who received GH treatment during childhood suggests that between 40% and 60% remain GH-deficient in adulthood and may therefore benefit from GH replacement therapy. A persistent GH-deficient state is more likely in those children who had organic pituitary disease and/or underwent radiotherapy in childhood. Surgical treatment of pituitary tumour significantly increases the risk of GH-deficiency.

Table 4. The etiologies and frequency of occurrence of adult GH-deficiency in 238 patients attending St.Thomas' Hospital (STH) between 1987 and 1997.

Clinical diagnosis	STH (n)	STH(%)
1. Tumours		
- Nonfunctional adenoma	69	29
- Cushing's syndrome	52	21
- Prolactinoma	18	7
- other functional adenoma	2	1
- Craniopharyngioma	32	13
- Dysgerminoma, pinealoma, ependynoma, glioma	10	4
- Menigioma	7	3.5
- Malignancies +/- radiotherapy (ENT carcinoma, brain tumours, metastatic disease)	2	1
2. Inflammation and Granuloma		
- Lymphocytic hypophysitis	2	1
- Histiocytosis X	2	1
- Tuberculous menigitis	-	-
3. Vascular disorders of the pituitary		
- Sheehan's syndrome	9	4
- Pituitary apoplexy	2	1
- Haemorrhage (vessel abnormality or Hanta-Virus)	-	-
4. Developmental disorders		
- empty sella and ectopic gland	4	2
5. Trauma	5	2
6. Others	7	3.5
7. Idiopathic	15	6

Despite the small number of reports on the occurrence of GH-deficiency in conditions other than pituitary disease, it appears that GH-deficiency is a common consequence of radiotherapy. The manifestation develops insidiously over many years following radiotherapy. The GH-secreting cell mass appears to be the most sensitive, followed by gonadotrophins, ACTH and TSH. Although prospective long-term studies are still lacking it appears that pituitary dysfunction after radiotherapy is dose-dependent: Patients who receive a total dose of 20 Gy or more to the hypothalamic-pituitary axis are at risk of hypopituitarism. This may be of particular importance for the long-term survivors of malignancies in whom endocrine disorder can be easily diagnosed and treated.

REFERENCES

1 De Boer H, Blok GJ, van der Veen VA. Clinical aspects of growth hormone deficiency in adults. *Endocrine Reviews* 1995; 16: 63-86.

2 Christ ER, Carroll PV, Russell-Jones DL, Sönksen PH. The consequences of growth hormone deficiency in adulthood, and the effects of growth hormone replacement. Schw Med. Wochenschrift 1997; 127:1440-1449

3 Frohman LA, Jansson JO. GHRH. *Endocrine Reviews* 1986; 7:223-253

4 Jones JI, Clemmons DR. Insulin-like growth factors and their binding proteins: biological actions. *Endocrine Reviews*. 1995; 16:3-34.

5 Kiess W, Kessler U, Schmitt S, Funk B. Growth hormone and insulin-like growth factor-I. In: *Growth hormone and IGF-I in human and experimental diabetes*. Published by John Wiley and Sons 1993.

6 Kelly PA, Djiane J, Postel-Vinay M-C, Edery M. The PRL/GH receptor family. *Endocrine Reviews* 1991; 12:235-251.

7 Godowski PJ, Leung DW, Meacham LR et al. Characterization of the human growth hormone receptor gene and demonstration of a partial gene deletion in two patients with Laron-type dwarfism. *Proceedings of the National Academy of Sciences of the U.S.A.* 1989; 86:8083-8087.

8 Laron Z, Anin S, Klipper-Aurbach Y, Klinger B. Effects of insulin-like growth factor on linear growth, head circumference and body fat in patients with Laron-type dwarfism. *Lancet.* 1992; 339:1258-1261.

9 Salomon F, Cuneo RD, Hesp R, Sönksen PH. The effects of treatment with recombinant human growth hormone on body composition and metabolism in adults with growth hormone deficiency. *New England Journal of Medicine* 1989; 321:1797-1803.

10 Jorgensen JOL, Pedersen SA, Thuesen L et al. Beneficial effects of growth hormone treatment in GH-deficient adults. *Lancet* 1989; i: 1221-25.

11 Rosen T, Bengtsson B-Å. Premature mortality due to cardiovascular disease in hypopituitarism. *Lancet* 1990; 336: 285-288

12 Bates AS, Van't Hoff W, Jones PJ, Clayton RN. The effect of hyopituitarism on life expectancy. *Journal of Clinical Endocrinology & Metabolism* 1996; 81: 1169-72

13 Bulow B, Hagmar, Mikozy Z, Nordstroem CH, Erfurth EM. Increased cerebrovascular mortality in patients with hypopituitarism. *Clinical Endocrinology* 1997; 46:75-81.

14 Attanasio AF, Lamberts SWJ, Matranga AMC, Birkett MA, Bates PC, Valk NK, Hilsted J, Bengtsson BA, Strasburger CJ. Adult growth hormone-deficient patients demonstrate heterogenity between childhood onset and adult onset before and during human GH treatment. *Journal of Clinical Endocrinology & Metabolism* 1997; 82: 82-88.

15 Toogood AA, Beardwell CG, Shalet SM. The severity of GH-deficiency in adults with pituitary disease is related to the degree of hypopituitarism. *Clinical Endocrinology*. 1994; 41: 511-516.

16 Cacciari E, Tassoni P, Parisi G et al. Pittfalls in diagnosing impaired GH secretion: Retesting after replacement therapy of 63 patients defined as GH deficient. *Journal of Clinical Endocrinology and Metabolism*. 1992; 74:1284-1289.

17 Clayton PE, Price DA, Shalet SM. Growth hormone state after completion of treatment with growth hormone. *Archives od Diseases in Childhood*. 1987; 62:222-226.

18 Nicolson A, Toogood AA, Rahim A, Shalet SM. The prevalence of severe GH-deficiency in adults who received GH replacement in childhood. *Clinical Endocrinology*. 1996; 44:311-316.

19 Weissberger AJ, Ho KK. Activation of the somatotrophic axis by testosterone an adult males: evidence for the role of aromatization. *Journal of Clinical Endocrinology and Metabolism*. 1993; 76:1407-1412

20 Bowes SB, Umpleby M, Cummings MH, Jackson NC et al. The effect of recombinant human GH on glucose and leucine metabolism in Cushing's syndrome. *Journal of Clinical Endocrinology and Metabolism*. 1996; 82:243-246.

21 Littley MD, Shalet SM, Beardwell CG et al. Hypopituitarism following external radiotherapy for pituitary tumours in adults. *Quarterly Journal of Medicine*. 1989; 70:145-160.

22 Tan BC, Kunaratnam N. Hypopituitary dwarfism following radiotherapy for nasopharyngeal carcinoma. *Clinical Radiology*. 1966; 17:302-304

23 Sheline GE, Wara WM, Smith V. Therapeutic irradiation and brain injury. *International Journal of Radiation Oncology Biology Physics*. 1980; 6:1215-122819. .

24 Ahmed SR, Shalet SM. Hypothalamic growth hormone releasing factor deficiency following cranial irradiation. *Clinical Endocrinology*. 1984; 21:483-488.

25 Shalet SM. Irradiation induced growth failure. *Clinics in Endocrinology and Metabolism*. 1986; 15:591-606.

26 Littley MD, Shalet SM, Beardwell CG. Radiation and hypothalamic-pituitary function. *Baillieres Clinical Endocrinology and Metabolism*. 1990; 4:147-175.

27 Lam KSL, Tse VKC, Wang C et al. Early effects of cranial irradiation on hypothalamic-pituitary function. *Journal of Clinical Endocrinology and Metabolism*. 1987; 64:418-424.

28 Littley MD, Shalet SM, Beardwell CG, Robinson EL, Sutton ML. Radiation induced hypopituitarism is dose-dependent. *Clinical Endocrinology.* 1989; 31:363-373.

29 Sharp PS, Fallon TJ, Brazier OJ et al. Long term follow-up of patients who underwent Yttrium-90 pituitary implantation for treatment of proliferative diabetic retinopathy. *Diabetologia.* 1987; 30:199-207.

30 Cogan JD, Phillips JA, Sakati, N, Frisch H, Schober E, Milner RDG. Heteerogenous growth hormone (GH) gene mutations in familial GH deficiency. *Journal of Clinical Endocrinology and Metabolism.* 1993; 76: 1224-1228.

31 Vimpani GV, Vimpani AF, Lidgard GP, Cameron EHD, Farquhar JW. Prevalence of severe growth hormone deficiency. *British Medical Journal.* 1977; 2: 427-430.

32 Perez-Jurado LA, Argente J. Molecular basis of familial growth hormone deficiency. *Hormone Research.* 1994; 42: 189-97.

33 Cohen LE, Wondisford FE, Salvatoni A, Maghnie M, Brucker-Davis F, Weintraub BD, Radovick S. *Journal of Clinical Endocrinology and Metabolism.* 1995; 80: 679-84.

34 Ozbey N, Inane S, Aral F. et al. Clinical and laboratory evaluation of 40 patients with Sheehan's syndrome. *Israel Journal of Medical Sciences.* 1994; 30:826-829.

35 Pfadenhauer K, Roesler A, herzog P, Grumme T. Adenomatous hypophyseal apoplexy. Clinical, diagnostic and therapeutic aspects of a frequently misdiagnosed emergency state. *Nervenarzt.* 1995; 66:288-292.

36 Frenandez-Real JM, Fernandez-castaner M, Villabona C et al. Giant intrasellar aneurysm presenting with panhypopituitarism and subarachnoidal hemorrhage:case report and literature review. *Clinical Investigator.* 1994; 72:302-306.

37 Suh DC, Park JS, Park SK, Lee HK, Chang KH. Pituitary hemorrhage as a complication of hantaviral disease. *American Journal of Neuroradiology.* 1995; 16:175-178.

38 Ruiz-Hernandez G, Chesa-Jimenez J, San Martin Ciges E. et al. Panhypopituitarism as a consequence of metastase of breast cancer in sella turcica. *Nuklearmedizin.* 1996; 35:143-145.

39 Ko JC, Yang PC, Huang TS, Yeh KH, Kuo SH, Luh KT. Panhypopituitarism caused by solitary parasellar metastasis from lung cancer. *Chest.* 1994; 105:951-953.

40 Lam KS, Sham MM, Tam SC, Ng. MM, Ma HT. Hypopituitarism after tuberculous menigitis in childhood. *Annales of Internal Medicine.* 1993; 118:701-706.

41 Patel MC, Guneratne N, Haq N, West TE, Weetham AP, Clayton RN. Paripartum hypopituitarism and lymphocytic hypophysitis. *Quarterly Journal of Medicine.* 1995; 88:571-580.

7 THE DIAGNOSIS OF ADULT GROWTH HORMONE DEFICIENCY

David M. Hoffman & Ken K. Y. Ho

Garvan Institute of Medical Research
St Vincent's Hospital
Sydney NSW 2010
Australia

1.Introduction

Recent clinical studies of growth hormone (GH) treatment in adults suspected of having GH deficiency have demonstrated beneficial effects on body composition, physical fitness, and psychological well-being (Jorgensen *et al.*, 1989; Salomon *et al.*, 1989). The evidence has been sufficiently convincing that replacement therapy for adults with GH deficiency has been approved for this indication in several countries. Accurate diagnosis of GH deficiency is important if we are to avoid treating adults who do not lack GH and for whom overall benefit awaits detailed evaluation.

Defining GH deficiency in the adult is difficult for a number of reasons. In contrast to children in whom auxological parameters provide a readily quantifiable index of GH deficiency, the clinical features of adult GH deficiency have poor discriminatory value. These features include increased adiposity and reduced lean body mass (Binnerts *et al.*, 1992; De Boer *et al.*, 1992; Rosen *et al.*, 1993; Hoffman *et al.*, 1995), reduced muscle strength and exercise capacity (Cuneo *et al.*, 1990; Rutherford *et al.*, 1991), and impaired psychological well-being (McGauley *et al.*, 1990; Rosen *et al.*, 1994). Some of these features are a normal accompaniment of aging, in particular the reduction in lean body mass and increase in fat mass (Forbes 1987). In view of the nonspecific clinical features of adult GH deficiency, specific biochemical criteria remain necessary for diagnosis. However, biochemical diagnosis may be confounded by a number of factors which impact on GH secretion, most notably the decline in GH secretion that occurs with aging and obesity. An ideal diagnostic test should be capable of distinguishing the healthy old and/or obese from GH deficiency arising from organic hypothalamic-pituitary disease. Finally, there is the issue of inter-laboratory assay standardisation which imposes a limitation in adopting and applying diagnostic criteria between laboratories using different assays.

In this chapter, the relative merits of different tests of GH deficiency are evaluated and recommendations are made on how these tests should be applied in an appropriate context to avoid misuse and unnecessary testing.

2. Evaluation of GH secretory status

There are three broad approaches to the assessment of GH secretory status. These include measuring (i) spontaneous GH secretion, (ii) peak GH response to provocative stimuli, and (iii) serum concentrations of GH-regulated proteins such as insulin-like growth factor-I (IGF-I) and IGF binding protein-3 (IGFBP-3). Paediatric experience report limited concordance between the various tests (Bercu *et al.*, 1986; Rose *et al.*, 1988; Lee *et al.*, 1990) and this has resulted in a long standing controversy as to which is the most accurate test for the diagnosis of GH deficiency.

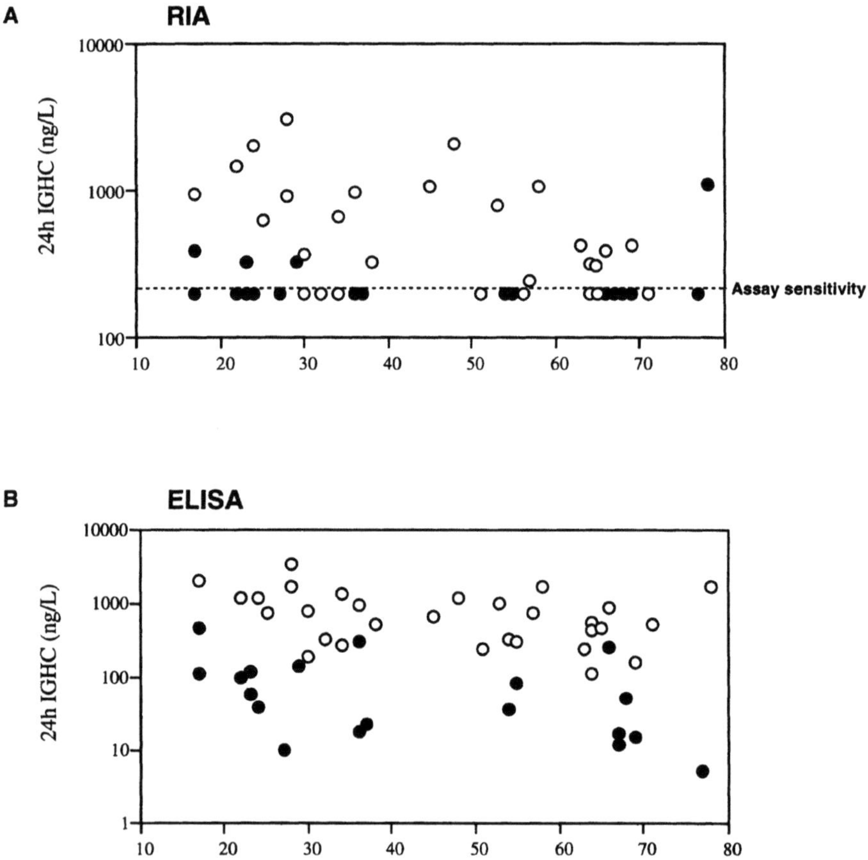

Figure 1. Mean 24 hour GH concentration (IGHC) as measured by (A) RIA (sensitivity 200 ng/L) and (B) ELISA (sensitivity 1 ng/L) in normal (●) and GH deficient (○) adults.

2.1 Spontaneous GH release

In view of the unpredictable, pulsatile nature of GH release, a single serum GH measurement gives very limited information about GH secretory status. A reliable estimate of the amount of GH secreted spontaneously can be obtained by measuring the mean GH concentration from samples pooled over a 24 hour period to give an "integrated growth hormone concentration" or IGHC. However, assessment of GH secretion by this technique is greatly limited by the resources required to undertake these labour intensive studies. Consideration must also be given to physiological modulators of GH secretion, in particular, age and adiposity. There are distinct age - related changes in IGHC with levels rising during puberty, peaking at mid- and late- puberty, and subsequently declining into old age (Rudman *et al.*, 1981; Zadik *et al.*, 1985; Ho *et al.*, 1987; Vermeulen 1987; Iranmanesh *et al.*, 1991; Rose *et al.*, 1991; Martha *et al.*, 1992). GH secretion beyond the 6th decade is less that a fifth of that during peak output in puberty. Spontaneous GH secretion is also reduced in obesity (Rudman *et al.*, 1981; Iranmanesh *et al.*, 1991; Veldhuis *et al.*, 1991; Rasmussen *et al.*, 1995a). Iranmanesh et al, 1991, using the mathematical technique of deconvolution analysis, showed that age and adiposity together accounted for more than 60% of the variability in 24h GH production rates in men (Iranmanesh *et al.*, 1991).

In an assessment of the diagnostic value of IGHC, we reported a significant overlap of values between normal and age, sex and adiposity- matched GH deficient subjects (Hoffman *et al.*, 1994). One-third of young normal subjects (age < 50 years) had values within the range of their GH deficient counterparts. Discrimination by IGHC was also limited by the sensitivity of the GH radioimmunoassay which was 0.2 ng/mL (0.6 mIU/L) since IGHC values were below the limit of detectability in 16% of normal subjects and 80% of GH deficient subjects (Figure 1A).

To determine whether IGHCs in patients with organic GH deficiency might be measurably lower than those of their matched normal counterparts, the samples were reassayed using a GH ELISA with a more than 100 fold greater sensitivity of 1 ng/L (Reutens *et al.*, 1995) (Figure 1B). All IGHCs from both groups fell within the dynamic range of the assay. GH deficient subjects clearly continue to secrete low but measurable amounts of GH. While discrimination between the two groups was improved, an overlap of values remained even after adjusting for age. Baum et al, 1996, have also reported good but incomplete separation of IGHC values between normal and GH deficient patients using an assay of comparable sensitivity. As it is known that intra-individual variation in 24 hour GH secretion may be as high as 50% (Saini *et al.*, 1991), it is possible that repeat testing would have improved discrimination. However, this is not practical for clinical diagnostic purposes

2.2 Provocative tests of GH secretion

A range of provocative tests have been developed over the past decades, primarily for the evaluation of GH reserve in short statured children. Several clinical trials in adults with GH deficiency have used different provocative tests to select patients for GH replacement treatment with diagnostic threshold criteria varying over a 10 fold range from 0.5 - 5 ng/mL (Jorgensen *et al.*, 1989; Salomon *et al.*, 1989; Binnerts *et al.*, 1992; Orme *et al.*, 1992; Whitehead *et al.*, 1992; Bengtsson *et al.*, 1993). These values appear to be either arbitrary or else based on values adopted from the paediatric literature. However, it is inappropriate to extrapolate threshold values obtained in paediatric studies since the GH response to provocative tests is markedly dependent on sexual maturation (Marin *et al.*, 1994). The consequences are evident in studies demonstrating normal GH responses in a substantial proportion of young adults who had impaired GH responses to provocative stimuli when previously tested in childhood (Tassoni *et al.*, 1990; Nicolson *et al.*, 1996).

The common practice of applying a single threshold value defining GH deficiency across the various stimulation tests should also be questioned. Since different stimulation tests modulate different components of the complex neuroendocrine control mechanism, the GH releasing potencies vary between the different stimuli (Lin & Tucci 1974; Tapanainen *et al.*, 1988; Koppeschaar *et al.*, 1992). From an evaluation of seven provocative tests of GH release in normal adult subjects, Lin & Tucci, 1974, reported differences in the peak GH levels to insulin tolerance test (ITT), L-DOPA, glucagon, ACTH, metyrapone, exercise and sleep with the ITT being the most potent. A recent study by Rahim et al, 1996, comparing 4 provocative tests reported the following order of potency: ITT > glucagon > arginine > clonidine with the latter response not different from placebo. These observations suggest that the threshold criteria for defining GH deficiency from provocative tests should be established for each test and that definitions based on extrapolations from other tests are invalid.

2.2.1 Insulin Tolerance Test

The ITT, first described as a diagnostic test for GH deficiency by Roth in 1963, is generally accepted as the standard against which other tests are compared (Frasier 1974). We compared the peak GH response to ITT in 21 patients with hypopituitarism to that in 25 normal subjects matched for sex, age and body mass index (Hoffman *et al.*, 1994). The patients with hypopituitarism were considered likely to be GH deficient because of coexistent organic hypothalamic pituitary disease requiring substitutive treatment for at least two pituitary hormone deficits. There was a clear separation of peak GH responses to ITT between normal subjects and patients with hypopituitarism. All normal subjects had peak GH responses = 5 ng/mL while all patients with organic hypopituitarism had peak GH responses = 3 ng/mL (Figure 2). This study confirms the earlier observation of Greenwood and colleagues (Greenwood *et al.*, 1966; Landon *et al.*, 1966), and suggests that the ITT is a simple and reliable test for GH deficiency in adults.

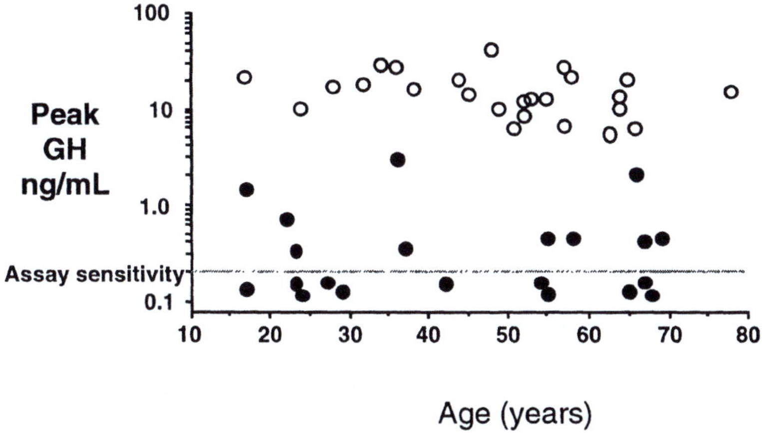

Figure 2. Peak GH response to ITT in GH deficient (●) and normal (○) adults.

Concerns have been expressed about the potential dangers of the ITT (Frasier 1974). In particular, there have been rare reports of major neurological sequelae and deaths in children undergoing the ITT (Shah *et al.*, 1992). However, the potential hazards of an ITT in adults may have been overemphasized. In a recent audit of 161 ITTs in adult subjects in an endocrinology unit in the United Kingdom, the only adverse event was a seizure in one subject who recovered without neurological deficit (Jones *et al.*, 1994). Furthermore, in an extensive literature review by Fish et al of over six thousand ITTs, there were just two reports of coma, two episodes of angina, and two impending seizures, all of which were abated with intravenous glucose (Fish *et al.*, 1986). The collective data suggest that the ITT is a safe test in adults and carries minimal risk of morbidity if performed under supervision in an experienced endocrine unit, and if patients with known seizure disorder or unstable coronary artery disease are excluded.

2.2.2 Other stimulation tests

The usefulness of arginine has been reported from 2 centres. In a study of the elderly, Toogood et al, 1996, found peak GH levels in normal subjects to be higher than, but not completely separated from patients with pituitary disease. Baum et al, 1996, studied a group of patients with pituitary disease who were selected for GH deficiency on the basis of failure to elicit any stimulatory response to arginine (< 0.5 ng/ml). Although they reported a virtually complete separation in arginine-stimulated GH responses between patients with organic pituitary disease and normal subjects, it is possible that there was a bias toward those with severe organic GH deficiency as a result of the selection process. The lower limit of peak GH response in normal subjects in the Baum and Toogood studies were relatively low (0.8 ng/mL and 1.6 ng/mL, respectively), indicating that arginine is not a particularly potent GH secretagogue and may impact on its specificity as a diagnostic test (see below).

Ghigo et al, 1996, have recently reported their experience with combined arginine and GH releasing hormone (GHRH) administration and shown this to be a more potent stimulator of GH release than arginine alone. There was an excellent separation in peak GH responses between normal subjects and patients with organic pituitary disease. The lowest peak GH response encountered in normal subjects was 16 ng/mL and the highest in hypopituitary patients was 9 ng/mL. The authors also analysed the responses to combined pyridostigmine and arginine administration and found that 5 of 27 hypopituitary patients had peak GH responses which overlapped those of normal subjects. Thus the available data suggest that combined arginine and GHRH testing offers a reliable alternative to the ITT as a diagnostic test of adult GH deficiency.

A number of studies have reported clonidine to be a weak and inconsistent stimulator of GH release in adults (Baum *et al.*, 1996; Rahim *et al.*, 1996). Baum et al, 1996, reported that 6 of 10 normal men showed no detectable responses to clonidine. The poor GH response indicates that this test has low specificity and likelihood of false positivity. Some normal subjects show no significant responses to L-DOPA and glucagon (Lin& Tucci 1974; Rahim *et al.*, 1996), although these tests have been less extensively evaluated.

2.2.3 Provocative tests in the elderly

There are varying reports on the effect of age on the peak GH response to provocative stimuli. In a study by Dudl et al, 1973, aging was not correlated with a diminished GH response to arginine. The response to GHRH has been found to be unchanged (Pavlov *et al.*, 1986) or reduced (Shibasaki *et al.*, 1984), and the response to the ITT has been variably reported as unchanged (Kalk *et al.*, 1973) or reduced (Muggeo *et al.*, 1975). These inconsistencies suggest that age may have a relatively minor impact on peak GH response to provocative tests and that provocative tests may be useful in distinguishing the functional hyposomatotropism of aging from that of organic GH deficiency. Indeed, we demonstrated no overlap in peak GH responses to ITT between normal and GH deficient patients in the subset of elderly patients (Hoffman *et al.*, 1994), (Figure 2). Similarly, Ghigo et al, 1996, showed no overlap of peak GH response to combined arginine/GHRH in their subset of elderly patients. Toogood et al, 1996, specifically investigated whether arginine stimulation can distinguish the hyposomatotropism of aging from GH deficiency. The mean GH response to arginine was significantly greater in normal elderly men compared to age-matched patients with pituitary disease although there was an overlap in responses. Since the group of pituitary patients were heterogenous with a range of pituitary hormone deficits (none to several), it is not clear whether the overlap reflected a limitation of the test or of inclusion of patients with minimal pituitary damage who were not GH deficient.

2.2.4 Provocative tests in obesity

Pharmacologically stimulated GH release is impaired in the presence of marked obesity. This has been demonstrated for a variety of stimuli including arginine (Dudl *et al.*, 1973), ITT (Williams *et al.*, 1984; Kopelman *et al.*, 1985; Rasmussen *et al.*, 1995b) and GHRH (Williams *et al.*, 1984; Kopelman *et al.*, 1985; Pavlov *et al.*,

1986). Furthermore, the response improves with weight reduction (Williams *et al.*, 1984; Gama *et al.*, 1990; Rasmussen *et al.*, 1995b). However, in distinguishing the functional hyposomatotropism of obesity from organic GH deficiency, the crucial consideration is whether the peak GH responses of obese individuals without pituitary disease fall within the range of GHD patients. We found the peak GH response to ITT in normal subjects was consistently above that of GHD patients matched for age and BMI, where the BMI in both groups ranged up to 39 kg/m^2 (Hoffman *et al.*, 1994), More recently, in an investigation of the peak GH response to ITT in a group of morbidly obese subjects (mean body mass index 40 kg/m^2), an excellent separation from those with organic GH deficiency has been reported (Hoffman *et al.*, 1997).

2.3 MARKERS OF GH ACTION

Serum IGF-I and IGFBP-3 are considered markers of the integrated action of GH because levels are markedly dependent on GH secretion and relatively constant throughout the day. The GH-dependence and simplicity of measurement of these markers by radioimmunoassay has led investigators to consider their usefulness in the diagnosis of GHD.

2.3.1. IGF-I

The changes in IGF-I levels throughout life mimic those of GH. With the onset of puberty there is a 2- to 3- fold rise in serum IGF-I concentrations, followed by a decline such that average adult levels are reached by the early twenties (Bala *et al.*, 1981; Juul *et al.*, 1994). There follows a gradual decline with advancing age (Rudman *et al.*, 1981; Florini *et al.*, 1985; Vermeulen 1987; Copeland *et al.*, 1990; Hoffman *et al.*, 1994). Unlike spontaneous GH secretion, IGF-I levels are little influenced by adiposity. Thus, age is a major consideration in the evaluation of IGF-I as a diagnostic test. It is also important to be aware of other conditions that may reduce IGF-I level; these include oral oestrogen administration (Weissberger *et al.*, 1991), malnutrition (Thissen *et al.*, 1994), hepatic disease (Cuneo *et al.*, 1995), poorly controlled diabetes (Tan & Baxter 1986) and hypothyroidism (Chernausek *et al.*, 1983).

Data on the reliability of IGF-I as a diagnostic indicator of adult GH deficiency are conflicting. We have reported that 60% of the IGF-I values in hypopituitary subjects were within the range of normal subjects even allowing for the effects of age (Hoffman *et al.*, 1994). Moreover, there was only a modest relationship between IGF-I and IGHC with GH accounting only for 30 - 40% of the variation in IGF-I (Reutens *et al.*, 1995). The significant dependency of IGF-I on non-GH related factors suggests that IGF-I may not be a robust marker of GH deficiency. The finding that IGF-I may be normal in a significant proportion of GH deficient subjects is supported by several recent studies (Bates *et al.*, 1995; Ghigo *et al.*, 1996; Svensson *et al.*, 1997).

In a study of patients with pituitary disease over 60 years of age, as many as 80% of patients had a normal IGF-I (Toogood *et al.*, 1996). In contrast, De Boer et al, 1994, reported in a study of GHD males under 40 years of age, that 96% had subnormal IGF-I values. An explanation for these conflicting observations may be provided by the observation that IGF-I levels fall with increasing age in normal subjects so that separation from GH deficiency would be expected to be greater in the young and reduced in the older age groups. Indeed, Ghigo et al, 1996, have reported that the overlap of IGF-I values between GH deficient and normal adults was 8.6% between 20 - 40 years, 50% between 41 - 60 years and 92% between 61 - 80 years. Thus sensitivity of IGF-I as a diagnostic marker of GH deficiency is age-dependent.

2.3.2. IGFBP-3

Data on IGFBP-3 in GH deficiency are limited. Although mean IGFBP-3 levels were lower, we found a substantial overlap in IGFBP-3 levels between hypopituitary and normal subjects (Hoffman *et al.,* 1994), an observation also made by Baum et al, 1996. From the report of De Boer, 1994, it would also appear that the sensitivity of IGFBP-3 as a diagnostic indicator of GH deficiency may be age-dependent. The limited data indicate that IGFBP-3 offers no advantage to IGF-I as a diagnostic indicator of adult GH deficiency.

3. Diagnostic Strategy

It is important to view the diagnostic utility of any test in the context of its practical application. According to Bayes' theorem, the value of a test is dependent on the prevalence or prior probability of the disease in the population being tested. Thus, the positive predictive value (PPV), being the probability that a positive test correctly identifies disease (true positive), increases as the prior probability of the disease approaches certainty. On the other hand, the negative predictive value (NPV), being the probability that a negative test correctly excludes disease (true negative), decreases as the prior probability approaches certainty and can only be offset by increasing the sensitivity. In other words, the diagnostic utility of any test of GH deficiency increases as the likelihood of GH deficiency in the target population increases.

Which patients should be targeted for tests of GH deficiency and what is the prior probability of GH deficiency in that target population? In practice, the target population is that group of patients with known or suspected pituitary disease. This should include patients who have had cranial irradiation where the prior probability of GH deficiency is at least 60% (Littley *et al.*, 1989; Lam *et al.*, 1991) as well as patients with a diagnosis of childhood onset GH deficiency who require confirmation of adult GH deficiency. From studies of patients attending teaching hospital endocrine services, Toogood et al, 1994, and Weissberger et al, 1994, independently showed that the estimated prior probability of GH deficiency is dependent on the number of other pituitary hormone deficiencies. Analysis of data from the study of Toogood et al showed that peak GH responses of less than 1 ng/ml and 5 ng/ml occurred, respectively in 7% and 52% of patients with otherwise intact pituitary function, and in 45% and 98% of patients with 2 or 3 additional hormone deficiencies. If GH deficiency is identified by the failure of peak GH to exceed 5 ng/mL, the combined data of Toogood et al and Weissberger et al, indicate an overall prevalence of GH deficiency amongst 293 patients of 80%.

Table 1 shows the relative diagnostic merits of the ITT, IGHC, IGF-I and IGFBP-3 based on a prior probability of GH deficiency of 80% with sensitivities and specificities of each of the tests derived from data in the literature. Data pertaining to GH responses to ITT in normal subjects were pooled from 10 published studies to determine specificity, which defines the normal GH threshold (Roth *et al.*, 1963; Greenwood *et al.*, 1966; Landon *et al.*, 1966; Penny *et al.*, 1969; Muggeo *et al.*, 1975; Kopelman *et al.*, 1985; Cordido *et al.*, 1990; Hoffman *et al.*, 1994). Of the 175 normal subjects, 97% responded with a peak GH greater than 5 ng/mL. Thus based on a specificity of 97%, and sensitivity of 100% (Hoffman *et al.*, 1994) the PPV of the ITT amongst patients with pituitary disease is 99%. In other words, applying a diagnostic cut-off response for a positive test as a peak GH response < 5 ng/mL, 99% of 'non-responding' pituitary patients will be truly GH deficient.

Table 1: Diagnostic merit of tests of GH deficiency in adults based on a prior probability of GH deficiency of 0.8 (Toogood *et al.,* 1994; Weissberger *et al.*, 1994).

Test	Sensitivity	Specificity	Predictive Value	
			Positive	Negative
ITT[1]	100	97	99	100
IGHC[2]	88	100	100	82
IGF-I[3]	58	100	100	57
IGFBP-3[4]	69	100	100	64

ITT = insulin tolerance test, IGHC = mean 24 hr GH concentration using supersensitive assays

Sensitivity and specificity derived from: 1. Roth *et al.*, 1963; Greenwood *et al.*, 1966; Landon *et al.*, 1966: Penny *et al.*, 1969: Muggeo *et al.*, 1975: Kopelman *et al.*, 1985: Cordido *et al.*, 1990; Hoffman *et al.*, 1994, 2. Reutens *et al.*, 1995; Baum *et al.*, 1996, 3. Jorgensen *et al.*, 1989; De Boer *et al.*, 1994; Hoffman *et al.*, 1994; Bates *et al.*, 1995; Baum *et al.*, 1996; Ghigo *et al.*, 1996; Svensson *et al.*, 1997, 4. De Boer *et al.*, 1994; Hoffman *et al.*, 1994; Baum *et al.*, 1996.

What happens to the predictive value of the ITT if the prior probability of GHD were less than 80%? Figure 3 illustrates the relationship between PPV and prior probability for the ITT. Here, it can be seen that the PPV is at least 95% for any condition where the prior probability of GH deficiency is above 35%. For example, patients who have undergone cranial irradiation have a high incidence of GH deficiency, varying from 60% of patients 5 years after cranial irradiation for nasopharyngeal carcinoma (Littley *et al.*, 1989) to almost 100% of patients 5 years following irradiation of pituitary tumours (Littley *et al.*, 1989). Thus, the PPV of an ITT in patients who have undergone cranial irradiation is at least 97%. Figure 3 also illustrates the problem of using an ITT to identify GH deficiency in an unselected group. Based on epidemiological data in Sweden that the prevalence of GH deficiency is 175 per million (Rosen & Bengtsson 1994) (prior probability of 0.02%), the PPV of an ITT is only 0.6%! Thus, the likelihood of a normal subject who fails to respond to an ITT having true GH deficiency is 6 in 1000. These theoretical considerations highlight the crucial importance of using a diagnostic test appropriately.

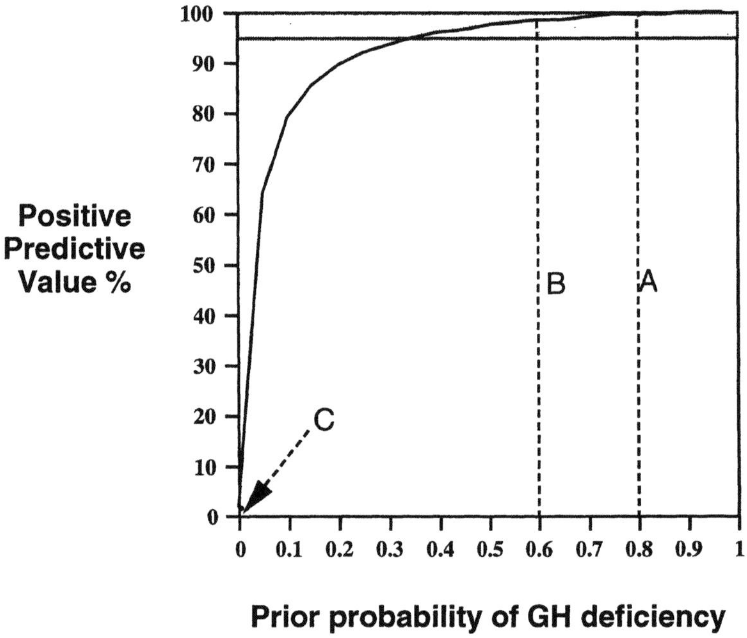

A: patients with pituitary disease and peak GH response to ITT <5 ng/ml
B: patients who have had cranial irradiation for nasopharyngeal carcinoma
C: general population (prior probability 0.0017, positively predictive value 0.6%)

Figure 3. Relationship between positive predictive value and prior probability for the ITT where sensitivity is 100% and specificity is 97%.

The diagnostic merit of IGF-I and IGFBP-3 can be evaluated in a similar way by setting the diagnostic threshold at the lower bounds of the respective ranges in normal subjects. When defined in this way, IGF-I and IGFBP-3 have a specificity of 100% (correctly identifies all non GH deficient subjects) and a PPV of 100% (correctly identifies GH deficiency from subnormal values). The collective data from the literature indicates that IGF-I and IGFBP-3 have moderate sensitivity (ability to identify patients who have GH deficiency). The data from 8 recent studies showed IGF-I levels to be subnormal in 198 of 340 GHD adults (Table 2). This gives a sensitivity of 58% which in turn confers a NPV of 57%. Thus there is only an approximate 57% probability that a patient with pituitary disease and a normal IGF-I does not have GH deficiency; put more simply, a normal IGF-I will miss GH deficiency in about 4 in 10 patients with pituitary disease. The NPV of IGFBP-3 is similar to that for IGF-I. Thus, for both IGF-I and IGFBP-3 the ability of a normal result to exclude disease is poor while the ability of a low result to indicate GH deficiency in a population with preexisting pituitary disease is excellent.

Table 2. Published studies of IGF-I values in adults with GH deficiency.

Study	**Criteria**	**number**	**low IGF-I**	**% low IGF-I**
Hoffman et al., 1994	ITT<5 ng/mL	23	9	39
Bates et al., 1995	ITT<1 ng/mL	40	21	53
De Boer et al., 1994	GHRH/ITT<7 ng/mL	50	48	96
Jorgensen et al., 1989	clonidine 5 ng/mL	22	14	64
Toogood et al., 1994	arginine < 1.6 ng/mL	24	4	17
Baum et al., 1996	arginine<0.5 ng/mL	23	20	87
Ghigo et al., 1996	GRHR+arginine<normal	54	31	57
Svensson et al., 1997	ITT<1.6 mL	104	51	49
Total		**340**	**198**	**58**

4. Assay Standardization

A major problem confronting investigators involved in the diagnosis of GH deficiency is the lack of standardisation of assays between laboratories. It is well documented that significant discrepancies in GH quantification occur between different immunoassays (Reiter *et al.*, 1988). The reasons are complex and include differences in antibody specificity, GH heterogeneity in serum, differences in standards and matrix. This observation suggests that cut-off values necessary to define the sensitivity or specificity of a particular GH provocative test cannot be readily adopted or extrapolated between laboratories even for the same test.

However, it is impractical and unreasonable for each laboratory to set its own cut off criteria from studies of normal subjects.

The problem can be approached in a number of ways. One approach is to undertake a systematic comparison between a number of major laboratories of a short list of commonly used GH assays for the purpose of deriving a conversion factor in relation to a designated reference assay. Another approach is to use provocative tests that are the most potent and consistent stimulators of GH release. Studies comparing the stimulatory potential of various test show the ITT, combined GHRH-arginine or GHRH-pyridostigmine are among the most potent, clonidine to be weak, and arginine or GHRH alone to be intermediate. It is evident that tests which are poor releasers of GH have poor specificity since a 'low' GH response to a poor stimulator has a lower probability of indicating GH deficiency than a similar response to a more potent stimulator. Thus a threshold could be set at a sufficiently low level that falls below a range that allows for measurement discrepancies between kits. This strategy may miss patients with mild GH deficiency but has the advantage of identifying the more severely affected, who are the ones most likely to benefit from replacement treatment.

IGF-I measurement has a valuable role both in the diagnosis and monitoring of treatment. Thus the issue of standardising IGF-I assays also requires consideration. There are a number of technical considerations unique to IGF-I assays which include antibody specificity to IGF-I and II, and effects of binding protein which may be variably affected by GH status. The potential impact of these factors and the variability of their effects on IGF-I quantification on adults with GH deficiency require clarification. Assays that measure 'free IGF-I' have recently been introduced (Frystyk *et al.*, 1996). Presently it is unclear whether there are advantages in the different assays and whether the differences are important in the context of diagnosis and monitoring treatment in adult GH deficiency.

4. Summary

In summary, the clinical features of GH deficiency are recognisable but are not sufficiently distinct to be useful in the diagnosis of GH deficiency. GH deficiency should be defined biochemically within an appropriate clinical context, that is, in patients with a history of pituitary disease, cranial irradiation, or a history of childhood onset GH deficiency. The diagnosis of adult GH deficiency is established by provocative testing of GH secretion. The ITT has been found to be the most reliable test of GH deficiency to date. When used in the appropriate setting, that is, in patients with known pituitary disease, a single ITT has a positive predictive value of greater than 95%. Serum IGF-I is not sufficiently sensitive to be a useful diagnostic test on its own, and serum IGFBP-3 offers no advantage over IGF-I. However, because of its high specificity, a single IGF-I level, in the absence of other pathological conditions associated with a low IGF-I, could be used as a screening test to determine the need to proceed to an ITT.

At a recent consensus meeting of the International Growth Hormone Research Society a number of recommendations were made with respect to the diagnosis of adult GH deficiency (GRS workshop 1997). The ITT was recommended as the test of choice with severe GH deficiency defined by a peak GH response to hypoglycaemia less than 3 μg/L, and a normal response greater than 5 μg/L. However, GH immunoassay results vary between different assay methods. The above recommendations for cut-off values for the insulin tolerance test are based on results obtained with polyclonal competitive RIAs calibrated against the pituitary derived preparation IRP 80/505 (1 mg = 2.6 IU). The Growth Hormone Research Society advocated future use of the recombinant hGH preparation IRP 88/624 (1 mg = 3.0 IU), and that further comparative studies were necessary both on a national and international level to achieve standardisation of GH assays (Attanasio *et al.,* 1997).

It is likely that other secretogogues will be established in the future as reliable diagnostic tools although these tests still require rigorous and systematic evaluation to define sensitivity and specificity. At present, the combined arginine/GHRH test (Ghigo *et al.*, 1996) is considered the most promising alternative to the ITT (Attanasio *et al.,* 1997).

ACKNOWLEDGMENTS
We are very grateful for the generous support of this work by the National Health and Medical Research Council of Australia.

References

GRS Workshop. Consensus guideline for the diagnosis and treatment of adults with growth hormone deficiency. J Clin Endocrinol Metab *In Press*.

Bala RM, Lopatka J, Leung A, McCoy E, McArthur RG. Serum immunoreactive somatomedin levels in normal adults, pregnant women at term, children at various ages, and children with constitutionally delayed growth. J Clin Endocrinol Metab 1981; 52: 508-512.

Bates AS, Evans AJ, Jones P, Clayton RN. Assessment of GH status in adults with GH deficiency using serum growth hormone, serum insulin-like growth factor-I and urinary growth hormone excretion. Clin Endocrinol 1995; 42: 425-430.

Baum HB, Biller BMK, Katznelson L, Oppenheim DS, Clemmons DR, Cannistraro KB, Schoenfeld DA, Best SA, Klibanski A. Assessment of growth hormone (GH) secretion in men with adult-onset GH deficiency compared with that in normal men-A clinical research center study. J Clin Endocrinol Metab 1996; 81: 84-92.

Bengtsson B-A, Eden S, Lonn L, Kvist H, Stokland A, Lindstedt G, Bosaeus I, Tolli J, Sjostrom L, Isaksson OGP. Treatment of adults with growth hormone (GH) deficiency with recombinant human GH. J Clin Endocrinol Metab 1993; 76: 309-317.

Bercu BB, Shulman D, Root AW, Spiliotis JE. Growth hormone (GH) provocative testing frequently does not reflect endogenous GH secretion. J Clin Endocrinol Metab 1986; 63: 709-716.

Binnerts A, Deurenberg P, Swart GR, Wilson JHP, Lamberts SWJ. Body composition in growth hormone-deficient adults. Am J Clin Nutr 1992; 55: 918-923.

Binnerts A, Swart GR, Wilson JHP, Hoogerbrugge N, Pols HAP, Birkenhager JC, Lamberts SWJ. The effect of growth hormone administration in growth hormone deficient adults on bone, protein, carbohydrate and lipid homeostasis, as well as on body composition. Clin Endocrinol 1992; 37: 79-87.

Chernausek SD, Underwood LE, Utiger RD, Van Wyk JJ. Growth hormone secretion and plasma somatomedin-C in primary hypothyroidism. Clin Endocrinol 1983; 19: 337-344.

Copeland KC, Colletti RB, Devlin JT, McAuliffe TL. The relationship between insulin-like growth factor-I, adiposity and aging. Metabolism 1990; 6: 584-587.

Cordido F, Dieguez C, Casanueva FF. Effect of central cholinergic neurotransmission enhancement by pyridostigmine on the growth hormone secretion elicited by clonidine, arginine, or hypoglycemia in normal and obese subjects. J Clin Endocrinol Metab 1990; 70: 1361-1370.

Cuneo RC, Hickman PE, Wallace JD, Teh BT, Ward G, Veldhuis JD, Waters MJ. Altered endogenous growth hormone secretory kinetics and diurnal GH-binding protein profiles in adults with chronic liver disease. Clin Endocrinol 1995; 43: 265-275.

Cuneo RC, Salomon F, Wiles CM, Sonksen PH. Skeletal muscle performance in adults with growth hormone deficiency. Horm Res 1990; 33 (Suppl 4): 55-60.

De Boer H, Blok G-J, Popp-Snijders C, van der Veen E. Diagnosis of growth hormone deficiency in adults. Lancet 1994; 343: 1645-1646.

De Boer H, Blok GJ, Voerman HJ, De Vries PMJ, Van Der Veen EA. Body composition in adult growth hormone-deficient men, assessed by anthropometry and bioimpedence analysis. J Clin Endocrinol Metab 1992; 75: 833-837.

Dudl RJ, Ensinck JW, Palmer HE, Williams RH. Effect of age on growth hormone secretion in man. J Clin Endocrinol Metab 1973; 37: 11-16.

Fish HR, Chernow B, O'Brian JT. Endocrine and neurophysiologic responses of the pituitary to insulin-induced hypoglycemia: a review. Metabolism 1986; 35: 763-780.

Florini JR, Prinz PN, Vitiello MV, Hintz RL. Somatomedin-C levels in healthy young and old men: relationship to peak and 24-hour integrated levels of growth hormone. J Gerentol 1985; 40: 2-7.

Forbes GB: Human body composition: growth, aging, nutrition and activity. 1987; Springer-Verlag, New York.

Frasier SDF. A review of growth hormone stimulation tests in children. Pediatr 1974; 53: 929-937.

Frystyk J, Vestbo E, Skjaerbaek C, Mogensen CE, Orskov H. Free insulin-like growth factors in human obesity. FEBS letters 1996; 348: 185-191.

Gama R, Teale JD, Marks V. The effect of synthetic very low calorie diets on the GH-IGF-1 axis in obese subjects. Clin Chim Acta 1990; 188: 31-38.

Ghigo E, Aimaretti G, Gianotti L, Belloni J, Arvat E, Camanni F. New approach to the diagnosis of growth hormone deficiency in adults. Eur J Endocrinol 1996; 134: 352-356.

Greenwood FC, Landon J, Stamp TCB. The plasma sugar, free fatty acid, cortisol, and growth hormone response to insulin. I. In control subjects. J Clin Invest 1966; 45: 429-436.

Ho KY, Evans WS, Blizzard RM, Veldhuis JD, Merriam GR, Samojlik E, Furlanetto R, Rogol AD, Kaiser DL, Thorner MO. Effects of sex and age on the 24-hour profile of growth hormone secretion in man: importance of endogenous estradiol concentrations. J Clin Endocrinol Metab 1987; 64: 51-58.

Hoffman DM, O'Sullivan A, Freund J, Ho KKY. Adults with growth hormone deficiency have abnormal body composition but normal metabolism. J Clin Endocrinol Metab 1995; 80: 72-77.

Hoffman DM, O'Sullivan AJ, Baxter RC, Ho KY. Diagnosis of growth hormone deficiency in adults. Lancet 1994; 343: 1064-68.

Hoffman DM, Rasmaussen MH, Leung K-c, Doyle N, Ho KKY. Can the insulin tolerance test (ITT) discriminate between the hyposomatotropism of obesity and organic growth hormone (GH) deficiency. 40th Annual Scientific Meeting of the Endocrine Society of Australia, Canberra, 1997; pp 94, Abstr 109.

Iranmanesh A, Lizarralde G, Veldhuis JD. Age and relative adiposity are specific negative determinants of the frequency and amplitude of growth hormone (GH) secretory bursts and the half-life of endogenous GH in healthy men. J Clin Endocrinol Metab 1991; 73: 1081-1088.

Jones SL, Trainer PJ, Perry L, Wass JAH, Besser GM, Grossman A. An audit of the insulin tolerance test in adult subjects in an acute investigation unit over one year. Clin Endocrinol 1994; 41: 123-128.

Jorgensen JOL, Thuesen L, Ingemann-Hansen T, Pederson SA, Jorgensen J, Skakkebaek NE, Christiansen JS. Beneficial effects of growth hormone treatment in GH-deficient adults. Lancet 1989; i: 1221-1225.

Juul A, Bang P, Hertel NT, Main K, Dalgaard P, Jorgensen K, Muller J, Hall K, Skakkebaek NE. Serum insulin-like growth factor-I in 1030 healthy children, adolescents, and adults: Relation to age, sex, stage of puberty, testicular size, and body mass index. J Clin Endocrinol Metab 1994; 78: 744-752.

Kalk WJ, Vinik AI, Pimstone BL, Jackson WPU. Growth hormone response to insulin hypoglycemia in the elderly. J Gerontol 1973; 28: 431-433.

Kopelman PG, Noonan K, Goulton R, Forrest AJ. Impaired growth hormone response to growth hormone releasing factor and insulin-hypoglycaemia in obesity. Clin Endocrinol 1985; 23: 87-94.

Koppeschaar HPF, ten Horn CD, Thijssen JHH, Page MD, Dieguez C, Scanlon MF. Differential effect of arginine on growth hormone releasing hormone and insulin induced growth hormone secretion. Clin Endocrinol 1992; 36: 487-490.

Lam KSL, Tse VKC, Wang C, Yeung RTT, Ho JHC. Effects of cranial irradiation on hypothalamo-pituitary function-a 5-year longitudinal study in patients with nasopharyngeal carcinoma. Q J Med 1991; 78: 165-176.

Landon J, Greenwood FC, Stamp TCB, Wynn V. The plasma sugar, free fatty acid, cortisol, and growth hormone response to insulin, and the comparison of this procedure with other tests of pituitary and adrenal function. II. In patients with hypothalamic or pituitary dysfunction or anorexia nervosa. J Clin Invest 1966; 45: 437-449.

Lee PDK, Wilson DM, Rountree L, Hintz RL, Rosenfeld RG. Efficacy of insulin-like growth factor I levels in predicting the response to provocative growth hormone testing. Pediatr Res 1990; 27: 45-51.

Lin T, Tucci JR. Provocative tests of growth-hormone release. A comparison of results with seven stimuli. Ann Intern Med 1974; 80: 464-469.

Littley MD, Shalet SM, Beardwell CG, Ahmed SR, Applegate G, Sutton ML. Hyopituitarism following external radiotherapy for pituitary tumours in adults. Q J Med 1989; 70: 145-160.

Marin G, Domene HM, Barnes KM, Blackwell BJ, Cassorla FG, Cutler Jr GB. The effects of estrogen priming and puberty on the growth hormone response to standardized treadmill exercise and arginine-insulin in normal girls and boys. J Clin Endocrinol Metab 1994; 79: 537-541.

Martha PM, Gorman KM, Blizzard RM, Rogol AD, Veldhuis JD. Endogenous growth hormone secretion and clearance rates in normal boys, as determined by deconvolution analysis: relationship to age, pubertal status, and body mass. J Clin Endocrinol Metab 1992; 74: 336-344.

McGauley GA, Cuneo RC, Salomon F, Sonksen PH. Psychological well-being before and after growth hormone treatment in adults with growth hormone deficiency. Horm Res 1990; 33: 52-54.

Muggeo M, Fedele D, Tiengo A, Molinari M, Crepaldi G. Human growth hormone and cortisol response to insulin stimulation in aging. J Gerontol 1975; 30: 546-551.

Nicolson A, Toogood AA, Rahim A, Shalet SM. The prevalence of severe growth hormone deficiency in adults who received growth hormone replacement in childhood. Clin Endocrinol 1996; 44: 311-316.

Orme SM, Sebastian JP, Oldroyd B, Stewart SP, Grant PJ, Stickland MH, Smith MA, Belchetz PE. Comparison of measures of body composition in a trial of low dose growth hormone replacement therapy. Clin Endocrinol 1992; 37: 453-459.

Pavlov EP, Harman SM, Merriam GR, Gelato MC, Blackman MR. Responses of growth hormone (GH) and somatomedin-C to GH-releasing hormone in healthy aging men. J Clin Endocrinol Metab 1986; 62: 595.

Penny R, Blizzard RM, Davis WT. Sequential arginine and insulin tolerance tests on the same day. J Clin Endocrinol Metab 1969; 29: 1499-1501.

Rahim A, Toogood AA, Shalet SM. The assessment of growth hormone status in normal young adult males using a variety of provocative agents. Clin Endocrinol 1996; 45: 557-562.

Rasmussen MH, Juul A, Kjems LL, Skakkebaek NE, Hilsted J. Lack of stimulation of 24-hour growth hormone release by hypocaloric diet in obesity. J Clin Endocrinol Metab 1995a; 80: 796-801.

Rasmussen MH, Hvidberg A, Juul A, Main KM, Gotfredsen A, Skakkebae NE, Hilsted J. Massive weight loss restores 24-hour growth hormone release profiles and serum insulin-like growth factor-I levels in obese subjects. J Clin Endocrinol Metab 1995b; 80:1407-1451.

Reiter EO, Morris AH, MacGillivray MH, Weber D. Variable estimates of serum growth hormone concentrations by different radioassay systems. J Clin Endocrinol Metab 1988; 66: 68-71.

Reutens AT, Hoffman DM, Leung K, Ho KKY. Evaluation and application of a highly sensitive assay for growth hormone (GH) in the study of adult GH deficiency. J Clin Endocrinol Metab 1995; 80: 480-485.

Rose SR, Municchi G, Barnes KM, Kamp GA, Uriarte MM, Ross JL, Cassorla F, Cutler GB. Spontaneous growth hormone secretion increases during puberty in normal girls and boys. J Clin Endocrinol Metab 1991; 73: 428-435.

Rose SR, Ross JL, Uriarte M, Barnes KM, Cassorla FG, Cutler GB. The advantage of measuring stimulated as compared with spontaneous growth hormone levels in the diagnosis of growth hormone deficiency. N Engl J Med 1988; 319: 201-207.

Rosen T, Bengtsson BA. Epidemiology of adult onset hypopituitarism in Goteborg, Sweden during 1956-1987. 18th International Symposium: Growth hormone and growth factors in endocrinology and metabolism, 1994; pp 60 (Abstr A3).

Rosen T, Bosaeus I, Tolli J, Lindstedt G, Bengtsson B. Increased body fat mass and decreased extracellular fluid volume in adults with growth hormone deficiency. Clin Endocrinol 1993; 38: 63-71.

Rosen T, Wiren L, Wilhelmsen L, Wiklund I, Bengtsson B-A. Decreased psychological well-being in adult patients with growth hormone deficiency. Clin Endocrinol 1994; 40: 111-116.

Roth J, Glick SM, Yalow RS, Berson SA. Hypoglycemia: a potent stimulus to secretion of growth hormone. Science 1963; 140: 987-988.

Rudman D, Kutner MH, Rogers CM, Lubin MF, Flemming GA, Bain RP. Impaired growth hormone secretion in the adult population. J Clin Invest 1981; 67: 1361-1369.

Rutherford OM, Jones DA, Round JM, Buchanan CR, Preece MA. Changes in skeletal muscle and body composition after discontinuation of growth hormone treatment in growth hormone deficient adults. Clin Endocrinol 1991; 34: 469-475.

Saini S, Hindmarsh PC, Matthews DR, Pringle PJ, Jones J, Preece MA, Brook CGD. Reproducibility of 24-hour growth hormone profiles in man. Clin Endocrinol 1991; 34: 455-462.

Salomon F, Cuneo RC, Hesp RC, Sonksen PH. The effects of treatment with recombinant human growth hormone on body composition and metabolism in adults with growth hormone deficiency. N Engl J Med 1989; 321: 1797-1803.

Shah A, Stanhope R, Matthew D. Hazards of pharmacological tests of growth hormone secretion in childhood. Brit Med J 1992; 304: 173-174.

Shibasaki T, Shizume K, Nakahar M, Masuda A, Jibiki K, Demura H, Wakabayashi I, Ling N. Age-related changes in plasma growth hormone response to growth hormone-releasing factor in man. J Clin Endocrinol Metab 1984; 58: 212.

Svensson J, Johannsson G, Bengtsson B-A. Insulin-like growth factor-I in growth hormone deficient adults: relationship to population based normal values, body composition and insulin tolerance test. Clin Endocrinol 1997; 46: 579-86.

Tan K, Baxter RC. Serum insulin-like growth factor I levels in adult diabetic patients: The effect of age. J Clin Endocrinol Metab 1986; 63: 651-655.

Tapanainen P, Knip M, Lautala P, Leppaluoto J. Variable plasma growth hormone (GH)-releasing hormone and GH responses to clonidine, L-Dopa, and insulin in normal men. J Clin Endocrinol Metab 1988; 67: 845-849.

Tassoni P, Cacciari E, Cau M, Colli C, Tosi M, Zucchini S, Cicognani A, Pirazzoli P, Salardi S, Balsamo A, Frejaville E, Cassio A, Zappulla F. Variability of growth hormone response to pharmacological and sleep tests performed twice in short children. J Clin Endocrinol Metab 1990; 71: 230-234.

Thissen J, Ketekslegers J, Underwood LE. Nutritional regulation of the insulin-like growth factors. Endocr Rev 1994; 15: 80-101.

Toogood AA, Beardwell CG, Shalet SM. The severity of growth hormone deficiency in adults with pituitary disease is related to the degree of hypopituitarism. Clin Endocrinol 1994; 41: 511-516.

Toogood AA, O'Neill PA, Shalet SM. Beyond the somatopause: Growth hormone deficiency in adults over the age of 60 years. J Clin Endocrinol Metab 1996; 81: 460-465.

Veldhuis JD, Iranmanesh A, Ho KKY, Waters MJ, Johnson ML, Lizarralde G. Dual defects in pulsatile growth hormone secretion and clearance subserve the hyposomatotropism of obesity in man. J Clin Endocrinol Metab 1991; 72: 51-59.

Vermeulen A. Nyctohemeral growth hormone profiles in young and aged men: correlation with somatomedin-C levels. J Clin Endocrinol Metab 1987; 64: 884-888.

Weissberger AJ, Ho KKY, Lazarus L. Contrasting effects of oral and transdermal routes of estrogen replacement therapy on 24-hour growth hormone (GH) secretion, insulin-like growth factor I, and GH-binding protein in postmenopausal women. J Clin Endocrinol Metab 1991; 72: 374-381.

Weissberger AJ, Verkikiou K, Sonksen PH. Diagnosis of growth hormone deficiency in adults. Lancet 1994; 344: 483.

Whitehead HM, Boreham C, McIlrath EM, Sheridan B, Kennedy L, Atkinson AB, Hadden DA. Growth hormone treatment of adults with growth hormone deficiency: results of a 13-month placebo controlled cross-over study. Clin Endocrinol 1992; 36: 45-52.

Williams T, Berelowitz M, Joffe SN, Thorner MO, Rivier J, Vale W, Frohman LA. Impaired growth hormone responses to growth hormone-releasing factor in obesity. N Engl J Med 1984; 311: 1403-1407.

Zadik Z, Chalew SA, Mccarter Jr RA, Meistas M, Kowarski AA. The influence of age on the 24-hour integrated concentration of growth hormone in normal individuals. J Clin Endocrinol Metab 1985; 60: 513-516.

8 ADULT GROWTH HORMONE DEFICIENCY - THE CLINICAL PICTURE

Jens Sandahl Christiansen
Troels Krarup Hansen
Steen Nielsen
Jens Otto Lunde Jörgensen

Medical Department M (Endocrinology and Diabetes)
Aarhus University Hospital, Denmark

Lacking or ceased GH secretion from the pituitary gland results in an abnormal metabolic condition where e.g. fasting and stress will be characterised by a relative increase in protein metabolism and reduction in lipid metabolism. Growth hormone deficiency (GHD) and the consequent reduced activity of the GH - IGF-I axis, is, however, not associated with quickly recognised and conspicuous symptoms as often seen in failures of other hormonal systems - e.g. in insulin, cortisol, thyroid or sex steroid deficiency.

Nevertheless endocrinologists more than 3 decades ago suggested the clinical importance of GH deficiency and partly described the clinical syndrome. Thus, Raben in 1962 [1] suggested that the metabolic actions of growth hormone was of importance for the normal maintenance of adult tissues - and also, as the first, reported on results of GH substitution therapy in a single patient [1]. The same year Falkheden described the endocrine aspects of hypophysectomy in man and observed a decline in basal metabolic rate despite replacement therapy with cortisone, thyroid hormone and sex hormones and suggested other pituitary factors to be responsible for this abnormality [2].

In children, GHD - like any other chronic disease - leads to a reduced growth rate which may be due to a combination of the above metabolic abnormalities and a lack of GH's specific stimulating properties as regards cartilage and bone formation.

In adults, long-lasting untreated GHD results in a number of symptoms of organic and psychosocial dysfunction. Individually the symptoms may be of less importance and as times go by they may be overshadowed by - or erroneously seen as - symptoms of inadequate replacement of the other hormonal axes in the complete or partly panhypopituitary condition which adult GHD most often is combined with.

In the following the symptoms and clinical findings in adults with GHD will be briefly outlined. It is important to realise, however, that adult GHD does not consist of one well defined patient group only. Thus it is obvious that the underlying disease leading to the hypopituitary state (e.g. trauma, tumours involving the optic chiasma) frequently will dominate the clinical picture. It is also evident, that the clinical condition of a patient suffering from isolated GHD is often different from the patient with multiple pituitary deficiency. However, most - if not all - of the abnormalities described in adult

GHD patients are found in as well isolated GHD as in multiple pituitary deficiency - and are equally alleviated by GH therapy in both conditions. These symptoms and signs are therefore thought to be due to GH deficiency per se. In addition it has been argued that the consequence of GHD in adults might be quite different if the disease was acquired in childhood as compared to adult onset GHD. Although there are merits to this point of view [3] there are at present no conclusive evidence from data obtained in groups of patient with comparable age and sex.

In the present short description, therefore, no distinction will be made between isolated GH deficiency and multiple pituitary deficiency and childhood versus adult onset GHD.

Psychosocial symptoms and signs

When applying validated and standardised quality of life questionnaires such as the Nothingham Health Profile and the Comprehensive Psychological Rating Scale it has been found that adult GHD patients compared to normal adults show signs of impaired psychological function. The symptoms, which are often quite suttle but can, however, be very serious, includes reduces energy levels, reduces vitality, increased anxiety, reduced emotional reaction as well as an increase in a number of symptoms of depression, hampered learning capacity and signs of social isolation [4, 5]. Although it could be argued that at least some of the psychosocial symptoms may be seen as secondary to the somatic consequences of GH deficiency it is on the other hand tempting to associate growth hormone with direct action on the brain because of the well known presence of GH receptors throughout the CNS system.

Muscular and adipose tissue

Although patients with long standing GHD are often obese it is important to realise that adult GHD patients are not necessarily overtly obese or have and increase in body mass index (BMI). When applying various methods (e.g. CT-scanning, bioimpedance, total body potassium, dexascan, skinfold thickness estimation), however, several independents studies have demonstrated a clear increase in fat mass, reduced lean body mass, reduced muscle volume and reduced muscle fat ratio in adults with untreated GHD [6, 7, 8, 9, 10, 11]. Muscle/fat ratio in untreated young GHD patients evaluated by a cross sectional CT-scan of the mid femoral region is approximately 64/36 compared to ratio of 78/22 in normal subjects of the same age [6]. Fat accumulation is seen in all parts of the body but in particular corresponding to the intraabdominal or visceral fat [12].

Muscle strength, exercise capacity and thermoregulation

The symptomatology of GHD in adults often includes complains of lack of strength and early fatigue. Isometric muscle strength corresponding to the quadriceps muscle is significantly reduced in GHD patients [7, 13]. The reduced muscle strength can easily be explained by the reduced muscle mass, and significant correlations between muscle mass and muscle strength have been demonstrated. In addition, as well muscle fibre density as muscle histology has been found to be normal [7]. Isometric force in muscle groups other than quadriceps have not shown striking differences in GHD patients compared to the normal population [7]. Overall, adults with GH deficiency have mild to moderate reduction in muscle strength, due to reductions in muscle mass.

In several studies the exercise capacity has been measured by means of an electromechanical bicycle-test with or without concurrent analysis of maximal oxygen uptake. In untreated GHD adults both exercise capacity (adjusted for weight and height), maximal oxygen uptake and power output are reduced considerably (20-30 %) compared to normal subjects [13, 14, 15].

In this context, it might be of potential interest that untreated adult GHD patients have been shown to suffer from a marked reduction in their sweat secretion capacity [16]. This reduction in sweat secretion might compromise thermoregulation [17] - a condition which very well might contribute to the reduced exercise capacity and power output.

Renal function and fluid homeostasis

It has been known for decades that acromegalic patients have an increased glomerular filtration rate (GFR), that traditionally substituted hypophysectomised patients have reduced GFR, and that GH treatment of normal subject increases GFR in the course of days [18]. Using the classical continuous infusion technique both GFR and renal plasma flow (RPF) adjusted for body surface area have been found to be approximately 20 % reduced in young adult GHD patients compared to normal subjects [6]. Standard clinical parameters of kidney function [e.g. s-creatinine), however, are normal. In addition untreated adult GHD patients are characterised by extracellular fluid depletion [11] which may not be readily recognised in the clinical situation but it is important to bear in mind during the early phases of replacement therapy, since correction of this fluid depletion takes place swiftly and might give rise to transient "dyshydration" which the patients experience as side effects.

Bone

Most authors agree, that adults GHD patients have a reduced bone mineral content (BMC) [19, 20, 21]. Although this reduced BMC might lead to clinical apparent osteoporosis and even an increase in osteoporotic fractures [21] the changes in bone tissue are presently considered to be of limited diagnostic and clinical relevance.

Cardiovascular aspect

It is presently an area of controversy whether the untreated adult growth hormone deficient state is associated with an increase in cardiovascular mortality and morbidity. Two relatively large independent studies from Sweden have demonstrated an almost two-fold increase in cardiovascular mortality [22, 23] while studies from the UK have failed to show similar results [24]. Cross-sectional investigations of adult patients with documented, untreated GHD have shown a reduced cardiac muscle mass corresponding to the left ventricle, reduced cardiac output (roughly estimated by the rate-pressure-product), reduced ejection rate and reduced ejection fraction [25, 26]. These ultrasonographics signs of reduced myocardial contractility various significantly with the degree of GHD as estimated by the IGF-I concentration in plasma [25]. Furthermore an ultrasound investigation of the large arteries has demonstrated an increased intima thickness and increased frequency of atheromatous plaques [27].

The absence of GH as such does not seem to influence blood pressure directly - but the gradual development of obesity and insulin resistance might lead to hypertension.

Carbohydrate metabolism

Fasting hypoglycaemia in adults with GHD appears to be less common than in GHD children [28]. In lean adults with GH deficiency fasting plasma glucose and insulin concentrations appear to be normal while in obese GHD adults hyperinsulinimia suggests a degree of insulin resistance [29]. In an English survey of 40 adults GHD patients 14 was found to have impaired glucose tolerance following an OGTT [30]. It is conceivable that the effects of GHD upon carbohydrate metabolism are secondary to the impact of body composition. Thus, patients with long standing GH deficiency who have developed an increase in fat mass - in particular an increase in intraabdominal fat - are most likely to show impaired glucose tolerance and even diabetes in the clinical setting.

Thyroid function

It was previously presumed that GH treatment could induce hypothyroidism, since decreasing plasma values of T4 was often observed in growth retarded GHD children after initiation of GH replacement. Recent studies, however, have revealed that GH stimulates the peripheral deiodination of T4 to T3 [31] - which give a full explanation of the above observations. In fact adults with untreated GHD have subnormal T3 values [31] - which might explain in part the significantly lower energy expenditure in these patients [8]. It could even been argued that part of the clinical picture of the untreated GHD adult is caused by of some degree of hypothyroidism.

Conclusion

As stated above the symptoms and signs of growth hormone deficiency in adults are not as obvious and even dramatic as might be seen in other endocrine deficient states. Taken together, however, the pleiotrophy of symptoms have distinct consequences for the individual patient. Thus, it has now been clearly demonstrated that untreated GHD patients - even with isolated GH deficiency - usually are not capable of living a normal life. Furthermore it must by now be considered a proven fact that GH replacement therapy results in a partial or often complete normalisation of all known abnormalities associated with the adult GHD syndrome. Thus recognition of the symptoms of GH deficiency have become of practical clinical importance.

REFERENCES

1. Raben MS. Growth hormone. N Eng J Med 1962; 266: 82-6.
2. Falkheden T, Norin T, Sjögren B, Skanse B. Thyroid function and basal metabolic rate following hypophysectomy in man. Acta Endocrinol 1962; 41: 457-473.
3. Attanasio AF, Lamberts SWJ, Matranga AMC, Birkett MA, Bates PC, Valk NK, Hilsted J, Bengtsson B-Å, Strasburger CJ and The Adult Growth Hormone Deficiency Study Group. Adult growth hormone (GH)-deficient patients demonstrate heterogeneity between childhood onset and adult onset before and during human GH treatment.
4. Björck S, Jönsson B, Westphal O, Levin J-E. Quality of life of adults with growth hormone deficiency: a controlled study. Acta Paediatr Scand 1989; suppl 356: 55-9.

5. McGauley GA, Cuneo RS, Salomon F, Sönksen PH. Psychological well-being before and after growth hormone treatment in adults with growth hormone deficiency. Horm Res 1990; 33 suppl. 4: 52-4.
6. Jörgensen JOL, Pedersen SA, Thuesen L, Jörgensen J, Ingemann-Hansen T, Skakkebaek NE, Christiansen JS. Beneficial effects of growth hormone treatment in GH-deficient adults. Lancet 1989; 1: 1221-25.
7. Cuneo RC, Salomon F, Wiles CM, Sönksen PH. Skeletal muscle performance in adults with growth hormone deficiency. Horm Res 1990; 33 suppl. 4: 55-60.
8. Salomon F, Coneo RC, Hesp R, Morris JF, Poston L, Sönksen PH. Basal metabolic rate in adults with growth hormone deficiency and in parients with acromegaly: relationship with lean body mass, plasma insulin level and leucocyte sodium pump activity. Clin Sci 1992; 83: 325-30.
9. Binnerts A, Deurenberg P, Swart GR, Wilson JHP, Lamberts SWJ. Body composition in growth hormone deficient adults. Am J Clin Nutr 1992; 55: 918-23.
10. De Boer H, Blok GJ, Voerman HJ, M De Vries PMJ, Van der Veen EA. Body composition in adult growth hormone deficient men, assessed by anthropometry and bioimpedance analysis. J Clin Endocrinol Metab 1992; 75, 883-7.
11. Rosén T, Boseaus I, Tölli J, Lindstedt G, Bengtsson B-Å. Increased body fat mass and decreased extracellular fluid volume in adults with growth hormone deficiency. Clin Endocrinol 1993; 38: 63-71.
12. Bengtsson B-Å, Edén S, Lönn L, Kvist H, Stokland A, Lindstedt G et al. Treatment of adults with growth hormone (GH) deficiency with recombinant human GH. J Clin Endocrinol Metab 1993; 76: 309-17.
13. Jörgensen JOL, Pedersen SA, Thuesen L, Jörgensen J, Möller J, Möller J et al. Longterm growth hormone treatment in growth hormone deficient adults. Acta Endocrinol (Copenh) 1991; 125: 449-53.
14. Whitehead HM, Boreham C, McIlrath EM, Sheridan B, Kennedy AL, Atkinson AB et al. Growth hormone treatment of adults with growth hormone deficiency: results of a 13-months placebo controlled cross-ova study. Clin Endocrinol 1992; 36: 45-52.
15. Cuneo RC, Salomon F, Wiles CM, Hesp R, Sönksen PH. Growth hormone treatment in growth hormone deficient adults. II. Effects on exercise performance. J Appl Physiol 1991; 70: 695-700.
16. Pedersen SA, Jörgensen JOL, Welling K, Michaelsen KF, Christiansen JS, Skakkebæk NE. Decreased sweating in adults with growth hormone deficiency: Beneficial effects of growth hormone therapy. Lancet 1989; 2: 681-2.
17. Juul A, Behrenscheer A, Tims T, Nielsen B, Halkjær-Kristensen J, Skakkebäk NE. Impaired thermoregulation in adults with growth hormone deficiency during heat exposure and exercise. Clin Endocrinol 1993; 38: 237-44.
18. Christiansen JS, Gammelgaard J, Örskov H, Andersen AR, Telmer S, Parving HH. Kidney function and size in normal subjects before and during growth hormone administration for one week. Eur J clin Invest 1981; 11: 487-90.

19. Kaufmann J-M, Taelman P, Vermeulen A, Vandeweghe M. Bone mineral status in growth hormone deficient males with isolated and multiple pituitary deficiencies of childhood onset. J Clin Endocrinol Metab 1992; 74: 118-23.
20. Rosén T, Hansson T, Granhed H, Szucs J, Bengtsson B-Å. Reduced bone mineral content in adult patients with growth hormone deficiency. Acta Endocrinol (Copenh) 1993; 129: 201-6.
21. Wüster C, Slenczka E, Ziegler R. Increased prevalence of osteoporosis and arteriosclerosis in patients with conventionally substituted pituitary insufficiency: is there a need for growth hormone substitution? Klin Wochenschr 1991; 69: 769-73.
22. Rosén T, Bengtsson B-Å. Premature cardiovascular mortality in hypopituitarism - a study of 333 consecutive patients. Lancet 1990; 336: 285-8.
23. Bülow B, Hagmart L, Mikoczy Z, Nordström C-H, Erfurth EM. Increased cerobrovascular mortality in patients with hypopituitarism. Clin Endocrinol 1997; 46: 75-81.
24. Bates AS, Van't Hoff W, Jones PL, Clayton RN. The effect of hypopituitarism on life expectancy. J Clin Endocrinol Metab 1996; 81: 1169-72.
25. Shahi M, Beshyah SA, Hackett D, Sharp PS, Johnston DG, Foale RA. Myocardial dysfunction in treated adult hypopituitarism: a possible explanation for increased cardiovascular mortality. Br Heart J 1992; 67: 92-6.
26. Merola B, Cittadini A, Colao A, Longobardi S, Fazio S, Sabatini D, Saccà L, Lombardi G. Cardiac structural and functional abnormalities in adult patients with growth hormone deficiency. J Clin Endocrinol Metab 1993; 77: 1658-61.
27. Markussis V, Beshyah SA, Fisher C, Sharp P, Nicolaides AN, Johnston DG. Detection of premature atherosclerosis by high-resolution ultrasonography in symptom-free hypopituitary adults. Lancet 1992; 340: 1188-92.
28. Degerblad M, Almkvist O, Grunditz R, Hall K, Kaijser L, Knutsson E et al. Physical and psychological capabilities during substitution therapy with recombinant growth hormone in adults with growth hormone deficiency. Acta Endocrinol (Copenh) 1990; 123: 185-93.
29. Salomon F, Cuneo RC, Umpleby AM, Sönksen PH. Glucose and fat metabolism in adults with growth hormone deficiency. Clin Endocrinol 1994; 41: 315-22.
30. Beshyah SA, Henderson A, Niththyananthan R, Johnston DG. Metabolic abnormalities in growth hormone deficient adults: II. Carbohydrate and lipid metabolism. Endocrinol Metab 1994; 1: 173-80.
31. Jörgensen JOL, Pedersen SA, Laurberg P, Weeke J, Skakkebæk NE, Christiansen JS. Effect of growth hormone therapy on thyroid function of growth hormone deficient adults with and without concomitant thyroxine substituted central hypothyroidism. J Clin Endocrinol Metab 1989; 69: 1127-31.

Growth hormone replacement therapy in adult GH deficient patients

9 EFFECTS OF GH ON BODY COMPOSITION

Thord Rosén

Research Centre of Endocrinology and Metabolism, Sahlgrenska University Hospital.

INTRODUCTION

Growth hormone (GH) has profound effects, not only on longitudinal bone growth, but also on body composition via its anabolic, lipolytic and anti-natriuretic actions. With the availability of new techniques there has been an increased interest in the assessment of body composition in various disorders including GH diseases. This paper will shortly describe the different methods for determining body composition, and the changes occurring secondary to GH treatment in GH deficient patients.

BODY COMPOSITION MODELS

Indirect methods of determining body composition are based on models, which divide the body into distinct compartments, in turn being chemically defined. The two-compartment model divides the body fat into a fat compartment (BF) and fat-free mass (FFM). Body fat, consisting of stored triglycerides and structural lipids, is assumed to be anhydrous, potassium-free and to have a density of 0.9 g/ml (Mendez et al, 1960). The chemically heterogenous fat-free mass compartment is supposed to have a water content of 73% (Moore et al, 1963; Streat et al, 1985), a potassium content of 66 mmol/kg (Forbes & Lewis, 1956; Smith et al, 1979) and a constant density of 1.1 g/ml at 37°C (Brozek et al, 1963). A more anatomically defined two-compartment model divides the body into adipose tissue (AT) and lean body mass (LBM). In this model AT also comprises the non-lipid stroma, cell membranes, cytoplasm and organelles of the adipocytes.

In the 1950s, a chemically defined four-compartment model was developed, which divided the body into water, protein, bone minerals and fat (Keys & Brozek, 1953). This model has been the basis for describing body composition in vivo. With the development of new techniques the components of the four-compartment model can now be determined by various methods, differing in technical complexity and accuracy.

In Göteborg, Sweden a four-compartment model comprising body fat, body cell mass (BCM), extracellular water (ECW) and the remaining fat-free extracellular solids has been developed and being extensively used. The input variables for this model are apart from body weight and body height, total body water (TBW) and total body potassium (TBK) (Bruce et al, 1980).

AVAILABLE TECHNIQUES

Anthropometry

Anthropometric data, such as body height, body weight, skinfold thickness, arm circumference and bone measurements are used in multiple regression equations to predict body density, body fat and fat-free mass. The accuracy of the skinfold thickness measurement is dependent on the investigator's skill and the site of the measurement. The body mass index (BMI) (Body weight/(Body height)^2 is the most widely used index of relative weight.

Densitometry

Whole body density is determined by measuring body weight in air and during submersion in water and applying to Archimede's principle. The assessment of body composition according to the two-compartment model assumes a constant density of body fat and fat-free mass compartment. The assumption of a constant density in turn implies a constant level of hydration, a constant proportion of bone mineral to muscle and of skin to FFM (Garrow et al, 1990), which is not always the case in GH-related disorders.

Isotope dilution

The dilution technique to estimate an unknown volume is an established chemical principle that is nowadays mainly used to determine total body water. By administer a known amount of a specific isotope (deuterium, tritiated water or ^{18}O) and by determine the isotope activity in a sample after an equilibrium time, the dilution volume of the isotope, and hence the total body volume, is calculated.

Total body potassium

Since a constant fraction (0.012%) of all natural potassium is the gamma emitter ^{40}K, total body potassium (TBK) can now be determined by counting the 1.46 MeV gamma radiation from ^{40}K in specially designed whole-body counters. The TBK value is then used to predict BCM in the 4-compartment model.

In Vitro neutron Activation Analysis

The specific electromagnetic radiation that emits from atomic nuclei being irradiated by neutrons from a suitable source, can be used to identify and quantitate the atoms with which the neutrons interacted. Thus, both hydrogen, carbon, nitrogen and mineral determinations can be achieved with this method. The estimation of total body nitrogen and calcium allows the use of multicompartment models in body composition, and furthermore in combination with TBK measurements, the nitrogen/potassium ratio in various states of disease will be provided.

Bioelectrical impedance
Impedance methods rely upon the difference in electrical conductivity of body fat and FFM, where, in general, body fluids are responsible for the electrical conduction and cell membranes determine capacitance. There are two techniques: bioelectrical impedance analysis (BIA) and total body electrical conductivity. BIA is used to predict fat-free mass- and thus body fat- and total body water in a two-compartment model on the basis of regression equations using anthropometric and BIA variables.

Photon beam absorptiometry
The original technique of photom beam absorptiometry (single and double photon absorptiometry) was designed for estimating bone mineral content, but with the current DEXA technique also body fat and lean body mass can be determined. The principle is that e.g. bone mineral content is assumed to be directly proportional to the amount of photon energy absorbed by the bone segment studied. In DEXA x-radiation two different energies is used.

Computed tomography (CT)
With CT it is possible to determine the area and the volume of different organs and structures because of their specific radiological attenuation, expressed in Hounsfield units. With the multicompartment technique volumes of different body compartments such as adipose tissue, muscle, skin and visceral organs, and their local subcompartments (legs, trunks, arms) can be determined.

Magnetic Resonance Image (MRI)
In concentional MRI, hydrogen nuclei are excited by electromagnetic radiation in a magnetic field, making the activated nuclei emit the absorbed energy during relaxation during their move towards the pre-excited stage. In summary, this novel technique can be used in body composition determinations, especially of adipose tissue. Although the MRI technique is reliable and accurate, it is slower and more expensive than the CT technique, and has not yet found its place in routine body composition determinations.

The accuracy of different body composition methods
Thus, all in-vivo body composition methods have underlying assumptions (such as constant hydration and intracellular potassium concentrations) and methodological errors, that might effect the result of the observations. It is beyond the scope of this chapter to review studies comparing the accuracy, benefits and limitations of the methods. In summary, there is no "golden standard" method, that in a superior way can determine all body composition fractions, although it seems that DEXA and BIA-methods are reliable and reproducible in daily routine situations.

PHYSIOLOGICAL EFFECTS OF GH

Anabolic effect

In the classic experiment by Evans and Long in 1921, it was shown that extracts from bovine pituitary glands promoted growth in normal female rats (Evans & Long, 1921). In 1944, Li et al. isolated purified GH from beef pituitaries, which was followed by the isolation of human GH in 1956 (Li, 1944; Li & Papkoff, 1956). Furthermore, in 1958 Ikkos et al. observed nitrogen retention in normal adult subjects after GH administration, thereby demonstrating an anabolic action by GH (Ikkos et al, 1958).

Hypophysectomy in the rat leads to a reduction in DNA level and cell division in muscle cells, which is restored by GH therapy (Beach & Kostyo, 1968; Goldspink & Goldberg, 1975). Pituitary dwarfs have lower muscle DNA content than normal individuals of the same age (Cheek & Hill, 1974). In contrast to insulin and IGF-I, GH increases protein synthesis without affecting proteolysis (Binnerts et al, 1992; Russell-Jones et al, 1993). GH stimulates muscle growth by increasing the number of cells rather than the cell size (Cheek & Hill, 1974). The effect of GH on muscle is probably a direct one, or mediated by a paracrine effect of IGF-I (Isgaard et al, 1989).

Lipolytic effect

The lipolytic action of GH was first demonstrated in the 1930s, when it was shown that pituitary extracts reduced carcass fat in rats (Lee & Schaffer, 1934). The lipolytic effect of GH has therafter been documented from several *in vivo* and *in vitro* studies. Following hypophysectomy in rats, there is an increase in body fat and a decrease in lean body mass, these changes are reversed by GH (Li et al, 1949; Scow, 1959). In 1959, GH-induced lipolysis was demonstrated in man (Raben & Hollenberg, 1959).

The effects of GH on the cellular level are complex. GH promotes the differentiation of preadipocytes to adipocytes (Green et al, 1985). GH opposes the actions of insulin on the adipocytes, except during the initial phase of 2-3 hours when a short transient insulin-like effect is exerted. Furthermore, GH treatment promotes a redistribution of body fat from an abdominal (android) to a more peripheral (gynoid) distribution (Zachman et al, 1980). These changes have been associated with GH-induced inhibitions in the anti-lipolytic effect of insulin, which differs in different adipose tissue regions (Rosenbaum et al, 1989).

Anti-natriuretic effect

In 1952, Whitney et al. demonstrated that GH administration resulted in sodium retention with a simultaneous expansion of the extracellular volume (Whitney et al, 1952). The mechanisms are still incompletely understood, although evidence for both a direct renal effect and an indirect effect via the renin angiotensin system is suggested. Thus, GH might increase the activity of the sodium pump (Herlitz et al, 1992) and furthermore, increases in both plasma renin activity and aldosterone concentrations were noticed in healthy volunteers given pharmacological GH doses (Möller et al, 1991).

BODY COMPOSITION IN ACROMEGALY

GH excess, as in acromegaly, leads to specific changes in body composition according to the known anabolic, lipolytic and anti-natriuretic effects of GH. In a study of 166 acromegalic patients where body composition was measured according to the four-compartment model the patients were found to be heavier (about 7kg), with an mean increase in body cell mass (4.7 kg) and extracellular water (7.1 kg) and a mean decrease in body fat (3.7 kg) compared to healthy subjects of the same body height (Bengtsson et al, 1989a). In response to treatment there is a decrease in extracellular water and an increase in fat mass towards normalization , but no normalization of body cell mass, although normal GH levels are obtained (Aloia et al, 1976; Strauch et al, 1977; Bengtsson et al, 1989b;).

BODY COMPOSITION IN ADULT GHD PATIENTS

Body composition is typically altered in adult patients with GHD. In summary, the patients are overweight due to an increased body fat mass, and furthermore a decrease in lean body mass and extracellular water is seen. These changes are identical with those seen in normal aging, indicating that relative hyposomatotropism might explain the body composition changes that typically appear in older age (Rudman et al, 1981).

INCREASED BODY WEIGHT AND BODY FAT MASS

Thus, the adult GHD patients are overweight for their height, secondary to increased body fat mass . In a study of 106 patients with GHD acquired in adulthood where body composition was determined in a 4-compartment model male patients were 7.5 kg and female patients 3.6 kg heavier than predicted for their length (Rosén et al, 1993). In a study comprising 24 adult GHD patients, the weight was 26.5±5.5% above the ideal body weight (Salomon et al, 1989).

The increased body fat in GHD patients has been observed in several studies (Salomon et al, 1989; Jörgensen et al, 1991; Binnerts et al, 1992; Rosén et al, 1993). In the study of 106 patients the body fat was increased by about 6 kg compared to that predicted for their weight (Rosén et al, 1993).This is in accordance with another study where the muscle to fat ratio, measured with CT scan of the thigh, was higher in adult GHD patients, compared with matched controls (Jörgensen et al, 1991). Furthermore, in a trial of 13 GH deficient adults there was a higher proportion of fat mass and a lower fat free mass, measured with BIA and total body water dilution technique, in the patients compared with normal subjects (Binnerts et al, 1992). The excess adipose tissue is characteristically placed around the abdomen rendering an increased waist/hip ratio to the patients (Salomon et al, 1989; Johansson et al, 1994).

REDUCED LEAN BODY MASS

Measurements of the lean body mass support the clinical impression that muscle mass is decreased and replaced by adipose tissue in adult GHD patients. Thus, in a study from London comprising 24 GH deficient adults, lean body mass was on average decreased to 92.6±2.0% of the predicted value (Salomon et al, 1989). Furthermore, as described above, the muscle to fat ratio in 21 GHD patients of childhood onset was significantly lower than in matched controls (Jörgensen et al, 1989).

DECREASED EXTRACELLULAR WATER

Dehydration is a typical sign of GH deficiency and low estimates of total body water and extracellular water are observed in several studies of adult GHD patients. In a study of 106 patients with adult onset GHD a decrease of extracellular water of about 15% were found in the patients compared with healthy controls using the 4-compartment model (Rosén et al, 1993). The ECV decrease was observed in both males and females. The low ECV value is in accordance with other studies on adult GHD patients (de Boer et al, 1992; Binnerts et al, 1992). In a study of seven adult GHD patients Möller et al. found a decrease in both extracellular volume and plasma volume in the patients compared to healthy controls, when the respective volumes were measured directly with isotope methods (Möller et al, 1996). However, interestingly in a recent Dutch study comprising ten adult GHD patients, where extracellular water and total body water were independently determined with deuterium and bromide dilution and by BIA, total body water and intracellular water were lower and ECW/ICW ratio higher in the patients compared with controls (van Marken Lichtenbelt et al, 1997). This indicates, in contrast to other studies that GHD adults have an enlarged extracellular water compartment related to intracellular water. Further studies are needed to confirm these results.

EFFECTS OF GH TREATMENT ON BODY COMPOSITION IN ADULT GHD PATIENTS

The first clinical effects of GH on body composition were observed in the late 1950s, when GH-deficient children were treated with GH prepared from human cadaveric pituitaries (hpGH). Loss of adipose tissue and increases in muscle and bone mass were readily seen with the initiation of GH therapy in these patients (Novak et al, 1972; Collipp et al, 1973). Due to limited supplies of hpGH, it was not until the advent of recombinant human growth hormone (rhGH) in the mid 1980s, that GH was available in such amounts that GH treatment to adult GHD patients became a possibility.

EARLY SHORT-TERM STUDIES

In 1989, the first two major studies of growth hormone therapy in GH-deficient adults were published (Jörgensen et al, 1989; Salomon et al, 1989). Both these short-term trials showed positive effects on body composition. Thus, Jörgensen *et al.* showed in a double-blind placebo-controlled cross-over study of 21 adult patients with GHD of childhood onset that 4 months of GH treatment resulted in a significant increase in muscle mass and a concomitant decrease in fat mass (7%) as compared to placebo treatment. Body composition was assessed with CT scan and skinfold measurements. Salomon *et al.* at St. Thomas' Hospital in London showed in a double-blind placebo-controlled parallel study of 24 adult onset GHD patients, where total body potassium methods were used, an increase in fat free mass of 5.5±1.1 kg (10%) and a decrease in fat mass of 5.7±0.9 kg (20%) after 6 months GH treatment compared with placebo. Essentially similar findings with a significant decrease in fat mass and an increase in fat-free mass were found in subsequent short-term studies of 4-6 months GH treatment (Whitehead et al., 1992; Degerblad et al., 1992; Orme et al., 1992; Binnerts et al., 1992b).

In a double-blind cross-over placebo controlled trial of 6 months, using the 4-compartment model, CT scan and BIA, a more detailed in formation of GH-induced changes in body composition was given (Bengtsson et al., 1993). With the 4-compartment model a decrease of body fat by 6kg and a 15% increase in muscle mass were observed (Fig 1). With the CT scan a redistribution of adipose tissue from visceral to subcutaneous depots was observed, in line with previous findings of GH treatment to GH-deficient children, when a redistribution of fat from central to peripheral regions was seen secondary to GH therapy (Collip et al., 1973). As discussed before, this has been associated with GH induced reductions of the anti-lipolytic effect of insulin, which differs in different regions (Rosenbaum et al., 1989).

Apart from the anabolic and lipolytic effects of GH, a natriuretic effect of GH was readily observed in the early studies. Thus, an increase in total body water of 6 kg was noticed after 6 weeks of GH treatment, but after 6 months the TBW increase was only 4 kg (Bengtsson et al., 1993). The increase in TBW is secondary to an expansion of ECW, according to observations using dilution techniques and BIA (Binnerts et al., 1992; De Boer et al., 1992; Bengtsson et al., 1993). The increase in ECW might render the patients fluid retention symtoms with swelling of hands and feet, but these symptoms often disappear with reduction of the GH dose.

FOLLOWING LONG-TERM STUDIES

Numerous long-term studies of different designs have in summary shown that the acute beneficial effects of GH treatment to adult GHD patients are sustained with continuation of the treatment.

In the Australian multicenter trial 176 adult GHD patients were studied in a 6 month placebo-controlled period, followed by a 6 month GH period for all patients (Cuneo et al., 1998). Body composition measured with DEXA, skinfold thickness and BIA, showed an increase in fat-free mass and total body water and a decrease in fat mass including truncal fat, skinfold thickness and waist-hip ratio in the first 6 months. Most changes were complete by 3 months of treatment and maintained to 12 months. With the target dose of 0.25 U/kg per week, however, a large proportion of the patients had fluid retention symtoms, suggesting that lower initial doses will be needed to avoid side effects and bad compliance.

Several studies with the same study period of about 12 months but with fewer patients included preceeded the Australian multicenter study. Thus, in a study including 25 adult GHD patients, an initial 6 month double-blind period was followed by an open treatment phase, ending when all patients had received rhGH for12 months (Rosén et al, 1994). According to the four-compartment model body fat decreased and total body water increased in the rhGH group after 6 months compared with placebo, but no change in body cell mass was observed. Between 6 and 12 months rhGH treatment there was no further change in body fat or total body water compared with baseline, but both body cell mass and total body nitrogen increased during the same period.

In a similar study of 22 hypopituitary GHD patients there was a reduction in waist/hip ratio, total fat and central fat measured with DEXA, during the initial 6 month placebo period; an effect that was sustained after further 6 months GH treatment (Weaver et al; 1995). Another study using DEXA showed a favourable increase in lean body mass and a decrease in fat mass after 12 months GH therapy in 13 GH deficient patients (Wallymahmed et al; 1997).

Body composition was studied according to the two-compartment model in a study with 40 GH deficient hypopituitary adults, of which 27 and 11 were followed for 12 and 18 months, respectively, after an intial 6 month placebo-controlled period (Beshyah et al., 1995). There was a continuous increase in LBM and a continuous decrease in percentage BF at 6, 12 and 18 months, respectively. Furthermore, also TBW increased at 6 and 12 months.

The Danish group has performed a double-blind, placebo-controlled study for one year in 29 adult GHD patients (Jörgensen et al., 1996; Hansen et al., 1995). The GH dose was 2.0 U/m^2. Twelve months of GH therapy induced a mean total fat mass reduction of 4.9 kg, with the reduction being most marked in the trunk (3.1 kg). Total lean soft tissue mass increased by 3.3 kg, with the increase being most prominent in the extremities, resulting in a normalization of the muscle:fat ratio of the midthigh.

Four major trials have studied the effects on body composition of GH therapy lasting 2-4 years. Johannsson *et al* presented in 1996 an open study of 34 adult GHD patients on GH therapy for 24 months (Johannsson et al., 1996). Body composition, measured by BIA and by the four-compartment model, normalized body composition with a 11% reduction in body fat and increments in body cell mass of 5% and 10%, respectively, after 2 years; however, most of the changes were present already after 6 months treatment. The response was highly variable, suggesting a considerable individual susceptibility to GH, not explained by different GH doses. Furthermore, the target dose of 0.25 U/kg/week had to be reduced by about 20% after two years because of side-effects such as fluid retention.

In a recent paper a multicenter study from Belgium published the body composition results of 2 years of GH therapy in 148 adult GHD patients, being placebo-controlled for the first 6 months (Verhelst et al., 1997). Body composition was measured every 3-6 months with BIA. The positive increase in LBM (5.1%) and TBW (5.4%) and decrease in body fat (10.9%) after 3 months of GH therapy were maintained after 12 months of GH therapy, but was slightly attenuated after 24 months. After two years almost 40% of the patients had stopped treatment because of insufficient subjective improvement or side effects, the last being secondary to the target dose of 0.25U/kg/week.

In 1994 Jorgensen *et al* reported data from three years of uninterrupted GH therapy in ten GH-deficient adults, who had all completed a previous double-blind study of 4 months (Jorgensen et al., 1994). The initial increase in thigh muscle volume recorded with DEXA was maintained after 3 years of GH therapy. Furthermore a slight increase in body weight and thigh fat volume was recorded.

Finally, thirteen adult GHD patients were followed during 4 years of GH treatment (Al-Shoumer et al, 1996). Body composition was evaluated at one years′ interval with TBK, BIA and DEXA. In summary, these data showed that 4 years of GH treatment in hypopituitary GHD adults is associated with sustained improvement in body composition including a increase in FFM and TBW and a decrease in percentage body fat and waist/hip ratio. Body weight and BMI changed minimally up to 3 years, but showed a slight significant increase at the fourth year.

INDIVIDUAL RESPONSIVENESS TO GH

Although all studies concerning GH treatment to adult GHD patients show a favourable change towards normalization of body composition, the clinical impression indicates a highly individual variability in GH responsiveness. Several factors such as gender, age, body mass index might influence the response. In a Swedish multi-center comprising 68 adult GHD patients, all patients had 12 months of GH treatment after an initial 6 month placebo period. (Johannsson et al., 1996). After 12 months of GH treatment, the individual changes in BF varied from -12.5 to 4.3 kg and from -4.5 to 10.1 kg in LBM. Age and and baseline GHBP level were inversely correlated with the increase in LBM. The GH-induced increment in IGF-1 and TBW was greater in men than in women, whereas the decreases in BF were similar in men and women. Thus, this study disclosed variability in responsiveness to GH administration in GH-deficient adults. The best response was obtained in younger patients with low GHBP levels. furthermore, men responded better than women. Gender differences were also found in another Swedish study comprising 36 GHD patients (Burman et al., 1997). These studies strengthen the current strategy when individualizing the dose in the treatment of GHD patients according to clinical effects, side-effects and IGF-1 concentrations, more than correcting the dose according to the patients body weight.

REFERENCES

Aloia JF, Petrak Z, Ellis K, Cohn AJ. Body composition and skeletal metabolism following pituitary irradiation in acromegaly. Am J Med 1976; 61:59-63.

Al-Shoumer KA, Page B, Thomas E, Murphy M, Beshyah SA, Johnston DG. Effects of four years′ treatment with biosynthetic human growth hormone (GH) on body composition in GH-deficient hypopituitary adults. Eur J Endocrinol 1996; 135(5):559-567.

Beach RK, Kostyo JL. Effect of growth hormone on the DNA content of muscles in young hypophysectomized rats. Endocrinology 1968; 82:882-884.

Bengtsson BÅ, Brummer RJM, Edén S, Bosaeus I. Body composition in acromegaly. Clin Endocrinol 1989a; 30:121-130.

Bengtsson BÅ, Brummer RJM, Edén S, Bosaeus I, Lindstedt G. Body composition in acromegaly:the effect of treatment. Clin Endocrinol 1989b; 31:481-490.

Bengtsson B-Å, Edén S, Lönn L, H Kvist, A Stokland, G Lindstedt et al. Treatment of adults with growth hormone deficiency with recombinant human growth hormone. J Clin Endocrinol Metab1993; 76:309-317.

Beshyah SA, Freemantle C, Shahi, M, Anyaoku V, Merson S, Lynch S, Skinner E, Sharp P, Foale R, Johnston DG. Replacement treatment with biosynthetic human growth hormone in growth hormone-deficient hypopituitary adults. Clin Endocrinol 1995; 42:73-84.

Binnerts A, Swart GR, Wilson JHP, Hoogerbrugge N, Pois HAP, Birkenhager JC, Lamberts WJ. The effect of growth hormone administration in growth hormone deficient adults on bone, protein, carbohydrate and lipid homeostasis, as well as on body composition. Clin Endocrinol 1992; 37:79-87.

Brozek J, Grande F, Anderson T, Keys A. Densitometric analysis of body composition: revisions of some quantitative assumptions. Ann N Y Acad Sci 1963; 110:113-140.

Bruce Å, Andersson M, Arvidssson B, Isaksson B. Body composition. Prediction of normal body potassium, body water and body fat in adults on the basis of body height, body weight and age. Scand J Clin Lab Invest 1980; 40:461-473.

Burman P, Johansson AG, Siegbahn A, Vessby B, Karlsson FA. Growth Hormone (GH)-deficient men are more responsive to GH replacement therapy than women. J Clin Endocrinol Metab 1997; 82:550-555.

Cheek DB, Hill DE. Effect of growth hormone on cell and somatic growth. In Handbook of Physiology 1974; Volume 4, Section 7 (eds R.O. Greep & E.B. Astwood), pp. 159-185. American Physiological Society, Washington D.C.

Collipp PJ, Curti V, Thomas J, Sharma RK, Maddaiah VT, Cohn SH. Body composition changes in children receiving human growth hormone. Metabolism 1973; 22:589-595.

Cuneo RC, Judd S, Wallace JD, Perry-Keene D, Burger H, Lim-Tio S, Strauss B, Stockigt J, Topliss D, Alford F, Hew L, Bode H, Conway A, Handelsman D, Dunn S, Boyages S, Cheung NW, Hurley D. The Australian multicenter trial of growth hormone (GH) treatment in GH-deficient adults. J Clin Endocrinol Metab 1998; 83:107-116.

De Boer H, Blok GJ, Voerman HJ, de Vries PMJM, van der Veen EA. Body composition in adult growth hormone-deficient men, assessed by anthropometry and bioimpedance analysis. J Clin Endocrinol Metab 1992; 75: 833-837.

Degerblad M, Elgindy N, Hall K, Sjöberg H-E, Thorén M. Potent effect of recombinant growth hormone on bone mineral density and body composition in adults with panhypopituitarism. Acta Endocrinol 1992; 126:387-393.

Evans HM, Long JA. The effect of the anterior lobe administered intraperitoneally upon growth maturity, and oestrus cycles of the rat. Anat Rec 1921; 21:62-63.

Forbes GB, Lewis AM. Total sodium, potassium and chloride in adult man. 1956; J Clin Invest 35:596-600.

Garrow JS, Garby L, Lammert O. Comparison of estimates of fat-free mass in normal and obese women from measurements of body potassium, body water and body density. Eur J Clin Nutr 1990; 44:213-217.

Goldspink DF, Goldberg AL. The influence of pituitary growth hormone on DNA synthesis in rat tissues. Am J Phys 1975; 228:302-309.

Green H, Morikawa M, Nixon T. A dual effector theory of growth-hormone action. Differentiation 1985; 29:195-198.

Hansen TB, Vahl N, Jorgensen JOL, Christiansen JS, Hagen C. Whole body and regional soft tissue changes in growth hormone deficient adults after one year of growth hormone treatment:a double-blind, randomized, placebo-controlled study. Clin Endocrinol 1995; 43:689-696.

Herlitz H, Jonsson O, Bengtsson BÅ. Relationship between plasma growth hormone concentration and cellular sodium transport in acromegaly. Acta Endocrinol 1992; 127:38-43.

Ikkos D, Luft R, Gemzell CA. The effect of human growth hormone in man. The Lancet 1958; i:720-721.

Isgaard J, Nilsson A, Vikman K, Isaksson OGP. Growth hormone regulates the level of insulin-like growth factor-I mRNA in rat skeletal muscle. J Endocrinol 1989; 120:107-112.

Johannsson G, Bjarnasson R, Bramnert M, Carlsson LMS, Degerblad M, Manhem P, Rosén T, Thorén M, Bengtsson B-Å. The individual responsiveness to growth hormone (GH) treatment in GH-deficient adults is dependent on the level of GH-binding protein, body mass index, age and gender. J Clin Endocrinol Metab 1996; 81:1575-1581.

Johannsson G, Rosén T, Lindstedt G, Bosaeus I, Bengtsson B-Å. Effects of 2 years of growth hormone treatment on body composition and cardiovascular risk factors in adults with growth hormone deficiency. Endocrinol Metab 1996); 3 (Suppl.A):3-12.

Johansson JO, Landin K, Tengborn L, Rosén T, Bengtsson BÅ. High fibrinogen and plasminogen activator inhibitor (PAI-1) activity in growth hormone deficient adults. Arterioscler Thromb 1994; 14:434-437.

Jörgensen JOL, Pedersen SA, Thuesen L, Jörgensen J, Ingeman-Hansen T, Skakkebaek NE, Christiansen JS: Beneficial effects of growth hormone treatment in GH-deficient adults. Lancet 1989; i:1221-1225.

Jörgensen JOL, Pedersen SA, Thuesen L, Jörgensen J, Möller J, Muller J et al. Long-term growth hormone treatment in growth hormone deficient adults. Acta Endocrinol 1991; 125:449-453.

Jörgensen JOL, Thuesen L, Müller J, Ovesen P, Skakkebaek NE, Christiansen JS. Three years of growth hormone treatment in growth hormone-deficient adults: near normalization of body composition and physical performance. Eur J Endocrinol 1994; 130 (3):224-228.

Jorgensen JOL, Vahl N, Hansen TB, Thuesen L, Hagen C, Christiansen JS. Growth hormone versus placebo treatment for one year in growth hormone deficient adults: increase in exercise capacity and normalization of body composition. Clin Endocrinol 1996; 45:681-688.

Keys A, Brozek J. Body fat in adult man. Physiol Rev 1953; 33:245-325.

Lee MO, Schaffer NK. Anterior pituitary growth hormone and the composition of growth. J Nutr 1934; 7:337-363.

Li CH. The isolation of pituitary growth hormone. Science 1944; 99:183-184.

Li CH, Simpson ME, Evans HM. Influence of growth and adrenocorticotropic hormones on the body composition of hypophysectomized rats. Endocrinology 1949; 44:71-75.

Li CH, Papkoff H. Preparation and properties of growth hormone from human and monkey pituitary glands. Science 1956; 124:1293-1294.

Mendez J, Keys A, Anderson JT, Grande F. Density of fat and bone mineral of mammalian body. Metabolism 1969; 9:472-477.

Moore FD, Olesen KH, McCurrey JD, Parker HV, Ball MR, Boyden CM. The body cell mass and its supporting environment.1963; Philadelphia:WB Saunders.

Möller J, Jörgensen JOL, Möller N, Hansen KW, Pedersen EB, Christiansen JS. Expansion of extracellular volume and suppression of atrial natriuretic peptide after growth hormone administration in normal man. J Clin Endocrinol Metab 1991; 72:768-772.

Möller J, Frandsen E, Fisker S, Jörgensen JOL, Christiansen JS. Decreased plasma and extracellular volume in growth hormone deficient adults and the acute and prolonged effects of GH administration:a controlled experimental study. Clin Endocrinol 1996); 44:533-539.

Novak LP, Hayles AB, Cloutier MD. Effect of HGH on body composition of hypopituitary dwarfs. Mayo Clin Proc 1972; 47:241-246.

Orme SM, Sebastian JP, Oldroyd B, Stewart SP, Grant PJ, Stickland MH, Smith MA, Belchetz PE. Comparison of measures of body composition in a trial of low dose growth hormone replacement therapy. Clin Endocrinol 1992; 37:453-459.

Raben MS, Hollenberg CH. Effect of growth hormone on plasma fatty acids. J Clin Invest 1959; 38:484-488.

Rosén T, Bosaeus I, Tölli J, Lindstedt G, Bengtsson B-Å.Increased body fat mass and decreased extracellular fluid volume in adults with growth hormone deficiency. Clin Endocrinol 1992; 38:63-71.

Rosén T, Edén S, Larson G, Wilhelmsen L, Bengtsson B-Å. Cardiovascular risk factors in adult patients with growth hormone deficiency. Acta Endocrinol1993b; 129:195-200.

Rosén T, Johannsson G, Hallgren P, Caidahl K, Bosaeus I, Bengtsson BÅ. Beneficial effects of 12 months replacement therapy with recombinant human growth hormone to growth hormone deficient adults. Endocrinol Metab 1994a; 1:55-66.

Rosenbaum M, Gertner JM, Leibel R. Effects of systemic growth hormone (GH) administration on regional adipose tissue distribution and metabolism in GH-deficient children. J Clin Endocrinol Metab 1989; 69:1274-1281.

Rudman D, Kutner MH, Rogers CM, Lubin MF, Flemming GA, Bain RP. Impaired growth hormone secretion in the adult population. J Clin Invest 1981;67: 1361-1369.

Russell-Jones DL, Weissberger AJ, Bowes SB, Kelly JM, Thomason M, Umpleby AM, Jones RH, Sönksen PH. The effects of growth hormone on protein metabolism in adult growth hormone deficient patients. Clin Endocrinol 1993; 38:427-31.

Salomon F, Cuneo RC, Hesp R, Sönksen PH. The effects of treatment with recombinant human growth hormone on body composition and metabolism in adults with growth hormone deficiency. N Engl J Med 1989; 321:1797-1803.

Scow RO. Effect of growth hormone and thyroxine on growth and chemical composition of muscle, bone and other tissues in thyroidectomized-hypophysectomized rats. Am J Phys 1959; 196:859-865.

Smith T, Hesp R, MacKenzie H. Total body potassium calibrations for normal and obese subjects in two types of whole body counter. Phys Biol Med 1979; 24:171-175.

Strauch G, Lego A, Therain F, Bricaire H. Reversible plasma and red cell volumes increases in acromegaly. Acta Endocrinol 1977; 85: 465-478.

Streat SJ, Beddoe AH, Hill GL. Measurement of body fat and hydration of the fat-free body in health and disease. Metabolism 1985; 34:509-518.

Verhelst J, Abs R, Vandeweghe M, Mockel J, Legros J-J, Copinschi G, Mahler C, Velkeniers B, Vanhaelst L, Van Aelst A, De Rijdt D, Stevenaert A, Beckers A. Two years of replacement therapy in adults with growth hormone deficiency. Clin Endocrinol 1997; 47:485-494.

Wallymahmed ME, Foy P, Shaw D, Hutcheon R, Edwards RHT, MacFarlane IA. Quality of life, body composition and muscle strength in adult growth hormone deficiency: the influence of growth hormone replacement therapy for up to 3 years. Clin Endocrinol 1997; 47:439-446.

Weaver JU, Monson JP, Noonan K, John WG, Edwards A, Evans KA, Cunningham J. The effect of low dose recombinant human growth hormone replacement on regional fat distribution, insulin sensitivity, and cardiovasular risk factors in hypopituitary adults. J Clin Endocrinol Metab 1995; 80:153-159.

Whitehead H, Boreham C, McIlrath EM, Sheridan B, Kennedy L, Atkinson AB et al. Growth hormone treatment of adults with growth hormone deficiency: results of a 13-month placebo controlled cross-over study. Clin Endocrinol 1992; 36:45-52.

Whitney JE, Bennet LL, Li CH. Reduction of urinary sodium and potassium produced by hypophyseal growth hormone in normal female rats. Proc Soc Exp Biol Med 1952; 79:584-587.

van Marken Lichtenbelt, Snel YEM, Brummer R-JM, Koppeschaar HPF. Deuterium and bromide dilution, and bioimpedance spectrometry independently show that growth hormone-deficient adults have an enlarged extracellular water compartment related to intracellular water. J Clin Endocrinol Metab 1997; 82:907-911.

Zachman M, Fernandez F, Tassinari D, Thakker R, Prader A. Anthropometric measurements in patients with growth hormone deficiency before treatment with human growth hormone. Eur J Ped 1980; 133:227-282.

10 EFFECTS OF GH ON LIPID METABOLISM

Gudmundur Johannsson

Research Centre of Endocrinology and Metabolism, Sahlgrenska University Hospital, Göteborg, Sweden

Introduction

Well-controlled clinical studies have shown that GH deficiency in adults is associated with a number of cardiovascular risk factors (Bengtsson *et al.* 1995). These patients have an increased amount of body fat with an abdominal preponderance (Salomon *et al.* 1989; Johansson *et al.* 1994), an increased prevalence of treated hypertension (Rosén *et al.* 1993), impaired fibrinolysis (Johansson *et al.* 1994), glucose intolerance (Beshyah *et al.* 1994), insulin resistance (Johansson *et al.* 1995) and premature atherosclerosis (Markussis *et al.* 1992). This cluster of cardiovascular risk factors together with the dyslipoproteinemia may therefore help to explain the doubling in cardiovascular mortality observed in hypopituitary patients with untreated GH deficiency (Rosén & Bengtsson 1990; Bülow *et al.* 1996).

The association between high continuos level of GH and increased prevalence of diabetes mellitus and hypertension is well known in acromegalic patients. This may in turn explain their doubled cardiovascular mortality (Wright *et al.* 1969; Bengtsson *et al.* 1988). Disturbances in the lipoprotein pattern is a common finding in acromegalic patients. Serum concentration of total cholesterol (TC) has been shown to be either increased or decreased, but a more consistent finding is increased serum level of triglycerides (TG) (Nikkilä & Pelkonen 1975; Takeda *et al.* 1982). The importance of GH on lipoprotein metabolism is demonstrated by the observation of increased serum levels of apolipoprotein (apo) A-1 and decreased serum levels of TG and lipoprotein (Lp)(a) in response to successful adenomectomy in acromegalic patients (Oscarsson *et al.* 1994).

GH deficiency and lipoproteins in adults

Studies in rodents have demonstrated the importance of GH on the lipoprotein metabolism. For example, hypophysectomy in the rat changes the lipoprotein pattern from a predominantly high-density lipoprotein (HDL) to one with a distinct low-

Table 1.

Lipoprotein pattern in adults with GH deficiency.

Study	*n (M/F)*	*TC*	*LDL-C*	*apoB*	*HDL-C*	*TG*
(Libber *et al.* 1990)	36(28/8)	↑	↑	nd	↑	↔
(Wüster *et al.* 1991)	122 (79/43)	↑	nd	nd	nd	↑
(Cuneo *et al.* 1993)	24 (16/8)	↑	↑	↑	↓	↑
(Rosén *et al.* 1993)	104 (66/34)	↔	nd	nd	↓	↑
(De Boer *et al.* 1994)	64 (64/0)	↑	↑	↑	↔	↔

↑ increased, ↓ decreased and ↔ not different from a reference population.
nd, not determined.

density lipoprotein (LDL) peak (Oscarsson *et al.* 1989), suggesting that the presence of GH is essential for maintaining a normal lipoprotein pattern.

In adult hypopituitary patients, hyperlipidemia was observed as early as the 1960s (Summers *et al.* 1967) and early 1970s (Merimee *et al.* 1972) when also the possible role of GH in the regulation of serum cholesterol was suggested in the hypophysectomized rat (Friedman *et al.* 1970). The presence of dyslipoproteinemia in adult GH-deficient patients has also been confirmed in more recent trials. Most studies have shown increased TC, LDL-C, apoB and serum TG concentrations and some investigators have also reported reduced HDL-C level (Table 1). One study also demonstrated normal serum concentration of serum Lp(a) as compared with apo(a) phenotype matched controls (Johannsson *et al.* 1995).

Effect of GH treatment on lipoproteins

The early treatment trials with human pituitary GH (Friedman *et al.* 1974; Merimee & Pulkkinen 1980) and the more recent ones using recombinant human GH in GH-deficient adults have produced varying results in terms of the effects on serum TC, LDL-C, HDL-C and apoB concentrations (Salomon *et al.* 1989; Binnerts *et al.* 1992; Whitehead *et al.* 1992; Johannsson *et al.* 1995). One consistent finding is, however, that serum TG and apo A-I concentrations are not significantly affected (Salomon *et al.* 1989; Whitehead *et al.* 1992; Johannsson *et al.* 1995) (Table 2). In addition serum apoE concentration increase in response to GH treatment (Johannsson *et al.* 1995; Oscarsson *et al.* 1996). More prolonged and open treatment trials with GH have demonstrated a sustained increase in HDL-C concentration and a reduction in TC and LDL-C after 18 and 24 months of treatment (Beshyah *et al.* 1995;

Johannsson *et al.* 1996), while a 36-month treatment trial only demonstrated a transient effect on these lipids (Garry *et al.* 1996). The last study demonstrated, however, a sustained increase in serum Lp(a) concentration.

GH is known to stimulate the release of free fatty acids (FFA) from fat cells through its lipolytic effects on adipose tissue. Through increased flux of FFA to the liver (Björntorp 1990) and a stimulatory effect of GH on the esterification of oleic acid into triglycerides and phospholipids hepatic secretion of very-low density lipoprotein (VLDL) is stimulated (Elam *et al.* 1992). Moreover, in response to GH the serum LDL-C and apoB concentrations decreases (Oscarsson *et al.* 1989), probably as a result of the increased clearance of these lipoproteins through increased hepatic LDL receptor activity (Rudling *et al.* 1992). This novel action of GH on the LDL receptor activity is further demonstrated by the observation that the hypocholesterolemic effect of estrogen is abolished in GH-deficient rats (Rudling *et al.* 1992) and that the maintaining of normal expression of the hepatic LDL receptor is achieved only with GH and not with IGF-I (Rudling *et al.* 1996).

Table 2.

Lipoprotein pattern in adults with GH deficiency during GH/placebo controlled trials.

Author	*Duration*	*TC*	*LDL*	*apoB*	*HDL*	*TG*	*Lp(a)*
(Edén *et al.* 1993)	6m	↔	↔	↔	↑	↔	↑
(Cuneo *et al.* 1993)	6m	↓	↓	↓	↔	↔	nd
(Russell-Jones *et al.* 1993)	2m	↓	↓	↔	↔	↔	
(Rosén *et al.* 1994)	6m	↔	nd	nd	↑	↔	nd
(Beshyah *et al.* 1995)	6m	↔	↔	↔	↔	↔	nd
(Weaver *et al.* 1995)	6m	↔	nd	↓	nd	↔	↔
(Snel *et al.* 1995)	6m	↓	↔	nd	↑	↔	nd
(O'Halloran *et al.* 1996)	6m	↔	↔	nd	↔	↔	↑
(Baum *et al.* 1996)	18m	↔	↔	nd	↔	↔	nd

↑ increased, ↓ decreased and ↔ not different from a reference population. nd, not determined.

The main influence of GH on lipoprotein metabolism may therefore be increased VLDL secretion from the liver and increased LDL-C catabolism. The overall effect of GH on lipoprotein metabolism is, however, much more complex with both direct and indirect actions. Thus, the impact of GH on body composition, thyroxine metabolism and glucose homeostasis may all have important indirect effects on the lipoprotein metabolism. Two clinical studies have addressed this question. One study could not demonstrate any significant association between changes in lipoproteins and other metabolic effects during 6 months of GH treatment (Cuneo *et al.* 1993). A more recent study using multiple regression analysis found that accompanying changes in serum thyroxine, insulin and IGF-I concentrations and body composition could explain at most 27% of the changes in lipoproteins during GH treatment (Johannsson *et al.* 1995). These findings suggests that the direct effect of GH is of primary importance in the regulation of lipoprotein metabolism in man.

Many of the effects of GH on lipoprotein metabolism may be explained by its effect on the hepatic lipoprotein metabolism (Angelin & Rudling 1994). Few studies have addressed the question of whether GH can influence the biliary lipid metabolism. One study demonstrated increased bile acid synthesis and chenodeoxycholate pool size in response to GH treatment in children with GH deficiency (Heubi *et al.* 1983), while biliary lipid metabolism was not affected by GH treatment in healthy male volunteers (Olivecrona *et al.* 1995). The influence of GH on biliary lipid metabolism is, however, indicated by markedly reduced activity of the rate-limiting enzymes in cholesterol and bile acid biosynthesis in the hypophysectomized rat, and enhanced hepatic microsomal cholesterol esterification, normalization of enzymatic activity of cholesterol 7alpha-hydroxylase and increased fecal output of bile acids in response to GH replacement (Rudling *et al.* 1997).

Lipoprotein (a)

GH plays a role in the regulation of Lp(a), producing increased Lp(a) concentration during GH administration in both healthy (Olivecrona *et al.* 1993) and GH-deficient adults (Edén *et al.* 1993), and decreased concentration in successfully treated patients with acromegaly (Oscarsson *et al.* 1994). Some studies have established a strong relationship between Lp(a) and coronary heart disease (Rosengren *et al.* 1990; Sandholzer *et al.* 1992). Lp(a) is a lipoprotein which is similar to LDL in terms of its lipid and apoB content, but which also contains a glycosylated protein, called apo(a), covalently linked to apoB. The serum level of Lp(a) is mainly genetically determined with an inverse relationship to the size polymorphism of apo(a) (Uterman 1989). Moreover, subjects with high molecular weight (HMW) and low molecular weight (LMW) isoforms of apo(a) have been shown to respond differently to disease and therapy (Dieplinger *et al.* 1993; Kronenberg *et al.* 1994). As in healthy subjects, adult GH-deficient patients with LMW apo(a) phenotypes tend to have higher Lp(a) concentrations than subjects with HMW phenotypes, but the response to treatment in terms of changes in lipoprotein concentrations is not influenced by the size polymorphism of the apo(a) phenotype (Johannsson *et al.* 1995).

The increased Lp(a) concentrations in response to GH treatment is a consistent findings in both short- and long-term studies (Edén *et al.* 1993; Garry *et al.* 1996; O'Halloran *et al.* 1996), but not in all (Russell-Jones *et al.* 1993). The finding of increased Lp(a) during GH treatment suggests an increase in hepatic Lp(a) secretion. This is in line with the observation that the different Lp(a) levels with similar size isoforms of apo(a) are a result of a variability in Lp(a) production rate and that the major pathway of Lp(a) catabolism is not the LDL receptor (White & Lanford 1995), as was previously suggested (Hofman *et al.* 1990). If the rate of Lp(a) particle assembly from apo(a) and apoB is dependent on the interaction of apoB-containing lipoprotein particles and apo(a) at the hepatocyte cell surface (Marcovina & Morrisett 1996), the increased secretion of apoB and VLDL by the liver induced by GH could explain the increased Lp(a) concentrations in response to GH treatment. The lipolytic action of GH with an increased flux of free fatty acids (FFA) to the liver might therefore be one explanation for the increase in Lp(a). This is supported by decreased levels of Lp(a) in response to nicotinic acid, a potent antilipolytic agent (Rader & Brewer 1992), and the stimulated formation of TG-rich apoB-apo(a) containing particles in cultured human hepatocytes by oleate (Edelstein *et al.* 1992). Because of the large impact of GH on Lp(a) levels, the similar Lp(a) concentration in untreated GH deficiency and matched controls remains unexplained. However, according to the above reasoning, low LDL receptor activity and visceral adiposity may be two plausible explanations.

A positive correlation is found both between the baseline levels of IGF-I and Lp(a) and between the increment in these proteins during treatment (Johannsson *et al.* 1995). These findings indicate a relationship between the GH regulation of IGF-I and Lp(a) in man. The effects of GH on Lp(a) could therefore be mediated by IGF-I. However, the finding that the IGF-I treatment of adults (Olivecrona *et al.* 1995; Oscarsson *et al.* 1995) decreases serum Lp(a) concentrations suggests that GH exerts direct effects on Lp(a) production. The observation that lower, individualized and probably more physiological doses of GH resulted in a similar increment in Lp(a) (Johannsson *et al.* 1997) suggests that both the pattern of sc. GH delivery (Oscarsson *et al.* 1996) and the total amount of GH administered is of importance to the increment in Lp(a).

Individual responsiveness

The reason for the inconsistent results relating to the effect of GH treatment on lipoproteins is not obvious. A large variation in the individual response (Snel *et al.* 1995; Johannsson *et al.* 1996), small cohorts of patients, different duration of hypopituitarism (Hyes *et al.* 1996), different baseline levels of lipoproteins (Johannsson *et al.* 1995) and genetic factors that control lipoprotein metabolism may all be of importance. The influence of two genetic factors, the apo(a) and apoE phenotype on the treatment response to GH by lipoproteins has been studied (Johannsson *et al.* 1995). ApoE is a ligand for the apoE and the LDL receptors. Three major isoforms of apoE appear in the population, E2, E3 and E4. They determine six different apoE phenotypes with varying affinity for the LDL receptor and therefore contribute significantly to the normal variation in serum lipoprotein levels within populations (Uterman 1987). The E2 and E4 isoforms have both been associated with dyslipoproteinemia when interacting with other genetic or environmental factors (Uterman 1987). During 12 months of GH replacement therapy in adults , the reduction in TC, LDL-C and apoB and the increase in apoE was affected by the patients' apoE phenotype, with a more marked reduction in TC and apoB and a more pronounced increment in apoE in subjects with the E2 phenotype, possibly reflecting a change towards normalization, as subjects with the E2 phenotype have lower TC and apoB concentrations and higher apoE concentrations compared with the E4 phenotypes (Uterman 1987). Furthermore, the baseline TG concentration and the response to GH were affected by the patients' apo(a) phenotype.

The influence of gonadal steroids on the treatment response to GH has been suggested. Men increase their Lp(a) concentration more than women in response to the GH treatment (Johannsson *et al.* 1995). Although it has been suggested that both androgens and estrogens decrease serum Lp(a) levels, it is conceivable that estrogens are more potent than androgens in lowering Lp(a) serum concentrations (Henriksson *et al.* 1992). Furthermore, there are indications that the reduction in TC, LDL-C and apoB and the increment in HDL-C is more marked in men while there is an increase in serum TG concentration in women during GH treatment (Johannsson *et al.* 1995; Burman *et al.* 1997). This is probably a result of the markedly different action of estrogen and androgen on the VLDL and apoB metabolism (Seed 1990; Sacks & Walsh 1994).

Measuring serum concentrations of lipoproteins may not reflect the overall impact of GH on the lipoprotein metabolism, as GH affects both production and secretion from the liver and the clearance of lipoproteins from the circulation (Angelin & Rudling 1994). GH thus enhances the turnover of VLDL and LDL-C, which may in turn explain the increase in HDL-C concentration (Nikkilä *et al.* 1987) and an increase in LDL particle diameter in response to GH treatment (Hew *et al.* 1996). The influence of GH and IGF-I on the interplay between mononuclear cells, macrophages and the endothelial lining (Hochberg *et al.* 1992) adds another perspective to the complex action of GH on lipid metabolism which cannot be measured exclusively by serum concentrations of lipoproteins.

Triglycerides

Serum concentrations of TG are high in GH-deficient adults. This may be associated with their increased abdominal adiposity (Snel *et al.* 1995) and insulin resistance (Johansson *et al.* 1995; Hew *et al.* 1996; Hew *et al.* 1996). However, although a dramatic reduction in visceral adipose tissue occurs in response to GH treatment, serum TG concentration is not reduced (Bengtsson *et al.* 1993; Snel *et al.* 1995; Jørgensen *et al.* 1996). The lipolytic action of GH treatment probably increases the flux of FFA to the liver (Björntorp 1990) and increases the synthesis and secretion of VLDL from the liver. The lipoprotein lipase activity in adipose tissue is attenuated (Ottoson *et al.* 1995) and the post-heparin plasma lipoprotein lipase is not affected by GH treatment (Oscarsson *et al.* 1996). As serum TG concentrations do not increase under conditions of increased VLDL secretion the peripheral catabolism must be enhanced. Increased lipoprotein lipase activity in other tissues such as muscle is therefore likely (Oscarsson *et al.* 1996). Furthermore, the strong association between glucose/insulin homeostasis and TG metabolism (Reaven 1995) might be reflected in the response to GH. The unaffected TG levels might thus be explained by essentially unchanged insulin sensitivity (Fowelin *et al.* 1993) and glucose tolerance (Beshyah *et al.* 1995; Jørgensen *et al.* 1996) during more prolonged GH treatment. The reduction in serum TG concentration in patients with high baseline levels (Edén *et al.* 1993; Johannsson *et al.* 1995) may, however, be a true response in a subgroup of patients with specific characteristics.

GH treatment of abdominally obese men

Striking similarities exist between Syndrome X (Reaven 1988) and untreated GH deficiency in adults (Bengtsson 1993). The most central findings in both these syndromes are abdominal/visceral obesity, insulin resistance, premature atherosclerosis and increased mortality from cardiovascular diseases (Salomon *et al.* 1989; Rosén & Bengtsson 1990; Markussis *et al.* 1992; Björntorp 1993; Johansson *et al.* 1994; Reaven 1995). A common finding in both conditions is high levels of serum TG and low HDL-C concentrations. This may be associated with their increased abdominal adiposity and insulin resistance. Because of these similarities between GH deficiency in adults and Syndrome X, undetectable and low levels of GH, respectively in these condition may be of importance for their metabolic consequences.

We have learned that GH can improve several of the aberrations associated with GH deficiency. With this background the effects of GH on the metabolic aberrations associated with abdominal/visceral obesity and Syndrome X was studied (Johannsson *et al.* 1997). Nine months of GH treatment in middle-aged men with abdominal/visceral obesity resulted in a specific and marked decrease in both abdominal subcutaneous and visceral adipose tissue. Moreover, insulin sensitivity improved and serum concentrations of TC and TG decreased. In healthy adults, short-term GH administration has been reported to increase serum TG concentrations (Marcus *et al.* 1990). In the above study, the serum TG concentration also displayed an initial increase in response to GH treatment. This could be an effect of enhanced the VLDL production from the liver. However, after 9 months of GH treatment, the serum TG concentration had decreased again, probably as an effect of the increased insulin-stimulated glucose uptake.

SUMMARY

The overall impact of GH treatment on lipoproteins may have important effects in reducing cardiovascular mortality in adults with GH deficiency. A reduction in TC and LDL-C reduces the incidence of cardiovascular disease in both men and women (Levine *et al.* 1995) and the increase in HDL-C should be favorable when it comes to restraining the progress of atherosclerosis (Manninen *et al.* 1988; Badimon *et al.* 1990). Not all prospective trials have found Lp(a) to be an independent risk factor for coronary heart disease (Rader *et al.* 1994). Moreover, in a prospective angiographic study Lp(a) was found not to be as important a predictor of the progression of coronary artery disease as LDL-C and apoB (Watts *et al.* 1994). This indicates that the possible increase in cardiovascular risk due to an increase in the level of Lp(a) is smaller than the decline in risk due to changes in TC, LDL-C and apoB concentrations. The observation that patients with high baseline LDL-C levels showed a more marked decrease in LDL-C and a less pronounced increase in Lp(a) concentrations (Johannsson *et al.* 1995) suggests a more favorable effect on the cardiovascular risk than can be predicted from the mean response to GH treatment. Only when effects of GH replacement on progress of atherosclerosis, cardiovascular morbidity and mortality are available will the final answer of the clinical effects of GH replacement therapy on lipoprotein metabolism be available.

References

Angelin, B. & Rudling, M. (1994). Growth hormone and hepatic lipoprotein metabolism. *Curr Opin Lipidol* **5**: 160-165.

Badimon, J.J., Badimon, L. & Fuster, V. (1990). Regression of atherosclerotic lesions by high density lipoprotein plasma fraction in the cholesterol-fed rabbit. *J Clin Invest* **85**: 1234-1241.

Baum, H.B., Biller, B.M.K., Finkelstein, J.S., Cannistraro, K.B., Oppenheim, D.S., Schoenfeld, A.D., Michel, T.H., Wittink, H. & Klibanski, A. (1996). Effects of physiological growth-hormone therapy on bone-density and body-composition in patients with adult-onset growth-hormone deficiency - a randomized, placebo-controlled trial. *Ann Intern Med* **125**: 883-890.

Bengtsson, B.-Å. (1993). The consequences of growth hormone deficiency in adults. *Acta Endocrinol* **128(Suppl 2)**: 2-5.

Bengtsson, B.-Å., Edén, S., Ernest, I., Odén, A. & Sjögren, B. (1988). Epidemiology and long-term survival in acromegaly: a study of 166 cases diagnosed between 1955 and 1984. *Acta Med Scand* **223**: 327-335.

Bengtsson, B.-Å., Edén, S., Lönn, L., Kvist, H., Stokland, A., Lindstedt, G., Bosaeus, I., Tölli, J., Sjöström, L. & Isaksson, O.G.P. (1993). Treatment of adults with growth hormone (GH) deficiency with recombinant human GH. *J Clin Endocrinol Metab* **76**: 309-317.

Bengtsson, B.-Å., Rosén, T., Johansson, J.-O., Johannsson, G., Oscarsson, J. & Landin-Wilhelmsen, K. (1995). Cardiovascular risk factors in adults with growth hormone deficiency. *Endocrinol Metab* **2(Suppl B)**: 29-35.

Beshyah, S.A., Henderson, A., Niththyananthan, R., Skinner, E., Anyaoku, V., Richmond, W., Sharp, P. & Johnston, D.G. (1995). The effects of short and long term growth hormone replacement therapy in hypopituitary adults on lipid metabolism and carbohydrate tolerance. *J Clin Endocrinol Metabol* **80**: 356-363.

Beshyah, S.A., Henderson, A., Niththyanathan, R., Sharp, P., Richmond, W. & Johnston, D.G. (1994). Metabolic abnormalities in growth hormone deficient adults: II. Carbohydrate tolerance and lipid metabolism. *Endocrinol Metab* **1**: 173-180.

Binnerts, A., Swart, G.R., Wilsson, J.H.P., Hoogerbrugge, N., Pois, H.A.P., Birkenhager, J.C. & Lamberts, S.W.J. (1992). The effects of growth hormone administration in growth hormone deficient adults on bone, protein, carbohydrate and lipid homeostasis, as well as on body composition. *Clin Endocrinol* **37**: 79-87.

Björntorp, B. (1990). "Portal" adipose tissue as a generator of risk factors for cardiovascular disease and diabetes. *Arteriosclerosis* **10**: 493-496.

Björntorp, P. (1993). Visceral obesity: A "Civilization Syndrome". *Obes Res* **1**: 206-222.

Burman, P., Johansson, A.G., Siegbahn, A., Vessby, B. & Karlsson, F.A. (1997). Growth hormone (GH)-deficient men are more responsive to GH replacement therapy than women. *J Clin Endocrinol Metab* **82**: 550-555.

Bülow, B., Hagmar, L., Mikozy, Z., Nordström, C.-H. & Erfurth, E.M. (1996). Increased cerebrovascular mortality in patients with hypopituitarism. *Clin Endocrinol* **46**: 75-81.

Cuneo, R.C., Salomon, F., Watts, G.F., Hesp, R. & Sönksen, P.H. (1993). Growth hormone treatment improves serum lipids and lipoproteins in adults with growth hormone deficiency. *Metabolism* **42**: 1519-1523.

De Boer, H., Voerman, H.J., Phillips, M. & Schouten, J.A. (1994). Serum lipid levels in growth hormone-deficient men. *Metabolism* **43**: 199-203.

Dieplinger, H., Lackner, C., Kronenberg, F., Sandholzer, C., Lhotta, K., Hoppichler, F., Graf, H. & König, P. (1993). Elevated plasma concentrations of lipoprotein(a) in patients with end-stage renal disease are not related to the size polymorphism of apolipoprotein(a). *J Clin Invest* **91**: 397-401.

Edelstein, C., Davidson, N.O. & Scanu, A.M. (1992). Oleate stimulates the formation of triglyceride-rich particles containing apoB100-apo(a) in long-term primary cultures of human hepatocytes. *Chem Phys Lipids* **67/68**: 135-143.

Edén, S., Wiklund, O., Oscarsson, J., Rosén, T. & Bengtsson, B.-Å. (1993). Growth hormone treatment of growth homone-deficient adults results in a marked increase in Lp(a) and HDL cholesterol concentrations. *Arterioscl Thromb Vasc Biol* **13**: 296-301.

Elam, M.B., Wilcox, H.G., Solomon, S.S. & Heimberg, M. (1992). In vivo growth hormone treatment stimulates secretion of very low density lipoproteins by the perfused rat liver. *Endocrinology* **131**: 2717-22.

Fowelin, J., Attvall, S., Lager, I. & Bengtsson, B.-Å. (1993). Effects of treatment with recombinant human growth hormone on insulin sensitivity and glucose metabolism in adults with growth hormone deficiency. *Metabolism* **42**: 1443-1447.

Friedman, M., Byers, S.O. & Elek, S.R. (1970). Pituitary growth hormone essential for regulation of serum cholesterol. *Nature* **225**: 464-467.

Friedman, M., Byers, S.O., Rosenman, R.H., Li, C.H. & Neuman, R. (1974). Effect of subacute administration of human growth hormone on various serum lipid and hormone levels of hypercholesterolemic and normocholesterolemic subjects. *Metabolism* **23**: 905-12.

Garry, P., Collins, P. & Devlin, J.G. (1996). An open 36-month study of lipid changes with growth hormone in adults: lipid changes following replacement of growth hormone in adult acquired growth hormone deficiency. *Eur J Endocrinol* **134**: 61-66.

Henriksson, P., Angelin, B. & Berglund, L. (1992). Hormonal regulation of serum Lp(a) levels. Opposite effects after estrogen treatment and orchidectomy in males with prostatic carcinoma. *J Clin Invest* **89**: 1166-1171.

Heubi, J.E., Burstein, S., Sperling, M.A., Gregg, D., Subbiah, M.T.R. & Matthews, D.E. (1983). The role of human growth hormone in the regulation of cholesterol and bile acid metabolism. *J Clin Endocrinol Metab* **57**: 885-891.

Hew, F.L., Alford, F.P., Christopher, M., Rantzau, C., Koschmann, M., O'Neal, D., Ward, G. & Best, J.D. (1996). Effects of growth hormone deficiency and therapy in adults on skeletal muscle glucose metabolism, lipid profiles and regional body composition. *Endocrinol Metab* **3**(suppl A): 55-60.

Hew, F.L., Koschmann, M., Christopher, M., Rantzau, C., Vaag, A., Ward, G., Beck-Nielsen, H. & Alford, F. (1996). Insulin resistance in growth hormone-deficient adults: defects in glucose utilization and glycogen synthase activity. *J Clin Endocrinol Metab* **81**: 555-564.

Hochberg, Z., Hertz, P., Maor, G., Oiknine, J. & Aviram, M. (1992). Growth hormone and insulin-like growth factor-I increase macrophage uptake and degradation of low density lipoprotein. *Endocrinology* **131**: 430-435.

Hofman, S.L., Eaton, D.L., Brown, M.S., McConathy, W.J., Goldstein, J.L. & Hammer, R.E. (1990). Overexpression of human low density lipoprotein receptors leads to accelerated catabolism of Lp(a) lipoprotein in transgenic mice. *J Clin Invest* **85**: 1542-7.

Hyes, F.J., Fiad, T.M. & McKenna, T.J. (1996). *Can baseline characteristics of growth hormone (GH)-deficient adults predict a favourable response to GH therapy?* 10th International Congress of Endocrinology, San Francisco.

Johannsson, G., Bjarnason, R., Bramnert, M., Carlsson, L.M.S., Degerblad, M., Manhem, P., Rosén, T., Thorén, M. & Bengtsson, B.-Å. (1996). The individual responsiveness to growth hormone (GH) treatment in GH-deficient adults is dependent on the level of GH binding protein, body mass index, age and gender. *J Clin Endocrinol Metab* **81**: 1575-1581.

Johannsson, G., Mårin, P., Lönn, L., Ottosson, M., Stenlöf, K., Björntorp, P., Sjöström, L. & Bengtsson, B.-Å. (1997). Growth hormone treatment of abdominally obese men reduces abdominal fat mass, improves glucose and lipoprotein metabolism, and reduces diastolic blood pressure. *J Clin Endocrinol Metab* **82**: 727-734.

Johannsson, G., Oscarsson, J., Rosén, T., Wiklund, O., Olsson, G., Wilhelmsen, L. & Bengtsson, B.-Å. (1995). Effects of 1 year of growth hormone therapy on serum lipoprotein levels in growth hormone-deficient adults: influence of gender and apo(a) and apoE phenotypes. *Arterioscler Thromb Vasc Biol* **15**: 2142-2150.

Johannsson, G., Rosén, T. & Bengtsson, B.-Å. (1997). Individualized dose titration of growth hormone (GH) during GH replacement in hypopituitary adults. *Clin Endocrinol* **47**: 571-581.

Johannsson, G., Rosén, T., Lindstedt, G., Bosaeus, I. & Bengtsson, B.-Å. (1996). Effects of 2 years of growth hormone treatment on body composition and cardiovascular risk factors in adults with growth hormone deficiency. *Endocrinol Metab* **3(Suppl. A)**: 3-12.

Johansson, J.-O., Fowelin, J., Landin, K., Lager, I. & Bengtsson, B.-Å. (1995). Growth hormone-deficient adults are insulin-resistant. *Metabolism* **44**: 1126-9.

Johansson, J.-O., Landin, K., Tengborn, L., Rosén, T. & Bengtsson, B.-Å. (1994). High fibrinogen and plasminogen activator inhibitor activity in growth hormone-deficient adults. *Arterioscler Thromb Vasc Biol* **14**: 434-437.

Jørgensen, J.O.L., Vahl, N., Hansen, T.B., Thuesen, L., Hagen, C. & Christiansen, J.S. (1996). Growth hormone versus placebo treatment for one year in growth hormone deficient adults: increase in exercise capacity and normalisation of body composition. *Clin Endocrinol* **45**: 681-688.

Jørgensen, J.O.L., Vahl, N., Nyholm, B., Juul, A., Müller, J., Møller, N., Schmitz, O., Skakkebæk, N.E. & Christiansen, J.S. (1996). Substrate metabolism and insulin sensitivity following long-term growth hormone (GH) replacement therapy in GH-deficient adults. *Endocrinol Metab* **3**: 281-286.

Kronenberg, F., König, P., Lhotta, K., Öfner, D., Sandholzer, C., Margreiter, R., Dosch, E., Uterman, G. & Dieplinger, H. (1994). Apolipoprotein(a) phenotype-associated decrease in lipoprotein(a) plasma concentrations after renal transplantation. *Arterioscler Thromb* **14**: 1399-1404.

Levine, G.N., Keaney, J.F. & Vita, J.A. (1995). Cholesterol reduction in cardiovascular disease. Clinical benefits and possible mechanisms. *N Engl J Med* **332**: 512-521.

Libber, S.M., Plotnick, L.P., Johanson, A.J., Blizzard, R.M., Kviterovich, P.O. & Migeon, C.J. (1990). Long-term follow-up of hypopituitary patients treated with human growth hormone. *Medicine* **69**: 46-55.

Manninen, V., Elo, M.O., Frick, M.H., Haapa, K., Heinonen, O.P., Heinsali, P., Helo, P., Huttunen, J.K., Kaitaniemi, P., Koskinen, P., Mäenpää, H., Mälkönen, M., Mänttäri, M., Norola, S., Pasternack, A., Pikkarainen, J., Romo, M., Sjöblom, T. & Nikkilä, E.A. (1988). Lipid alterations and decline in the incidence of coronary heart disease in the Helsinki Heart Study. *JAMA* **260**: 641-651.

Marcovina, S.M. & Morrisett, J.D. (1996). Structure and metabolism of lipoprotein (a). *Curr Opin Lipidol* **6**: 136-145.

Marcus, R., Butterfield, G., Holloway, L., Gilliland, L., Baylink, D.J., Hintz, R.L. & Sherman, B.M. (1990). Effects of short-term administration of recombinant human growth hormone to elderly people. *J Clin Endocrinol Metab* **70**: 519-27.

Markussis, V., Beshyam, S.A., Fisher, C., Sharp, P., Nicolaides, A.N. & Johnston, D.G. (1992). Detection of premature atherosclerosis by high-resolution ultrasonography in symptom-free hypopituitary adults. *Lancet* **340**: 1188-1192.

Merimee, T.J., Hollander, W. & Fineberg, S.E. (1972). Studies of hyperlipidemia in the HGH-deficient state. *Metabolism* **21**: 1053-1061.

Merimee, T.J. & Pulkkinen, A. (1980). Familial combined hyperlipoproteinemia: evidence for a role of growth hormone deficiency in effecting its manifestation. *J Clin Invest* **65**: 829-35.

Nikkilä, E.A. & Pelkonen, R. (1975). Serumlipids in acromegaly. *Metabolism* **24**: 829-838.

Nikkilä, E.A., Taskinen, M.-R. & Sane, T. (1987). Plasma high-density lipoprotein concentration and subfraction distribution in relation to triglyceride metabolism. *Am Heart J* **113**: 543-548.

O'Halloran, D.J., Wieringa, G., Tsatsoulis, A. & Shalet, S.M. (1996). Increased serum lipoprotein(a) concentrations after growth hormone (GH) treatment in patients with isolated GH deficiency. *Ann Clin Biochem* **33**: 330-334.

Olivecrona, H., Ericsson, S. & Angelin, B. (1995). Growth hormone treatment does not alter biliary lipid metabolism in healthy adult men. *J Clin Endocrinol Metab* **80**: 1113-1117.

Olivecrona, H., Ericsson, S., Berglund, L. & Angelin, B. (1993). Increased concentrations of serum lipoprotein(a) in response to growth hormone treatment. *Br Med J* **306**: 1726-1727.

Olivecrona, H., Johansson, A.G., Lindh, E., Ljunghall, S., Berglund, L. & Angelin, B. (1995). Hormonal regulation of serum lipoprotein(a) levels. Contrasting effects of growth hormone and insulin-like growth factor-I. *Arterioscl Thromb Vasc Biol* **15**: 847-849.

Oscarsson, J., Lundstam, U., Gustafsson, B., Wilton, P., Edén, S. & Wiklund, O. (1995). Recombinant human insulin-like growth factor-I decreases serum lipoprotein(a) concentrations in normal adult men. *Clin Endocrinol* **42**: 673-676.

Oscarsson, J., Olofsson, S.O., Bondjers, G. & Edén, S. (1989). Differential effects of continuous versus intermittent administration of growth hormone to hypophysectomized female rats on serum lipoproteins and their apoproteins. *Endocrinology* **125**: 1638-49.

Oscarsson, J., Ottosson, M., Johansson, J.-O., Wiklund, O., Mårin, P., Björntorp, P. & Bengtsson, B.-Å. (1996). Two weeks of daily injections and continuous infusion of recombinant human growth hormone (GH) in GH-deficient adults: II. Effects on serum lipoproteins and lipoprotein and hepatic lipase activity. *Metabolism* **45**: 370-377.

Oscarsson, J., Wiklund, O., Jakobsson, K.-E., Petrusson, B. & Bengtsson, B.-Å. (1994). Serum lipoproteins in acromegaly before and after 6-15 months after transsphenoidal adenomectomy. *Clin Endocrinol* **41**: 603-608.

Ottoson, M., Vikman-Adolfsson, K., Enerbäck, S., Elander, A., Björntorp, P. & Edén, S. (1995). Growth hormone inhibits lipoprotein lipase activity in human adipose tissue. *J Clin Endocrinol Metab* **80**: 936-941.

Rader, D.J. & Brewer, H.B. (1992). Lipoprotein(a): clinical approach to a unique atherogenic lipoprotein. *JAMA* **267**: 1109-1112.

Rader, D.J., Hoeg, J.M. & Brewer, H.B. (1994). Quantitation of plasma apolipoproteins in the primary and secondary prevention of coronary artery disease. *Ann Int Med* **120**: 1012-1025.

Reaven, G.H. (1988). Role of insulin resistance in human disease. *Diabetes* **37**: 1595-1607.

Reaven, G.M. (1995). Pathophysiology of insulin resistance in human disease. *Physiol Rev* **75**: 473-486.

Rosén, T. & Bengtsson, B.-Å. (1990). Premature mortality due to cardiovascular diseases in hypopituitarism. *Lancet* **336**: 285-288.

Rosén, T., Edén, S., Larsson, G., Wilhelmsen, L. & Bengtsson, B.-Å. (1993). Cardiovascular risk factors in adult patients with growth hormone deficiency. *Acta Endocrinol* **129**: 195-200.

Rosén, T., Johannsson, G., Hallgren, P., Caidahl, C., Bosaeus, I. & Bengtsson, B.-Å. (1994). Beneficial effects of 12 months replacement therapy with recombinant human growth hormone to growth hormone deficient adults. *Endocrinol Metab* **1**: 55-66.

Rosengren, A., Wilhelmsen, L., Eriksson, E., Risberg, B. & Wedel, H. (1990). Lipoprotein(a) and coronary heart disease. A prospective case-control study in a general population sample of middle aged men. *Br Med J* **301**: 1248-1251.

Rudling, M., Norstedt, G., Olivecrona, H., Reihnér, E., Gustafsson, J.-Å. & Angelin, B. (1992). Importance of growth hormone for the induction of hepatic low density lipoprotein receptors. *Proc Natl Acad Sci USA* **89**: 6983-6987.

Rudling, M., Olivecrona, H., Eggertsen, G. & Angelin, B. (1996). Regulation of rat hepatic low density lipoprotein receptors. In vivo stimulation by growth hormone is not mediated by insulin-like growth factor I. *J Clin Invest* **97**: 292-299.

Rudling, M., Parini, P. & Angelin, B. (1997). Growth hormone and bile acid synthesis. Key role for the activity of hepatic microsomal cholesterol 7 alpha-hydroxylase in the rat. *J Clin Invest* **99**: 2239-2245.

Russell-Jones, D.L., Watts, G.F., Weissberger, A., Naoumova, R., Myers, J., Thompson, G.R. & Sönksen, P.H. (1993). The effect of growth hormone replacement on serum lipids, lipoproteins, apolipoproteins and cholesterol precursors in adult growth hormone deficient patients. *Clin Endocrinol* **41**: 345-350.

Sacks, F.M. & Walsh, B.W. (1994). Sex hormones and lipoprotein metabolism. *Curr Opin Lipidol* **5**: 236-240.

Salomon, F., Cuneo, R.C., Hesp, R. & Sönksen, P.H. (1989). The effects of treatment with recombinant human growth hormone on body composition and metabolism in adults with growth hormone deficiency. *N Engl J Med* **321**: 1797-1803.

Sandholzer, C., Saha, N., Kark, J.D., Rees, A., Jaross, W., Dieplinger, H., Hoppichler, F., Boerwinkle, E. & Utermann, G. (1992). Apo(a) isoforms predict risk for coronary heart disease: a study in six populations. *Arterioscler Thromb* **12**: 1214-1226.

Seed, M. (1990). Sex hormones, lipoproteins, and cardiovascular risk. *Atherosclerosis* **90**: 1-7.

Snel, Y.E.M., Doerga, M.E., Brummer, R.-J.M., Zelissen, P.M.J. & Koppeschaar, H.P.F. (1995). Magnetic resonance imaging-assessed adipose tissue and serum lipid and insulin concentrations in growth hormone-deficient adults: Effect of growth hormone replacement. *Arterioscler Thromb Vasc Biol* **15**: 1543-1548.

Summers, V.K., Hipkin, L.J. & Davis, J.C. (1967). Serum lipids in disease of the pituitary. *Metabolism* **12**: 1106-1113.

Takeda, R., Tatami, R., Ueda, K., Sagara, H., Nakabayashie, H. & Mabushi, H. (1982). The incidence and pathogenesis of hyperlipidaemia in 16 consequtive acromegalic patients. *Acta Endocrinol* **100**: 358-362.

Uterman, G. (1987). Apolipoprotein E polymorphism in health and disease. *Am Heart J* **113**: 433-440.

Uterman, G. (1989). The mysteries of lipoprotein(a). *Science* **246**: 904-910.

Watts, G., Mandalia, S., Slavin, B.M., Brunt, J.N.H., Coltart, D.J. & Lewis, B. (1994). Metabolic determinants of the course of coronary artery disease in men. *Clin Chem* **40**: 2240-2246.

Weaver, J.U., Monsson, J.P., Noonan, K., John, W.G., Edwards, A., Evans, K.A. & Cunningham, J. (1995). The effect of low dose recombinant human growth hormone replacement on regional fat distribution, insulin sensitivity, and cardiovascular risk factors in hypopituitary adults. *J Clin Endocrinol Metab* **80**: 153-159.

White, A.L. & Lanford, R.E. (1995). Biosynthesis and metabolism of lipoprotein(a). *Curr Opin Lipidol* **6**: 75-80.

Whitehead, H., Boreman, C., McIlrath, E.M., Sheridan, B., Kennedy, L., Atkinson, A.B. & Hadden, D.R. (1992). Growth hormone treatment of adults with growth hormone deficiency: results of a 13-month placebo controlled cross-over study. *Clin Endocrinol* **36**: 45-52.

Wright, A.D., Hill, D.M., Lowry, C. & Russell Fraser, T. (1969). Motality in acromegaly. *Q J Med* **34**: 1-16.

Wüster, C.H.R., Slenczka, E. & Ziegler, R. (1991). Arteriosklerose bei konventionell substituierter hypophysenvorderlappeninsuffizienz: bedarf einer zusätzlichen Wachstumshormonsubstitution? *Klin Wochenschr* **69**: 769-773.

11 THE EFFECTS OF GROWTH HORMONE ON GLUCOSE & PROTEIN METABOLISM

D. L. RUSSELL-JONES AND A. M. UMPLEBY

UMDS Guy's and St Thomas's Medical and Dental School
Division of Medicine, Department of Endocrinology
Diabetes & Metabolis Medicine
St Thomas's Hospital, London, UK

GLUCOSE METABOLISM

Houssay was the first to observe that hypophysectomy reduced the hyperglycaemia of diabetes (1) in animals. Following these observations the metabolic properties of pituitary extracts were studied in detail (2). Young was the first to show that the diabetogenic factor in the pituitary co-extracted with the growth promoting activity (3) and subsequently short term and long term GH administration was shown to induce transient or permanent diabetes respectively (4 , 5) . Following the full characterisation and identification of growth hormone (GH) a number of studies were performed that demonstrated profound effects on glucose metabolism. Acromegaly "natures" experiment of GH excess is associated with a spectrum of impaired glucose tolerance and diabetes (6). Mintz elegantly demonstrated the diabetogenic effects of endogenous GH in a study in which glucose tolerance was measured after insulin induced hypoglycaemic GH release. (7). Glucose levels were much higher in the presence of elevated GH. When high doses of GH are administered to normal subjects the insulin antagonistic effect is seen after a lag of 3-4 hours (8,9). Although the mechanisms are poorly understood GH appears to reduce the sensitivity to insulin at both the liver and peripheral tissues (fat and muscle) (10, 11, 12, 13). GH increases lipolysis and FFA release and this has been proposed as a possible mechanism for the reduced insulin sensitivity. FFA delivery to the liver promotes gluconeogenesis and increased availability of FFA in the periphery competes with glucose as an oxidative substrate (14). There is also some evidence to suggest that GH has a direct effect on the pancreatic islets leading to increased insulin secretion (15).

In addition to these "diabetogenic" effects of GH, paradoxical transient insulin-like effects have been observed. GH increases glucose oxidation and uptake in isolated muscle, kidney and adipose tissue within 20-60 minutes of exposure (16). In normal and hypohysectomised rats a similar acute effect is seen (17). These findings have been confirmed in humans (18), but in the continued presence of GH tissues become refractory (19), their physiological relevance is therefore questionable.

GLUCOSE METABOLISM IN ADULT GHD PATIENTS AND THE RESPONSE TO TREATMENT

The importance of GH in glucose homeostasis in humans can be demonstrated by the common finding of hypoglycaemia in children with GH deficiency.(20) Studies have shown that lower than expected hepatic glucose production combined with increased glucose (21) utilisation are responsible for this fasting hypoglycaemia. In lean adults with GH deficiency fasting blood glucose levels are normal, however the glucose response to prolonged (4 day) fasting is altered with decreased blood glucose and plasma insulin levels compared with control subjects.(22). Thus although frank hypoglycaemia is not seen abnormalities on profound fasting become apparent. It is possible to explain these observations as glucagon and adrenaline play a more prominent and GH plays a less prominent gluco-regulatory role in the adult compared to children (23). Conversely during oral glucose tolerance tests GH deficient patients have been shown to have higher glucose levels and increased prevalence of impaired glucose tolerance in comparison to normal controls (24, 25). Adult GH deficient patients also have altered body composition which contributes to the observed insulin resistance with reduced skeletal muscle and lean tissue combined with increased visceral and total fat mass (24, 26).

A study of 24 adults with GH-deficiency demonstrated fasting insulin levels above the normal reference range and a significant positive correlation between fasting plasma insulin and both fat mass and waist girth (26). Johansson et al (27) showed using a hyperinsulinaemic clamp that GH deficient patients were relatively insulin resistant (fig 1). This was studied in greater detail in 14 adult GH deficient patients and 12 controls by Hew et al (28) using hyperinsulinaemic euglycaemic clamp (55 mU/min), combined with measurement of glycogen synthetic rate by muscle biopsy. Both groups had similar basal rates of glucose utilisation, but a reduction (64%) in insulin stimulated glucose utilisation occurred in the GH-deficient adults compared with control subjects. In addition the insulin sensitivity index of glucose disposal and glycogen synthesis were reduced in the patients.. These data confirm that patients with GH-deficiency have insulin resistance, with inhibition of the glycogen synthase pathway. Salomon et al (26) assessed fasting glucose metabolism in 24 GH deficient adults using D-[3-^{3}H] glucose and indirect calorimetry. In these studies glucose turnover was low when expressed in relation to body weight but normal when corrected for LBM. Glucose turnover correlated positively with fat oxidation. These changes are consistent with those observed in normal adults during a prolonged fast and suggest that adults with GH deficiency have reduced stores of glycogen.

GH replacement therapy has also been shown to affect glucose homeostasis in a number of studies. In a double-blind placebo-controlled cross-over study Fowelin et al (29) assessed the effects of 6 months GH replacement on glucose metabolism in 9 adults with GH-deficiency. Using D-[3-^{3}H] glucose tracer and a hyperinsulinaemic euglycaemic clamp they investigated basal glucose and insulin stimulated glucose metabolism after 6 weeks and 6 months GH treatment. After 6

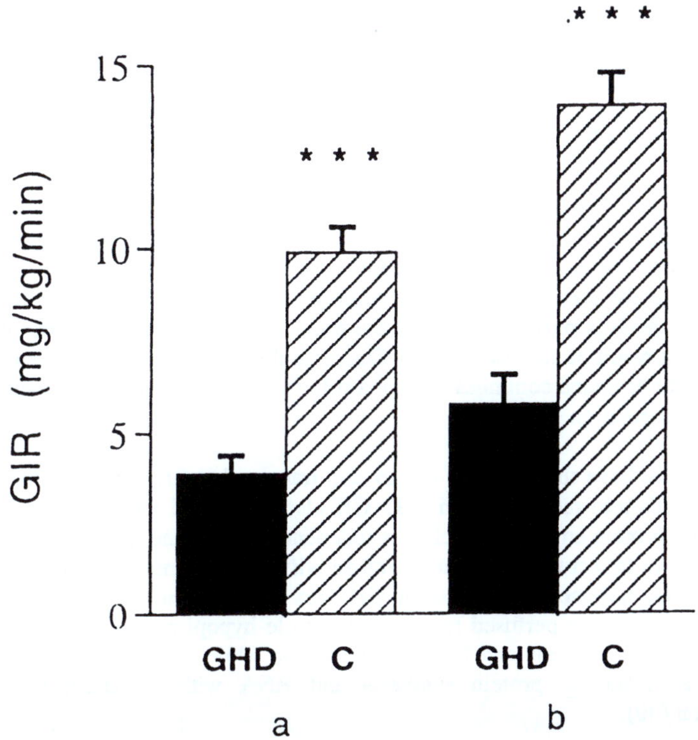

Figure 1: Glucose infusion rate (GIR) in adult patients with growth hormone deficiency (GHD) compared to matched controls (C) (a) GIR/body weight (b) GIR/lean body mass.

weeks GH treatment, fasting glucose levels and plasma insulin concentrations were elevated but both had returned to baseline at 26 weeks. The amount of glucose required to maintain euglycaemia during an insulin clamp was significantly lower after 6 weeks of therapy but there was no difference above baseline at 26 weeks. These changes were interpreted as a short term reduction of insulin stimulated glucose utilisation, mediated by GH, with reversal of these changes with time, perhaps as a result of altered body composition. Using an intravenous glucose tolerance test (IVGTT) combined with Bergman's minimal model O'Neal et al (30) demonstrated increased fasting glucose, insulin and C-peptide as well as an increase in both the first and second phase of insulin secretion following 1 week of GH therapy. Insulin-mediated glucose disposal and glucose decay were decreased but the fractional clearance rate of insulin was unchanged. Following 3 months GH, the fasting plasma insulin and C-peptide concentrations remained elevated but all other variables had returned to baseline values.

In summary, GH-deficient adults are insulin resistant which may be related to their altered body composition and visceral adiposity. Following GH replacement glucose and insulin levels rise transiently and insulin resistance measured by clamp studies worsens. In the longer term insulin resistance returns to baseline values. It is thought that the likeliest mechanism for this late change is an improvement in body composition with more prolonged GH replacement.

PROTEIN METABOLISM

In 1957 human GH was isolated (31) and was shown to be effective in children not only as a stimulator of growth but also to significantly affect body composition increasing lean tissue and reducing fat mass (32). However normal growth occurs over a relatively short time period while growth hormone secretion continues throughout life. It is only recently that the importance of GH in adults has become apparent as a result of studies in GH deficient adults (33, 34). It is now clear that in adults growth hormone continues to play a role in the control of protein metabolism and body composition.

In vitro studies have shown, using amino acid tracers, that GH stimulates the uptake of amino acids into rat diaphragm (38). GH also stimulated the uptake of labelled glycine into protein an effect which was independent of amino acid uptake (36,37). GH was shown to increase the number and activity of ribosomes in rat muscle and to increase RNA polymerase suggesting GH may control synthesis at the level of translation (38,39). In perfused rat skeletal muscle hypophysectomy was shown to decrease protein synthesis and RNA concentration while treatment for 5 days with GH increased muscle protein synthesis and RNA with no effect on protein degradation (40).

The effect of GH on protein metabolism in adults is clearly demonstrated by its effect to increase lean body mass (LBM). Studies of body composition in adult GHD have demonstrated a mean reduction in LBM of 7-8% corresponding to approximately 4 kg of lean tissue in these patients compared to an age and sex matched control group (33). In the first clinical trials of GH LBM was shown to increase by 4 kg following 0.07U/kg/day GH treatment for 1 month (33). All subsequent studies of GH replacement in adult GH deficient patients have also shown an increase in lean body mass of a mean of 2-5.5 kg after 6 months despite the use of a range of different measurement techniques (44, 45, 46, 47, 48). Open studies conducted over longer periods of GH treatment (up to 3 years) suggest that the restoration of lean body mass is maintained over longer periods (46). GH treatment has also been shown to significantly increase thigh muscle cross-sectional area demonstrating that skeletal muscle, a major component of lean body mass increases in a parallel manner (46).

While studies of LBM can provide an insight into the effects of GH on protein metabolism only net effects can be determined. Using isotopic techniques several studies have investigated the effect of GH on protein synthesis and degradation. In normal subjects the effect of GH on protein metabolism at pharmacological doses (0.1mg/kg/day) was investigated using ^{13}C leucine as a tracer.

The study showed that GH increased whole body protein synthesis but had no effect on proteolysis (47). Forearm protein metabolism has also been studied in normal subjects using ^{3}H phenylalanine as a tracer. GH infused locally into the brachial artery to raise forearm but not systemic GH levels stimulated forearm protein synthesis by 66% within 6h after beginning the infusion without altering proteolysis. However GH also increased forearm blood flow (48).

The effects of GH on protein metabolism in GHD adults have produced conflicting data. In a 6 month open study of GH replacement (0.5 IU/kg/day) in 8 adult GH deficient adults Binnerts et al (44) investigated the effects of GH on protein metabolism using ^{15}N glycine. Protein synthesis was increased at 1 month, with no alteration in protein degradation. This effect was unsustained and was no longer evident following 3, 6 or 9 months GH therapy. When rates of protein metabolism were expressed in terms of lean body mass (LBM) no significant differences were detectable throughout the entire study duration. In a study of 16 adults with GH deficiency compared with 20 matched controls Beshyah et al (49) used L-[1-^{13}C] leucine to study protein metabolism. Compared with a control group the patients had decreased leucine flux, oxidation and protein synthesis. Seven of these patients underwent an identical study after 6 months GH therapy (0.18-0.35 IU/kg/week) which demonstrated a non-significant increase in protein flux, oxidation and synthesis.

In contrast to these findings two other studies have shown GH to have a significant effect on protein synthesis in adult GHD after 1 and 2 months respectively. Salomon et al (50) demonstrated an increase in protein synthesis in a study of 24 patients in a double blind study of GH replacement of 0.07U/kg/day for 6 months. Using a bolus injection of L-[^{14}C leucine] to measure protein metabolism an increase in protein synthesis expressed in terms of LBM was demonstrated after 1 month of GH treatment but not after 6 months of treatment. In a randomised double blind trial of GH treatment for 2 months (0.018U/kg/day for 1 month followed by 0.036U/kg/day for 1 month) protein metabolism was measured using an infusion of 13C leucine. GH treatment was shown to result in an increase in non oxidative leucine disappearance rate expressed in terms of lean body mass (a measure of protein synthesis) and had an additional effect to reduce leucine oxidation rate. Although proteolysis per kg body weight was increased there was no significant effect on the rate of proteolysis when expressed in terms of lean body mass (51).

The major alterations in body composition occur within 1 month of commencement of GH therapy (33,34) and it is likely that any significant alteration in protein metabolism would occur over this period. This may account for the results of the studies of protein metabolism, which have demonstrated an increase in protein synthesis up to 2 months (51) after commencement of GH, but little alteration above baseline after 6 months therapy (44,50) most likely as a result of achieving a new baseline rate of metabolism.

The studies described above were performed in the postabsorptive state. However most protein gain occurs in the postprandial state when circulating amino acids are elevated. Russell-Jones et al also studied the effect of GH treatment (0.018U/kg/day for 1 month followed by 0.036U/kg/day for 1 month) on postprandial protein metabolism in a double blind placebo study of 12 GHD adults (52) using ^{13}C leucine. The study demonstrated that in the GH treated group the increase in protein synthesis in the postabsorptive state was maintained during the meal resulting in a greater increase in post-prandial protein synthesis than in the placebo group. This study suggests that the GH induced increase in LBM in these patients is due to increased protein synthesis in both the postprandial and postabsorptive state.

Thus growth hormone seems to produce its anabolic effect by stimulating protein synthesis without affecting proteolysis in contrast to insulin (53) which predominantly inhibits proteolysis. These effects of GH on protein metabolism provide a mechanism for the changes in body composition and increased lean body and muscle mass observed in all studies of GH replacement in adult GH deficient adults.

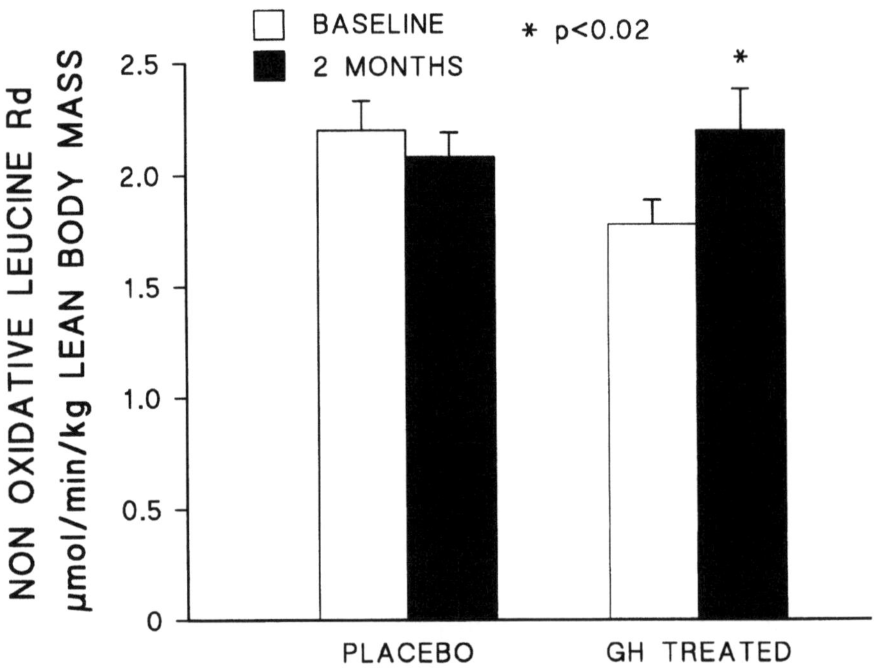

Figure 2: The effect of GH on leucine production rate (Ra; a measure of protein degradation) and non-oxidative leucine disposal rate (NOLD; a measure of protein synthesis), measured following an infusion of 1-^{13}C leucine, in 18 growth hormone

deficient adults. The study was a double blind placebo controlled trial of recombinant GH (0.18U/kg/day for 1 month followed by 0.036U/kg/day for 1 month). Nine patients received GH treatment and 9 patients received placebo. Patients were studied at 0 and 2 months (94). * significantly different from baseline, $p<0.05$.

REFERENCES

1. Houssay BA. The hypophysis and metabolism. N Eng J Med 1936; 214: 961-986.
2. Evans HM, Meyer K, Simpson ME, Reichart FL. Disturbance of carbohydrate metabolism in normal dogs injected with the hypophyseal growth hormone. Proc Soc Exp Biol N.Y. 1931-2; 29: 857-858.
3.Young FG. Anterior pituitary fraction and carbohydrate metabolism I; the preparation and properties of diabetogenic extracts. J Endocrinol 1939; 1: 339-354.
4.Young GF. The growth hormone and diabetes. Recent Prof Horm Res 1953; 8: 471-510.
5.Young FG. Growth and the diabetogenic action of anterior pituitary preparations. Br Med J 1941; (vol 2)4225: 897-901.
6.Sönksen PH, Greenwood FC, Ellis JP, Lowy C, Rutherford A, Nabbaro JDN. Changes of carbohydrate tolerance in acromegaly with progress of the disease and in response to treatment. J Clin Endocrinol. 1967; 27: 1418-1430.
7. Mintz DH, Finster JL, Taylor AL, Fefer A. Hormonal genesis of glucose intolerance following hypoglycaemia. Am J Med. 1968; 45: 187-197.
8. Raben MS, Hollenberg CH. Effect of growth hormone on plasma fatty acids. J Clin Invest. 1959; 38: 484-488.
9. Rosenfeld RG, Wilson DM, Dollar LA, Bennett A, Hintz RL. Both human pituitary growth hormone and recombinant DNA-derived human growth hormone cause insulin resistance at a postreceptor site. J Clin Endocrinol Metab 1982; 54: 1033-1038.
10. Bratusch-Marrain PR, Smith D, DeFronzo RA. The effect of growth hormone on glucose metabolism and insulin secretion in man. J Clin Endocrinol Metab. 1982; 55: 973-982.
11. Fowelin J, Attval S, Lager I, Bengtsson B-A. Effects of treatment with recombinant human growth hormone on insulin sensitivity and glucose metabolism in adults with growth hormone deficiency. Metabolism 1993; 42: 1443-1447.
12. Moller N, Jorgensen JOL, Schmitz O, Moller J, Christiansen JS, Alberti KGMM, Orskov H. Effects of a growth hormone pulse on total and forearm substrate fluxes in humans. Am J Physiol. 1990; 258 (1:1): E86-E91.
13. Rizza RA, Mandarino LJ, Gerich JE. Effects of growth hormone on insuin action in man. Mechanisms of insulin resistance, impaired suppression of glucose production and impaired stimulation of glucose utilization. Diabetes. 1982; 31: 663-669.
14. Davidson MB. Effect of growth hormone on carbohydrate and lipid metabolism. Endocr Rev 1987; 8: 115-131.

15. Nielsen JH. Effects of growth hormone, prolactin and placental lactogen on insulin content and release, and deoxyribonucleic acid synthesis in cultured pancreatic islets. Endocrinology. 1982; 110: 600-606.
16. Maler RJ, Szego C. Early insulin synergistic activity of GH. Diabetes. 1969; 550-555.
17. MacGorman LR, Rizza RA, Gerich JE. Physiological concentrations of growth hormone exert insulin-like and insulin antagonistic effects on both hepatic and extrahepatic tissues in man. J Clin Endocrinol Metab. 1981; 53: 556-559.
18. Goodman HM. Early and late effects of GH on the metabolism of glucose in adipose tissue. Endocrinology. 1965; 76: 1134-1140.
19. Goodman HG, Grumbach MM, Kaplan SL. Growth and growth hormone. II. A comparison of isolated growth-hormone deficiency and multiple pituitary-hormone deficiencies in 35 patients with idiopathic hypopituitary dwarfism. N Eng J Med. 1968; 278; 57-68.
20.Brasel JA, Wright JC, Wilkins L, Blizzard RM. An evaluation of seventy-five patients with hypopituitariam beginning in childhood. Am J Med. 1965; 38: 484-498.
21. Bougnères P-F, Artavia-Loria E, Ferre P, Chaussain J-L, Job J-C. Effects of hypopituitarism and growth hormone replacement therapy on the production and utilization of glucose in childhood. J Clin Endocrinol & Metab. 1985; 61: 1152-1157.
22. Merimée TJ, Felig P, Marliss E, Fineberg SE, Cahill Jr., GG. Glucose and lipid homeostasis in the absence of human growth hormone. J Clin Invest. 1971; 50:574-582.
23. Cryer PE. Glucose homeostasis and hypoglycemia. In Williams `Textbook of Endocrinology', ed. JD Wilson & DW Foster, 8th edn: 1223-1253. Philadelphia: Saunders, 1992.
24. Markussis V, Beshyah SA, Fisher C, Sharp P, Nicolaides AN & Johnston DG. Detection of premature atherosclerosis by high-resolution ultrasonography in symptom-free hypopituitary adults. Lancet. 1992; 340: 1188-1192.
25. Merimée TJ, Rabinowitz D, Rimoin DL & McKusick VA. Isolated human growth hormone deficiency. III Insulin secretion in sexual ateliotic dwarfism. Metabolism. 1968; 17: 1005-1011.
26. Salomon F, Cuneo RC, Umpleby AM et al. Interactions of body fat and muscle mass with substrate concentrations and fasting insulin levels in adults with growth hormone deficiency. Clinical Science 1994; 87: 201-06.
27. Johannsson G, Bengtsson BA, Andersson B, Isgaard J, Caidahl K. Long term cardiovascular effects of growth hormone treatment in GH-deficient adults. Preliminary data in a small group of patients. Clinical Endocrinology 1996; 45:305-14.
28. Hew FL, Koschmann M, Christopher M, Rantzau C, Vaag A, Ward G, Beck-Nielsen H, Alford F. Insulin resistance in growth hormone-deficient adults: defects in glucose utilization and glycogen synthase activity. J. Clin. Endocrinol. Metab. 1996; 81: 555-564.
29. Fowelin J, Attvall S, Lager I, Bengtsson B-A. Effects of treatment with recombinant human growth hormone on insulin sensitivity and glucose metabolism in adults with growth hormone deficiency. Metabolism. 1993; 42: 1443-1447.

30. O'Neal DN, Kalfas A, Dunning PL, Christopher MJ, Sawyer SD, Ward GM, Alford FP. The effect of 3 months of recombinant human growth hormone (GH) therapy on insulin and glucose-mediated glucose disposal and insulin secretion in GH-deficient adults: a minimal model analysis. J. Clin. Endocrinol. Metab. 1994; 79: 975-983.
31. Beck JC, McGarry EE, Dyrenfurth I, Venning EH. Metabolic effects of human and monkey growth hormone in man. Science 1957; 125:884-886.
32. Tanner JM, Whitehouse RH. The effect of human growth hormone on subcutaneous fat thickness in hyposomatotrophic and panhypopituitary dwarfs. J Endocrinol 1967; 39: 263-70.
33. Salomon F, Cuneo RC, Hesp R, Sönksen PH. The effects of treatment with recombinant human growth hormone on body composition and metabolism in adults with growth hormone deficiency. N Engl J Med 1989; 321: 1797-1803.
34. Jorgensen JOL, Pedersen SA, Thuesen L et al. Beneficial effects of growth hormone treatment in GH-deficient adults. *Lancet* 1989; i: 1221-25.
35. Kostyo JL. Rapid effects of growth hormone on amino acid transport and protein synthesis. Ann NY Acad Sci 1968; 148:389-407
36. Kosyto J.: Separation of the effects of growth hormone on muscle amino acid transport and protein sythesis. Endocrinology 1968; 75: 113-119
37. Kostyo JL, Knobil E. The effect of growth hormone on the in vitro incorporation of leucine-2-14C into the protein of rat diaphragm. J Endocrinol 1959; 65:395-401
38. Florini JR, Breuer CB. Effects of growth hormone on muscle ribosomes and RNA polymerase. Biochemistry 1966; 5: 870-1876.
39. Kostyo JL & Rillema JA. In vitro effects of GH on the number and activity of ribosomes engaged in protein synthesis in the isolated rat diaphragm. Endocrinology 1971; 88: 1054-1060.
40. Flaim KE, Li JB, Jefferson LS. Protein turnover in rat skeletal muscle: effects of hypophsectomy and growth hormone. Am J Physiol 1978; 234:E38-43.
41. Bengtsson B-A, Eden S, Lonn L, Kvist H, Stokland A, Linstedt G et al. Treatment of adults with growth hormone (GH) deficiency with recombinant human GH. J Clin Endocrinol Metab 1993; 76: 309-317.
42. Amato G, Carella C, Fazio S, La Montagna G, Cittadini A, Sabatini D et al Body composition, bone metabolism, and heart structure and function in growth hormone (GH)-deficient adults before and after GH replacement therapy at low doses. J Clin Endocrinol Metab 1993; 77: 1671-76.
43. Whitehead HM, Boreham C, McIlrath EM, Sheridan B, Kennedy L, Atkinson AB, Hadden DR. Growth hormone treatment of adults with growth hormone deficiency: results of a 13-month placebo controlled cross-over study. Clinical Endocrinology 1992; 36:45-52.
44. Binnerts A, Swart GR, Wilson JHP, Hoogerbrugge N, Pois HAP, Birkenhager JC, Lamberts WJ. The effect of growth hormone administration in growth hormone deficient adults on bone, protein, carbohydrate and lipid homeostasis, as well as body composition. Clinical Endocrinology 1992; 37:79-87.
45. Chong PKK, Jung RT, Scrimgeour CM, Rennie MJ, Paterson CR. Energy expenditure and body composition in growth hormone deficient adults on exogenous growth hormone. Clinical Endocrinology 1994; 40: 103-110.

46. Jórgensen JOL, Thuesen L, Muller J, Ovesen P, Skakkebaek NE, Christiansen JS. Three years of growth hormone treatment in growth hormone-deficient adults: near normalization of body composition and physical performance. Eur J Endocrinol 1994; 130: 224-28.
47. Horber FF & Haymond MW. Human growth hormone prevents the protein catabolic side effects of prednisolone in humans. J Clin Invest 1990; 86: 265-272.
48. Fryburg DA, Gelfand RA, Barrett EJ. Growth hormone acutely stimulates forearm muscle protein synthesis in normal humans. Am J Physiol 1991; 260: E499-E504.
49. Beshyah SA, Sharp PS, Gelding SV et al. Whole-body leucine turnover in adults on conventional treatment for hypopituitarism. Acta Endocrinologica 1993;129: 58-64.
50. Salomon F, Umpleby AM, Cuneo RC, Sonksen PH. Protein, fat and glucose metabolism during treatment with recombinant human growth hormone in adults with growth hormone deficiency. Endocrinology and Metabolism 1977; 4:121-128.
51. Russell-Jones DL, Weissberger AJ, Bowes SB, Kelly JM, Thomason M, Umpleby AM, et al. The effects of growth hormone on protein metabolism in adult growth hormone deficient patients. Clin Endocrinol 1993; 38: 427-31.
52. Russell-Jones DL, Bowes SB, Rees SE, Jackson NC, Weissberger AJ, Hovorka R, Sönksen PH & Umpleby AM. The effect of growth hormone treatment on post-prandial protein metabolism in growth hormone deficient adults. Am. J. Physiol. (IN PRESS)
53. Umpleby AM, Boroujerdi MA, Brown PM, Carson ER and Sönksen PH. The effect of metabolic control on leucine metabolism in type 1 insulin-dependent diabetic patients. Diabetologia 1986; 29: 131-141.

12 EFFECTS OF GROWTH HORMONE ON BLOOD COAGULATION AND FIBRINOLYSIS

Kerstin Landin-Wilhelmsen

Department of Medicin, Sahlgrenska University Hospital, Göteborg, Sweden

Introduction

Both growth hormone deficiency (GHD) and growth hormone excess (as in acromegaly) are associated with increased risk for cardiovascular disease (1-5). Increased prevalence of lipid disturbances and hypertension are reported in both diseases as well as insulin resistance in GHD and high prevalence of diabetes in acromegaly (6-12).

Disturbed lipid and glucose metabolism as well as hypertension are factors which have been described as a "metabolic" or "insulin resistance syndrome" by Reaven (13). The syndrome has later been extended to include increased fibrinogen and plasminogen activator inhibitor-1 (PAI-1), factors involved in coagulation and fibrinolysis (14-18), thereby linking atherogenesis with thrombogenesis.

GHD patients have an increased abdominal fat distribution, higher fibrinogen and PAI-1 concentrations (19). It was recently shown that acromegalic patients had markedly increased fibrinogen concentrations compared to controls (20).
Insulin like growth factor-1 (IGF-1), which mainly reflects the growth hormone secretion, correlated also positively with fibrinogen independently of age in a random population sample (21).

This chapter will review the role of growth hormone on coagulation and fibrinolysis, mainly focused on fibrinogen and PAI-1.

Coagulation and fibrinolysis

A cascade of factors contributes to a normal hemostasis and a balance between the coagulation, the protein C dependent system and the fibrinolytic system is maintained in healthy individuals. A defect in either system can lead to bleeding disorders, or the opposite, thrombus formation.
All factors are intimately involved in blood rheology which is also depending on platelets and the von Willebrand factor (vWf), which later is needed for normal platelet adhesion and is a carrier of coagulation factor VIII.

t-PA

↓ – PAI-1

Plasminogen ------------------→ Plasmin

↓ – a_2-antiplasmin

Fibrinogen------------------→ Fibrin degradation products

Figure 1. *Schematically the fibrinolytic system. t-PA = tissue plasminogen activator, PAI-1 = plasminogen activator inhibitor type 1, – = inhibition.*

The fibrinolytic system is schematically outlined in **figure 1**. Plasminogen activator inhibitor type 1 (PAI-1) rapidly inhibits the activity of the important key factor; tissue plasminogen activator (t-PA). t-PA degrades plasminogen to the active enzyme plasmin which, in turn, proteolytically degrades fibrinogen to fibrin(ogen) degradation products.

PAI-1 is a member of the serine protease inhibitor gene family and is synthesized in platelets, endothelial cells and hepatocytes and forms a t-PA/PAI-1 complex in the circulation (22-24). The t-PA antigen concentration reflects the t-PA/PAI-1 mass concentration.

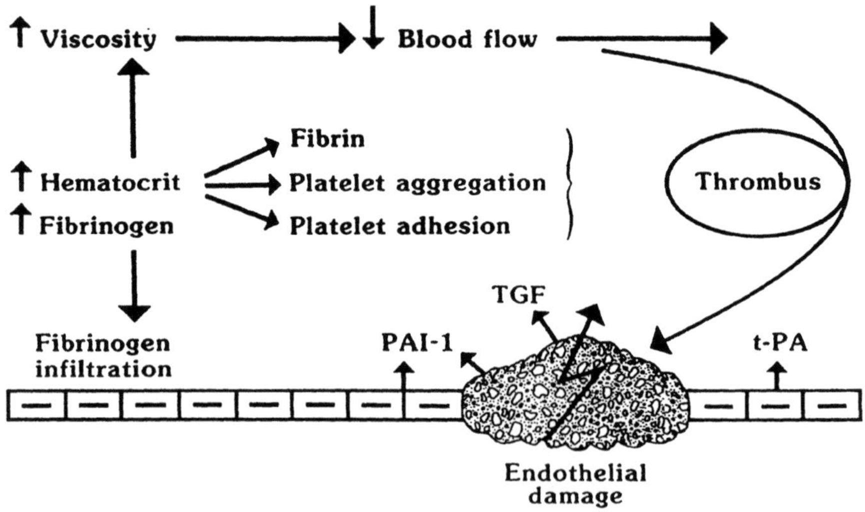

Figure 2. *Transsection of a vessel wall with a fibrin-platelet thrombus, schematically showing the release of plasminogen activator inhibitor type 1 (PAI-*

1) and its role in vascular fibrinolysis. t-PA = tissue plasminogen activator, TGF = transforming growth factor.

Furthermore, the vessel morphology and a number of circulating substances such as vitamines, hormones, blood lipids and other growth and metabolic factors, tumour necrosis factor, endotoxin, interleukins etc. are involved in blood rheology.

The age of the individual and the overall health status influence on the hemostasis as many of the fibrinolytic factors are acute phase reactants. PAI-1 secretion is stimulated by inflammatory reactions as well as damaged endothelium as in the arteriosclerotic process, **figure 2**.

Coagulation and fibrinolysis in cardiovascular disease

Fibrinogen is an independent risk factor for myocardial infarction and stroke. The risk was additive with increasing levels of blood pressure and cholesterol, respectively. Fibrinogen is strongly related to smoking (25-26). Fibrinogen is also increased in obesity, especially abdominally distributed obesity (16, 27), **figure 3**.

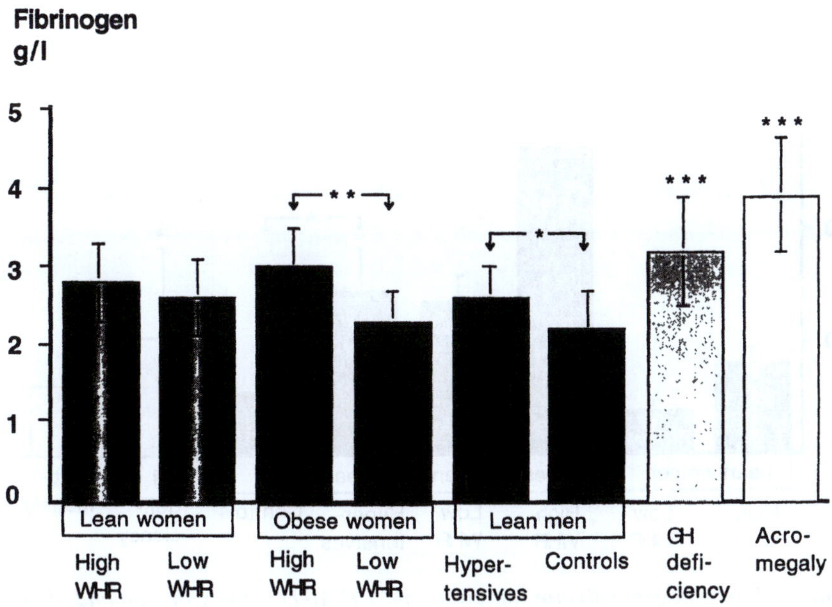

Figure 3. *Plasma fibrinogen in lean and obese healthy subjects with a high (>0.80) and a low (<-0.80) waist/hip ratio, indicative of abdominal and femoral fat distribution, respectively, men with and without hypertension, growth hormone (GH)*

Increased factor VII and factor VIII have been associated with increased risk of myocardial infarction (28-29), but factor VII was not elevated in young postinfarction patients (30). It has been speculated whether vWf is involved in the arteriosclerotic process; studies have indicated that vWf is a risk factor for ischemic heart disease (31).

PAI-1 has been considered as a risk factor for recurrent myocardial infarction in young age (32), and high levels have been found in patients with angina pectoris (33-34).

PAI-1 is strongly related to body weight, waist/hip circumference ratio, blood pressure, triglycerides, plasma insulin and degree of insulin resistance measured with the clamp technique (14, 16-18), factors included in the insulin resistance syndrome (13), which has been extended to include also fibrinogen and PAI-1 (14, 16-18). All are factors involved in the arteriosclerotic and thrombotic process.

Levels of fibrinogen and PAI-1 in some groups of subjects and patients are exemplified in **figures 3 and 4**.

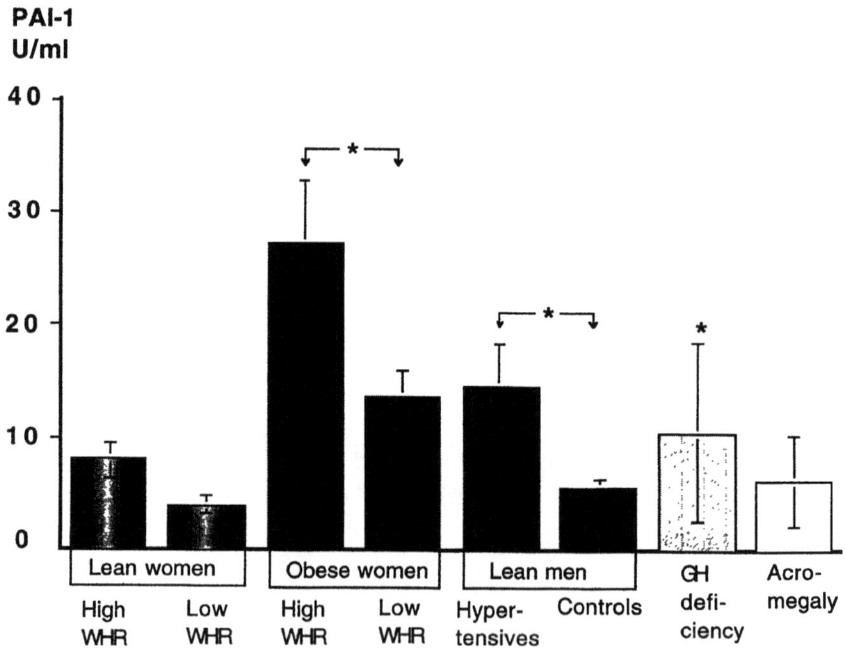

Figure 4. *Plasminogen activator inhibitor type 1 (PAI-1) activity in lean and obese healthy subjects with a high (>0.80) and a low (<-0.80) waist/hip ratio, indivative of abdominal and femoral fat distribution, respectively, men with and without hypertension, growth hormone (GH) deficient adults and in acromegaly (16-17, 19-20). Significances against age- and weight-matched controls, * = p<0.05*

Growth hormone, coagulation and fibrinolysis in healthy subjects

Sarji *et al* showed that the growth hormone secretion pattern imitated the secretion of vWf during sleep and at oral glucose suppression (35). An increase in both vWf and growth hormone during physical exercise has been reported in normal subjects (36).

An epidemiological study of 1140 healthy individuals aged between 20 and 59 years showed that PAI-1 increased and t-PA activity decreased with increasing age in men and after the menopause also in women (37). The growth hormone secretion decreases with increasing age and was negatively correlated to PAI-1 concentrations also when adjusted for age, body mass index, blood glucose and lipids in men, but not in women (37).

IGF-1 decreases (21, 38-41) and fibrinogen increases with increasing age (27). However, when adjustment for age was done, IGF-1 correlated positively with fibrinogen in a random population sample of men aged 25-64, but not in women (21).
There was no correlation between growth hormone and PAI-1 or between IGF-1 and fibrinogen in premenopausal women, which might be due to the effect of estradiol (21, 37). The decline in PAI-1 and IGF-1 was accentuated after the menopause (21, 37). Estrogen replacement therapy causes lower fibrinogen and PAI-1 concentrations (27, 42).

When growth hormone was given during one week to eight healthy men the fibrinogen concentration increased but there was no effect on t-PA antigen (43). Hence, it seems as growth hormone has different effects on t-PA/PAI-1 and fibrinogen concentrations in healthy men.

Coagulation and fibrinolysis in GHD

Pituitary deficient adults substituted with thyroxin, cortisone and sex steroids, but not growth hormone, have an increased risk for cardiovascular disease (1-2). Increased arteriosclerosis in carotic and femoral arteries among pituitary deficient patients, including GHD, in comparison with controls was demonstrated by Markussis *et al* (44). Serum cholesterol was higher, but there was no difference in plasma fibrinogen compared to controls (44).

Sarji *et al* showed that the growth hormone peaks during sleep were associated with peaks of von Willebrand activity and that one subject with isolated GHD had no peaks of either (35). Lowe *et al* could not find any differences in factor VIII concentrations between three adults with GHD and controls (45).
Jörgensen *et al* reported concentrations of factor VIII, vWf, fibrinogen and t-PA antigen within the normal range in patients with GHD (46). Beshyah *et al* also found normal hemostatic mechanisms in hypopituitary adults (47).

However, Johansson *et al* demonstrated higher plasma fibrinogen and PAI-1 activity in 20 adults with GHD in comparison with age- and weight-matched controls (19), **figures 3 and 4**. Smoking was less common among the GHD patients and could not explain the differences in hemostatic factors (19).

The reason for lack of differences between coagulation and fibrinolytic factors in the GHD patients and controls reported above (35, 45-47) might be the small number of patients studied, poorly defined control groups, and fairly high fibrinogen concentrations among controls (>3 g/l in the study by Markussis *et al* (44)).

Effects of growth hormone on coagulation and fibrinolysis in GHD

Growth hormone infusion and injections during three to six hours lead to rised factor VIII in eight young patients with GHD (48), but no rise in factor VIII in another study comprising three adolescent boys with GHD (45) and increased vWf in three adults with GHD (35).

Padayatty *et al* did not see any change in PAI-1, t-PA antigen or euglobulin clot lysis time in eight pituitary deficient adults treated with growth hormone during two months (49), **table 1**. Neither did Jörgensen *et al* concerning coagulation factors, t-PA antigen or capillary fragility in 22 similar patients after four months of growth hormone treatment in a placebo-controlled study (46), **table 1**.

Table 1. *Published studies concerning the effect of growth hormone treatment on tissue plasminogen activator (t-PA) and plasminogen activator inhibitor-1 (PAI-1) in growth hormone deficient adults. act.=activity, ag=antigen, mo.=months, p=placebo controlled.*
↔ = no effect, ↓ = decrease, nd = not determined

Author	Year	n	Duration	PAI-1 act.	PAI-1 ag	t-PA ag
Jörgensen *et al* (46)	1990	22	4 mo. p	nd	nd	↔
Padayatty *et al* (49)	1993	8	2 mo.	↔	↔	↔
Johansson *et al* (50)	1996	17	24 mo.	↓	↓	↓
Burman *et al* (51)	1997	36	9 mo. p	↓ men	nd	nd

Johansson *et al* found decreased PAI-1 activity and antigen as well as decreased t-PA antigen during two years of growth hormone treatment (50), **table 1**. Protein C and alfa$_2$-antiplasmin, the latter an inhibitor of plasmin, decreased also during that period (50).

Fibrinogen, however, did not change during growth hormone treatment for two years (50). The decrease in PAI-1 activity was confirmed in men but not in women by Burman *et al* in a nine months placebo-controlled trial (51), **table 1**.

Taken together, it seems that the effect of growth hormone therapy on the t-PA/PAI-1 complex is dependent on the duration of treatment as studies during nine and 24 months decreased t-PA/PAI-1 (50-51), but not the shorter studies (46, 49), **table 1**.

Possible mechanisms for the growth hormone effect on PAI-1 in GHD

Johansson *et al* found that PAI-1 activity correlated positively with plasma insulin, blood glucose and triglycerides in GHD as in healthy adults, as well as in insulin resistant obese and hypertensive patients (14, 16-19). Abdominal obesity is associated with both higher fibrinogen and PAI-1 levels (16, 27), **figures 3 and 4**. Hence, growth hormone deficient adults have many similarities with the "insulin resistance syndrome" (13, 52).

The decrease in PAI-1 and t-PA antigen during growth hormone treatment was not, however, followed by any decrease in glucose, insulin, blood lipids, body weight or waist/hip ratio. The only variable that changed favourably was lean body mass which increased (50), **table 2**.

Table 2. *Body composition and metabolic variables in growth hormone deficient adults in relation to controls and effects of two years growth hormone (GH) treatment (19, 50).*
↓ = lower or decrease, ↑ = higher or increase, ↔ = no difference, nd = not determined

Variable	Compared with controls	After 2 years GH treatment
Body weight	↔	↑
Waist/hip ratio	↑	↔
Body fat	nd	↔
Lean body mass	nd	↑
Plasma insulin	↔	↑
Blood glucose	↓	↔
Serum cholesterol	↔	↔
Serum triglycerides	↑	↔
IGF-1	↓	↑
Plasma fibrinogen	↑	↔
PAI-1 actvity	↑	↓

Adults with GHD are highly insulin resistant, (8). Insulin sensitivity initially worsened but improved then up to six months of growth hormone treatment (53). PAI-1 activity is inversely correlated to insulin sensitivity (16-17). Hence, it cannot

be excluded that the beneficial effect of long-term growth hormone therapy on PAI-1 activity is mediated via improved insulin sensitivity.
In another study men lost more body fat than women and got higher lean body mass in parallel with a greater decrease in PAI-1 on growth hormone treatment (51). Lean body mass, or skeletal muscle tissue, is the main target tissue for peripheral insulin mediated glucose uptake (54).

Insulin sensitivity might be the major factor responsible for PAI-1 regulation rather than plasma insulin per se. This could be exemplified with the fact that during a two hours supraphysiological infusion of insulin at the hyperinsulinemic, euglycemic clamp in healthy men, PAI-1 activity, PAI-1 antigen as well as t-PA antigen decreased together with increasing glucose uptake (55). During this experiment, the insulin antagonistic growth hormone did not change and could not be resonsible for the acute PAI-1 decrease during the clamp (55).
In obese subjects PAI-1 antigen was not either influenced by growth hormone changes during oral glucose or insulin tolerance tests, respectively (56).

Effects of growth hormone on lipoprotein (a) and its role in fibrinolysis

Lipoprotein (a) has structural similarities with plasminogen (57-58) and competes with plasminogen on its binding receptor (59-60), **figure 1**. Growth hormone treatment for 6 to 12 months in GHD adults increased lipoprotein (a) concentrations which were similar to controls at baseline (61-62). It cannot be excluded that lipoprotein (a) interferes with fibrinolysis.

Lipoprotein (a) might be genetically determined (63) and gene mutations are also associated with acromegaly (64). Lipoprotein (a) was higher in acromegaly when compared with a population sample from the same region (65-66). Furthermore, lipoprotein (a) decreased 6 to 15 months after successful transsphenoidal adenomectomy in acromegaly (65).

Lipoprotein (a), as such, is an independent risk factor for coronary heart disease (66-68) and might act both in the arteriosclerotic process and in the thrombogenesis by competing with plasminogen receptor sites.

Effects of growth hormone and IGF-1 on coagulation and fibrinolysis in experimental studies

Experimental studies in female rats showed that hypophysectomy lead to a decrease in all vitamin K-dependent coagulation factors, which increased again after growth hormone substitution (69). The opposite was found in rats treated with growth hormone which decreased vitamin K-dependent coagulation factors while the growth hormone inhibitor somatostatin increased them (70).
In vitro studies have shown that IGF-1 increases PAI-1 secretion and PAI-1 mRNA in hepatoma (Hep G2) cells (49, 71-72), in contrast to human studies where IGF-1 administration had no effect on PAI-1 concentrations, see below (49).

Growth hormone excess, coagulation and fibrinolysis

Growth hormone excess has been shown to play a role in precipitation of diabetic angiopathy in man and hypophysectomy inhibited the progression of the diabetic retinopathy (73). Furthermore, growth hormone deficient dwarfs did not develop diabetic angiopathy (74). Lowe *et al* found normal factor VIII in both acromegalic and GHD patients (45). However, diabetics had higher factor VIII concentrations but normal growth hormone levels (45).

When type II diabetics were given recombinant human IGF-1 leading to supraphysiological serum IGF-1 concentrations, from a mean of 123 ng/ml at start to a mean of 823 ng/ml after treatment during five days no effects were seen on either PAI-1 concentrations or euglobulin clot lysis time (49).

PAI-1 and fibrinogen in acromegaly before and after treatment

Acromegaly, with hypersecretion of growth hormone and consequently high IGF-1 concentrations, is associated with increased morbidity and mortality from cardiovascular disease (3-5).
Glucose intolerance, diabetes, lipid disturbances and hypertension are commonly found in this disease (9-12) and similar to the "insulin resistance syndrome" (13).

Markedly elevated fibrinogen levels were also found in acromegaly as compared with healthy controls (20), **figure 3**. Fibrinogen was also higher than the mean fibrinogen from a random population sample of similar ages and this difference was not due to differences in smoking habits (20). Interestingly, PAI-1 activity was normal in spite of deteriorated glucose and lipid metabolism in the patients, **figure 4**. Lean body mass was high and body fat low in acromegaly (20), **table 3**.

One year after cure of acromegaly i.e. mean growth hormone levels < 5 mU/l, either with transsphenoidal surgery or with octreotide treatment, plasma fibrinogen had decreased together with improved glucose and lipid metabolism (20), **table 3**. However, fibrinogen was still higher than in controls matched for weight and smoking as well as when compared with a random population sample. PAI-1 activity and lean body mass did not change on therapy.

Fibrinogen correlated positively with IGF-1 in acromegalics as in the population sample and the highest fibrinogen level among the acromegalics was found in a young women with complete anterior pituitary deficiency (except for growth hormone).

The lowest fibrinogen concentrations among the acromegalics were found in two women on estrogen hormone replacement therapy. However, their mean fibrinogen levels were still above the mean of the controls and the population mean where women on estrogen replacement tended to have lower fibrinogen concentration than age-matched women without estrogen replacement (20, 27).

Table 3. *Body composition and metabolic variables in acromegaly compared with controls and one year after normalization of growth hormone in acromegaly (20).* *↓ = lower or decrease, ↑ = higher or increase, ↔ = no difference.*

Variable	Compared with controls	1 year after cure
Body weight	↔	↔
Body fat	↓	↔
Lean body mass	↑	↔
Plasma insulin	↑	↓
Blood glucose	↔	↓
Serum cholesterol	↔	↓
Serum triglycerides	↑	↔
Plasma fibrinogen	↑	↓
PAI-1 activity	↔	↔

Beneficial effects on fibrinogen, glucose and lipid metabolism were seen after cure of the disease and may well contribute to the improved survival reported after normalization of growth hormone concentrations in acromegaly (20, 75).

U-shaped curve for the relation between IGF-1 and fibrinogen

Figure 3 shows that both growth hormone deficient adults and acromegalics had higher fibrinogen concentration than all of the other studied groups.
It is also interesting to see that IGF-1 concentrations after growth hormone treatment in the growth hormone deficient adults reached similar levels as the acromegalics after cure, 312 and 345 ug/l respectively, together with similar fibrinogen concentrations, 3.0 and 3.2 g/l, respectively (20, 50). These values are still elevated as compared with controls of similar age who had mean IGF-1 around 200 ug/l and mean fibrinogen of 2.4 g/l (20).

The relation between IGF-1 and fibrinogen seemed to be U-shaped. Such U-shape could not be seen within the normal range in the random population sample of 392 men and women (20).
As fibrinogen is a strong independent risk factor for cardiovascular disease, and stroke in particular (25), the recent finding of markedly elevated fibrinogen might play a role for the increased thrombotic propensity in acromegaly.

Summary

Growth hormone seems to be beneficial for PAI-1 concentrations in man. The positive correlations between growth hormone and IGF-1 and fibrinogen both in healthy individuals and in acromegaly must, however, be considered with caution when modifying the dose of the hormone both for substitution and in new fields of applications for growth hormone use (76).

References

1. Rosén T, Bengtsson B-Å. Premature mortality due to cardiovascular disease in hypopituitarism. Lancet 1990; 336: 285-288
2. Bulow B, Hagmar L, Erfurth EM. Increased cerebrovascular mortality in patients with hypopituitarism. Clin Endocrinol 1997; 46(1): 75-82.
3. Wright AD, Hill DM, Lowy C, Russell Fraser T. Mortality in acromegaly. Quarterly J Med 1970; 53: 1-16.
4. McGuffin Jr WL, Sherman BM, Roth J, Gorden P, Kahn R, Roberts WC, Frommer PL. Acromegaly and cardiovasular disorders. A prospective study. Annals Intern Med 1974; 81: 11-18.
5. Bengtsson B-Å, Edén S, Ernest I, Odén A, Sjögren B. Epidemiology and long-term survival in acromegaly. A study of 166 patients diagnosed 1955-1984. Acta Med Scand 1988; 223: 327-335.
6. Merimee TJ, Hollander W, Fineberg SE. Studies of hyperlipidemia in the HGH-deficient state. Metabolism 1972; 21: 1053-1061.
7. Rosén T, Edén S, Larsson G, Wilhelmsen L, Bengtsson B-Å. Cardiovascular risk factors in adult patients with growth hormone deficiency. Acta Endocrinol 1993; 129: 195-200.
8. Johansson J-O, Fowelin J, Landin K, Lager I, Rosén T, Bengtsson B-Å. Growth hormone deficient adults are insulin resistant. Metabolism 1995; 44: 1126-1129.
9. Davidoff L, Cushing H. Studies in acromegaly. IV The disturbance of carbohydrate metabolism. Arch Int Med 1927; 39: 751-779.
10. Nikkilä EA, Pelkonen R. Serum lipids in acromegaly. Metabolism 1975; 24: 829-838.
11. Nabarro JD. Acromegaly - a review. Clin Endocrinol 1987; 26: 481-512.
12. Lam KSL, Pang RWC, Janus ED, Kung AWC, Wang CCL. Serum apolipoprotein(a) correlates with growth hormone levels in Chinese patients with acromegaly. Atherosclerosis 1993; 104: 183-188.
13. Reaven GM. Role of insulin resistance in human disease. Diabetes 1988; 37: 1595-1607.
14. Vague P, Juhan-Vague I, Aillaud MF, Badier C, Viard R, Alessi MC, Collen D. Correlation between blood fibrinolytic activity, plasminogen activator inhibitor level, plasma insulin level, and relative body weight in normal and obse subjects. Metabolism 1986; 35: 250-253.
15. Juhan-Vague I, Roul C, Alessi MC, Ardissone JP, Heim M, Vague P. Increased plasminogen activator inhibitor activity in non insulin dependent

diabetic patients - relationship with plasma insulin. Thromb Haemost 1989; 61: 370-373.
16. Landin K, Stigendal L, Eriksson E, Krotkiewski M, Risberg B, Tengborn L, Smith U. Abdominal obesity is associated with an impaired fibrinolytic activity and elevated plasminogen activator inhibitor-1. Metabolism 1990; 39: 1044-1048.
17. Landin K, Tengborn L, Smith U. Elevated fibrinogen and plasminogen activator inhibitor (PAI-1) in hypertension are related to metabolic risk factors for cardiovascular disease. J Intern Med 1990; 227: 273-278.
18. Jansson JH, Johansson B, Boman K, Nilsson TK. Hypofibrinolysis in patients with hypertension and elevated cholesterol. J Intern Med 1991; 229: 309-316.
19. Johansson J-O, Landin K, Tengborn L, Rosén T, Bengtsson B-Å. High fibrinogen and plasminogen activator inhibitor activity in growth hormone-deficient adults. Arterioscler Thromb 1994; 14: 434-437.
20. Landin-Wilhelmsen K, Tengborn L, Wilhelmsen L, Bengtsson B-Å. Elevated fibrinogen levels decrease following treatment of acromegaly. Clin Endocrinol 1997; 46: 69-74.
21. Landin-Wilhelmsen K, Wilhelmsen L, Lappas G, Rosén T, Lindstedt G, Lundberg P-A, Bengtsson B-Å. Serum insulin-like growth factor I in a random population sample of men and women: relation to age, sex, smoking habits, coffee consumption and physical activity, blood pressure and concentrations of plasma lidids, fibrinogen, parathyroid hormone and osteocalcin. Clin Endocrinol 1994; 41: 351-357.
22. Chmielewska J, Rånby M, Wiman B. Evidence for a rapid inhibitor to tissue plasminogen activator in plasma. Thromb Res 1983; 31: 427-36.
23. Sprengers ED, Kluft C. Plasminogen activator inhibitors. Blood 1987; 69: 381-387.
24. Schleef RR, Loskutoff DJ. Fibrinolytic system of vascular endothelial cells. Role of plasminogen activator inhibitors. Haemostasis 1988; 18: 328-341.
25. Wilhelmsen L, Svärdsudd K, Korsan-Bengtsen K, Larsson B, Welin L, Tibblin G. Fibrinogen as a risk factor for stroke and myocardial infarction. N Engl J Med 1984; 311: 501-505.
26. Meade TW, Mellows S. Brozovic M, Miller GH, Chakrabarti RR, North WRS, Haines AP, Stirling Y, Imeson JD, Thompson SG. Haemostatic function and ischaemic heart disease: Principal results of the Northwick Park Heart Study. Lancet 1996; 2: 533-537.
27. Dotevall A, Johansson S, Wilhelmsen L. Association between fibrinogen and other risk factors for cardiovascular disease in men and women. Results form the Göteborg MONICA survey 1985. Annals of Epidemiology 1994; 4: 369-374.
28. Korsan-Bengtsen K, Wilhelmsen L, Tibblin G. Blood coagulation and fibrinolysis in a random sample of 788 men 54 years old. II. Relations of the variables to "risk factors" for myocardial infarction. Thromb Diathes haemorrh 1972; 28: 99-108.
29. Meade TW, Chakrabarti R, Haines AP, North WRS, Stirling Y, Thompson SG. Haemostatic function and cardiovascular death: early results of a prospective study. Lancet 1980; 1: 1050-1054.

30. Moor E, Silveira A, van't Hooft F, Suontaka AM, Eriksson P, Blombäck M, Hamsten A. Coagulation factor VII mass and activity in young men with myocardial infarction at a young age. Role of plasma lipoproteins and factor VII genotype. Arterioscler Thromb Vasc Biol 1995; 15: 655-664.
31. Jansson J-H, Nilsson TK, Johnson O. von Willebrand factor in plasma: a novel risk factor for recurrent myocardial infarction and death. Br Heart J 1991; 66: 351-355.
32. Hamsten A, deFaire U, Walldius G, Dahlén G, Szamosi A, Landou C, Blombäck M, Wiman B. Plasminogen activator inhibitor in plasma; risk factor for recurrent myocardial infarction. Lancet 1987; 2: 3-9.
33. ECAT Angina Pectoris Study Group. ECAT angina pectoris study: baseline associations of haemostatic factors with extent of coronary arteriosclerosis and other coronary risk factors in 3000 patients with angina pectoris undergoing coronary angiography. Eur Heart J 1993; 14: 8-17.
34. Jansson JH, Olofsson Bo, Nilsson TK. Predictive value of tissue plasminogen activator mass concentration on long-term mortality in patients with coronary artery disease. A 7-year follow-up. Circulation 1993; 88: 2030-2034.
35. Sarji KE, Levine JH, Nair RMG, Sagel J, Colwell JA. Relation between growth hormone levels and von Willebrand factor activity. J Clin Endocrinol Metab 1977; 45: 853-856.
36. Sarji KE, Graves JM, Colwell JA. von Willebrand factor activity in normal subjects: sex difference and variability. Thromb Res 1975; 7: 885-895.
37. Heinrich J, Dirkes-Kersting A, Schulte H, Assmann G. Effects of age, menopause and human growth hormone on variables of fibrinolysis. Fibrinolysis 1992; 6, Suppl 3: 33-35.
38. Rudman D, Kutner MH, Rogers CM, Lubin MF, Fleming GH, Bain RP. Impaired growth hormone secretion in the adult population. Relation to age and adiposity. J Clin Invest 1981; 67: 1361-1369.
39. Kelijman M. Age-related alterations of the growth hormone/insuline-like-growth-factor I axis. J Amer Geriatr Soc 1991; 39: 295-307.
40. Yamamoto H, Sohmiya M, Oka N, Kato Y. Effects of aging and sex on plasma insulin-like growth factor I (IGF-I) levels in normal adults. Acta Endocrinol 1991; 124: 497-500.
41. Corpas E, Harman SM, Blackman MR. Human growth hormone and human ageing. Endocrine Reviews 1993; 14: 20-39.
42. Koh KK, Mincemoyer R, Bui MN, Csako G, Pucino F, Guetta V, Waclawiw M, Cannon RO. Effects of hormone-replacement therapy on fibrinolysis in postmenopausal women. N Engl J Med 1997; 336: 683-390.
43. Ingerslev J, Dahl Christensen P, Stenbjerg S, Möller N, Örskov H, Christiansen JS. Diabetes-like alterations in hemostatic parameters after growth hormone administration for one week in normal man. J Diab Compl 1989; 3: 2: 103-106.
44. Markussis V, Beshyah SA, Fisher C, Sharp P, Nicolaides AN, Johnston DG. Detection of premature atherosclerosis by high-resolution ultrasonography in symptom-free hypopituitary adults. Lancet 1992; 340: 1188-1192.
45. Lowe GDO, Middleton SM, Hay ID, Beastall GH, Singaraveloo M, Cohen HN, Thomson JA, Forbes CD, Prentice CRM. Effects of growth hormone,

glucose and insulin on the factor VIII complex. Thromb Res 1982; 28: 201-211.
46. Jörgensen JOL, Pedersen SA, Ingerslev J, Möller J, Skakkebaek NE, Christiansen JS. Growth hormone (GH) therapy in GH-deficient patients, the plasma Factor VIII - von Willebrand factor complex, and capillary fragility. A double-blind, placebo-controlled crossover study. Scand J Clin Lab Invest 1990; 50: 417-420.
47. Beshyah SA, Markussis V, Harbourne T, Nicolaides AN, Johnston DG. Haemostatic mechanisms are normal despite increased vascular mortality in hypopituitary adults. Horm Metab Res 1993; 25: 449-450.
48. Borkenstein M, Muntean W. Effects of growth hormone on the factor VIII complex in patients with growth hormone deficiency. Metabolism 1984; 33: 1065-1067.
49. Padayatty SJ, Orme S, Zenobi PD, Stickland MH, Belchetz PE, Grant PJ. The effects of insulin-like growth factor-1 on plasminogen activator inhibitor-1 synthesis and secretion: results from in vitro and in vivo studies. Thromb Haemost 1993; 70: 1009-1013.
50. Johansson J-O, Landin K, Johannsson G, Tengborn L, Bengtsson B-Å. Long-term treatment with growth hormone decreases plasminogen activator inhibitor-1 and tissue plasminogen activator in growth hormone-deficient adults. Thromb Haemost 1996; 76: 422-428.
51. Burman P, Johansson AG, Siegbahn A, Vessby B, Karlsson FA. Growth hormone (GH)-deficient men are more responsive to GH replacement therapy than women. J Clin Endocrinol Metab 1997; 82: 550-555.
52. Bengtsson B-Å, Rosén T, Johansson J-O, Johannsson G, Landin-Wilhelmsen K. Cardiovascular risk factors in adults with growth hormone deficiency. Endocrinol Metab 1995: 2 (Suppl B): 29-35.
53. Fowelin J, Attvall S, Lager I, Bengtsson B-Å. Effects of treatment with recombinant human growth hormone on insulin-sensitivity and glucose-metabolism in adults with growth hormone deficiency. Metabolism 1993; 42: 1443-1447.
54. DeFronzo RA, Tobin JD, Andres R. Glucose clamp technique: a method for quantifying insulin secretion and resistance. Am J Physiol 1979; 237: E214-223.
55. Landin K, Tengborn L, Smith U. Effects of metformin and metoprolol CR on hormones and fibrinolytic variables during a hyperinsulinemic, euglycemic clamp in man. Thromb Haemost 1994; 71: 783-787.
56. McCormack L, Stickland MH, Grant PJ. Plasminogen activator inhibitor-1 antigen concentrations during insulin and oral glucose tolerance tests in obese man. Fibrinolysis 1993; 7: 225-228.
57. Eaton DL, Fless GM, Kohr WJ, McLean JW, Xu QT, Miller CG, Lawn RM, Scanu AM. Partial amino acid sequence of apolipoprotein(a) shows that it is homologous to plasminogen. Proc Natl Acad Sci USA 1987; 84: 3224-3228.
58. McLean JW, Tomlinson JE, Kuang WJ, Eaton DL, Chen EY, Fless GM, Scanu AM, Lawn RM. cDNA sequence of human apolipoprotein(a) is homologous to plasminogen. Nature 1987; 330: 132-137.

59. Edelberg JM, Gonzalez-Gronow M, Pizzo SV. Lipoprotein a inhibits streptokinase-mediated activation of human plasminogen. Biochemistry 1989; 28: 2370-2374.
60. Gonzalez-Gronow M, Edelberg JM, Pizzo SV. Further characteriazation of the cellular plasminogen binding site: evidence that plasminogen 2 and lipoprotein(a) compete for the same site. Biochemistry 1989: 28: 2374-2377.
61. Edén S, Oscarsson J, Rosén T,Wiklund O, Bengtsson B-Å. Growth hormone treatment of growth hormone deficient adults results in a marked increase in Lp(a) and HDL cholesterol concentrations. Arterioscler Thromb 1993; 13: 296-301.
62. Johannsson G, Oscarsson J, Rosén T, Wiklund O, Olsson G, Wilhelmsen L, Bengtsson B-Å. Effects of one year of growth hormone therapy on serum lipoprotein levels in growth hormone-deficient adults. Influence of gender, apo(a) and apo E phenotypes. Arterioscler Thromb Vasc Biol 1995; 15: 2142-2150.
63. Utermann G, Menzel HJ, Kraft HG, Duba HC, Kemmler HG, Seitz C. Lp(a) glycoprotein phenotypes. Inheritance and relation to Lp(a)-lipoprotein concentrations in plasma. J Clin Invest 1987; 80: 458-465.
64. Melmed S. Acromegaly. N Engl J Med 1990; 322: 966-977.
65. Oscarsson J, Wiklund O, Jakobsson KE, Petruson B, Bengtsson B-Å. Serum lipoproteins in acromegaly before and 6-15 months after transsphenoidal adenomectomy. Clin Endocrinol 1994; 41: 603-608.
66. Rosengren A, Wilhelmsen L, Eriksson E, Risberg B, Wedel H. Lipoprotein(a) and coronary heart disease: a prospective case-control study in a general population sample of middle aged men. Brit Med J 1990; 301: 1248-1251.
67. Kostner GM, Avogaro P, Cazzolato G, Marth E, Bittolo G, Quinci GB. Lipoprotein Lp(a) and the risk for myocardial infarction. Atherosclerosis 1981; 38: 51-61.
68. Dahlén GH, Guyton JR, Attar M, Farmer JA, Kautz JA, Gotto AM. Association of levels of lipoprotein(a), plasma lipids, and other lipoproteins with coronary artery disease documented by angiography. Circulation 1986; 74: 758-765.
69. Sävendahl LSG, Grankvist K, Engström KG. Growth hormone deficiency impairs blood clotting and reduces factor VII coagulant activity in rat. Thromb Haemost 1995; 73: 626-629.
70. Negrev N, Nyagolov Y, Stanchewa E. Somatotropin and somatostatin effects on vitamin K-dependent plasma coagulation factors. Eur J Pharmacol 1995; 277: 145-149.
71. Schneider DJ, Sobel BE. Augmentation of synthesis of plasminogen activator inhibitor type 1 by insulin and insulin-like growth factor type 1: Implications for vascular disease in hyperinsulinaemic states. Proc Natl Acad Sci 1991; 88: 9959-9963.
72. Anfosso F, Alessi MC, Nalbone G, Chomiki N, Henry M, Juhan-Vague I. Up-regulated expression of plasminogen activator inhibitor-1 in Hep G2 cells: interrelationship between insulin and insulin-like growth factor 1. Thromb Haemost 1995; 73: 268-274.

73. Lundbaek K, Christensen NJ, Jensen VA, Johansen K, Steen Olsen T, Prange Hansen Å, Örskov H, Österby R. Diabetes, diabetic angiopathy, and growth hormone. Lancet 1970; 131-133.
74. Merimee TJ, Fineberg S, Hollander W. Vascular disease in the chronic hGH-deficient state. Diabetes 1973: 22: 813-819.
75. Bates AS, Van't Hoff W, Jones JM, Clayton RN. Does treatment of acromegaly affect life expectancy? Metabolism 1995; 44: 1-5.
76. Gomberg-Maitland M, Frishman WH. Recombinant growth hormone: A new cardiovascular drug therapy. Am Heart J 1996; 132: 1244-1262.

13 GROWTH HORMONE AND THE IMMUNE SYSTEM

Christoph J. Auernhammer
Martin Bidlingmaier
and Christian J. Strasburger

Klinikum Großhadern and Klinikum Innenstadt
Ludwig-Maximilians-Universität
Munich, Germany

INTRODUCTION

Growth hormone (GH) and insulin like growth factor I (IGF-I) play an important role in the neuro-endocrine-immune crosstalk. GH and IGF-I receptors are expressed on the surface of immune cells. GH and IGF-I stimulate the proliferation of immunocompetent cells and modulate several humoral and cellular immune functions. In addition, GH and IGF-I are also synthesized and secreted by various immunocompetent cells. Increasing evidence suggests, that the "immune GH-IGF-I axis" acting in a paracrine/autocrine system between the immunocompetent cells has an important function, and at least in part may exert this function autonomously independent of the endocrine system.

Besides GH, prolactin (PRL) plays an important role in the neuro-endocrine-immune crosstalk. PRL as well as PRL receptors are also abundantly expressed in immunocompetent cells. GH and PRL receptors belong to the same family of receptors (Bazan JF 1989; Postel-Vinay MC 1996). Human GH (hGH) can bind and activate the PRL receptor, thus mediating lactotrophic effects. On the other hand, PRL mediates some effects on immune cells by stimulating the secretion of IGF-I (Lin B 1997b). Therefore, a close interaction of the GH-PRL-IGF-I system exists. This review will only emphasize the effects of GH and IGF-I on the immune system. Several reviews on the interactions between GH and PRL on the immune system have recently been published (Auernhammer CJ 1995; Ben-Jonathan N 1996; Weigent DA 1996; Kooijman R 1996a; Yu-Lee LY 1997; Clark R 1997).

GH-IGF-I AXIS IN THE IMMUNE SYSTEM

GH and IGF-I receptors on immune cells

GH receptor mRNA has been detected in the human B lymphoblastoid IM-9 cell line and human peripheral B- and T-cells (Yang Y 1993; Hansen B 1993; Badolato R 1994). Human IM-9 lymphocytes and human peripheral blood mononuclear (PBMN) cells have been shown to bind specifically 125I-GH with an affinity constant of 1.3 -1.5 x 10-9 M-1. The calculated number of receptors per cell was 4,000 for IM-9 cells (Lesniak MA 1976) and about 7,000 for PBMN cells (Kiess W 1988).

The IM-9 cell is frequently used as a model to study GH and GH receptor interaction and its use has substantially contributed to our understanding of the GH receptor signalling mechanism (for review see: Argetsinger LS 1996 and Postel-Vinay MC 1996). The cDNA sequence coding for the 246-amino acid long N-terminal extracellular portion of the hGH receptor in IM-9 cells was shown to be identical to that reported for human liver and placenta, with an alternative splicing in exon 3 (Hansen B 1993). The GH receptor on IM-9 cells has been demonstrated to be functional in activating the signalling cascade. GH binding causes a change of conformation and dimerization of the GH receptor (Ilondo MM 1994; Frank SJ 1994a; Mellado M 1997), resulting in subsequent tyrosine phosphorylation of Jak-1, Jak-2 and STAT-1, -3 and -5 (Silva CM 1993a and b; Frank SJ 1994b; Silva CM 1994; Clayton PE 1994; Silva CM 1996). However, in the IM-9 cell model, IGF-I mRNA transcription is not increased after GH treatment, indicating the signalling pathway being not linked to the IGF-I gene (Clayton PE 1994).

Flow cytometry studies using a FITC labelled monoclonal antibody against the human growth hormone receptor (FITC-mAb 263) or FITC-labelled GH confirmed the existence of hGH-receptors on IM-9 cells and human peripheral blood mononuclear cells (Badolatao R 1994; Rapaport R 1995a; Valerio G 1997a). The hGH-receptor is expressed in a relatively high number on CD20+ B-cells (Badolatao R 1994; Rapaport R 1995a, Valerio G 1997a) and CD14+ monocytes (Rapaport R 1995a), while CD2+ T-lymphocytes and CD16+ NK cells show a far lower expression of hGH receptors (Badolatao R 1994; Rapaport R 1995a, Valerio G 1997a). For human peripheral B-cells and monocytes about 6,000 receptors per cell have been calculated (Rapaport R 1995a). Therefore, B-cells and monocytes appear to account for most of the GH binding observed in peripheral blood mononuclear cells (Kiess W 1988). Neither the percentage of CD20+ B-cells staining positive for the growth hormone receptor nore the mean fluorescence intensity are significantly different between GH-deficient patients, patients with Turner syndrome and normal controls (Rapaport R 1995a; Valerio G 1997a). In contrast, an enhanced binding of GH to CD16+ NK cells was observed in GH-deficient adults (Bidlingmaier M 1997), and CD2+ T-cells were found to show an increased GH receptor expression in short children (Valerio G 1997a). On the other hand, a significant downregulation of GH receptor expression after 6 months of GH treatment was observed in CD20+ B-cells but not in CD2+ T-cells (Valerio G 1997a). This downregulation of GH receptors in B-cells by GH therapy is consistent with the observation of GH-induced GH receptor downregulation in IM-9 cells (Ilondo MM 1994; Smal J 1985). In normal subjects, a correlation was suggested between serum levels of growth hormone binding protein and the percentage of B lymphocytes, expressing the GH receptor (Rapaport R 1995), thus implying, that the immune system may be a significant source of GHBP in circulation. Also in sheep the serum levels of GHBP were positively correlated with the specific binding of GH to peripheral blood mononuclear cells (Khosraviani M 1996), and human IM-9 cells have been demonstrated to release GHBP (Hansen BS 1993; Saito Y 1995).

IGF-I receptors are expressed abundantly on several human myeloid, B- and T- cell lines (Yang Y 1993; Cross RJ 1995; Li YM 1997), human peripheral blood mononuclear cells (Stuart CA 1991; Kooijman R 1992; Kooijman 1995a; Hattori Y 1996) and human thymocytes (Kooijman 1995b). One study reported 97% of monocytes and 88% of B-lymphocytes to express IGF-I-receptors, while only about 2% of T-lymphocytes expressed IGF-I-receptors (Stuart CA 1991). In contrast, another study found IGF-I receptors in every lymphocyte subpopulation, with relatively high numbers of receptors on monocytes, natural killer cells and CD 4+ T-helper cells, an intermediate number of receptors on CD 8+ suppressor/cytotoxic T-cells, and a relatively low number of receptors on B-lymphocytes (Kooijman R 1992). Scatchard analysis of 125I-IGF-I binding to purified subpopulations revealed about 730±430 receptors / monocyte and 230±50 receptors / T-cell (Kooijman 1992). Radioligand binding studies in mixed populations of unstimulated or mitogen-stimulated T-lymphocytes observed an increased number of IGF-I receptors on mixed T-cells after mitogen-stimulation (Kozak RW 1987; Tapson VF 1988; Xu X 1995). However, flow cytometry analysis revealed, that the expression of the IGF-I receptor in human T-cells shows a differential expression in different stages of T-cell development (Kooijman 1995 a). In vivo and in vitro activated (HLA-DR+) T-cells express IGF-I receptors in a far lower percentage than non-activated (HLA-DR-) T-cells (Koojiman 1995a). While about 75% of naive (CD45RA+) T-cells express IGF-I receptors, only 37% of memory (CD45R0+) T-cells were found positive for the IGF-I receptor.

GH, IGF-I and IGFBPs - synthesis and secretion by immune cells

GH synthesis and secretion could be demonstrated in rodent bone marrow, spleen and thymus cells (Weigent DA 1991a and b; Weigent DA 1992; Kooijman R 1997a). In spleen cells of the rat, synthesis and secretion of GH is primarily seen in B-cells, T-helper cells and macrophages, while the T-suppressor cells and NK cells produce only minor amounts of GH (Weigent 1991b). GH synthesis and secretion has also been found in human peripheral blood mononuclear (PBMN) cells (Weigent DA 1988; Hattori N 1993; Hattori N 1994; Varma S 1993; Rapaport R 1996; Melen L 1997), the human B-cell lymphoma cell line IM-9 and the Burkitt-lymphoma cell line sfRamos (Kao TL 1992; Lytras A 1993).

The transcription factor Pit-1 regulates the expression of pituitary-derived growth hormone gene GH-N (Camper SA 1990; Cohen LE 1996). Whether the placenta-derived growth hormone gene GH-V requires Pit-1 has not yet been proven, although Pit-1 is expressed in the placenta (Bamberger AM 1995). Pit-1 has also been demonstrated to be expressed in murine hematopoietic and lymphoid tissues and human neutrophils (Delhase M 1993, Kooijman 1997b). Snell-Bagg dwarf mice bear a mutation in the Pit-1 gene, causing a deficiency in GH, PRL and thyroxine (Camper SA 1990; Cohen LE 1996). However, GH expression in bone marrow, spleen and thymus cells is not significantly decreased in Snell-Bagg dwarf mice in comparison to normal control mice, suggesting a Pit-1 independent regulation of GH expression in immune cells (Weigent DA 1994; Kooijman R 1997a). The GH cDNA derived from rat leukocytes was confirmed by sequencing to be identical to the pituitary GH-N gene, encoding a 22kD protein (Rohn WM 1995). In human neutrophils only GH-N, but no GH-V expression could be demonstrated (Kooijman R 1997b). In contrast, another study found GH-V and GH-N transcription in human peripheral blood mononuclear (PBMN) cells of normal men and women (Melen L 1997). In a Pit-1-defective female patient the PBMN cells were found to express only the GH-V transcript but not the GH-N transcript (Melen L 1997), indicating that the expression of the GH-N transcript but not the GH-V transcript is dependent on Pit-1. A Pit-1 independent GH transcription has also been observed in immune cells of Snell-Bagg (dw) mice (Weigent DA 1994; Kooijman R 1997a). One hypothetic mechanism might be a compensatory up-regulation of the non Pit-1 dependent GH-V expression in Pit-1-defective immune cells. However in mice different GH transcripts, corresponding to the human GH-N and GH-V, have not been discriminated in these studies (Weigent DA 1994; Kooijman R 1997a). Isolated PBMN cells of GH-deficient children have been shown to produce normal amounts of GH in comparison to controls (Rapaport R 1996) .

GHRH mRNA transcripts and secretion of immunoreactive and biologically active GHRH peptide have been demonstrated in rodent and human leukocytes (Weigent DA 1990; Weigent DA 1991b; Stephanou A 1991). Specific binding of 125I-GHRH has been reported in rodent leukocytes (Guarcello V 1991), but, to our knowledge, has not yet been investigated in human immune cells. By in vitro stimulation of PBMN cells with GHRH, stimulating effects (Guarcello V 1991; Kao TL 1992), as well as a lack of any effect on GH secretion (Hattori N 1994), were reported. By the incubation of leukocytes with IGF-I a decrease of GH secretion in rat cells (Baxter JB 1991b; Weigent DA 1992), as well as no negative feedback in human cells (Hattori N 1993) have been described. Presence of exogenous GH was reported to stimulate endogenous GH secretion of human PHA-stimulated PBMN cells (Hattori N 1993). Some of these discrepant data indicate, that GH synthesis and secretion in immunocompetenT-cells might be regulated by mechanisms different from the pituitary system.

In several studies, in vitro stimulation of rodent and human leukocytes with the T-cell mitogens phytohemagglutinin (PHA) and concanavalin A (Con A) produced a marked rise in GH secretion (Weigent DA 1992; Varma S 1993; Hattori N 1993 ; Hattori N 1994), caused not only by an increase of hGH secreted per single cell, but also by an increase of the percentage of cells secreting hGH (Weigent DA 1992; Varma S 1993). In contrast, another study observed no effect of these T-cell mitogens on GH secretion of human PBMC (Rapaport R 1996). In vitro, the B-cell mitogens pokeweed mitogen (PWM) and lipopolysaccharid (LPS) cause only a modest or no increase of GH secretion, respectively (Varma S 1993,: Rapaport R 1996). However the intraperitoneal application of LPS and Freund's complete adjuvans in rats caused an increase in GH production of leucocytes from spleen, thymus and peritoneum in vivo (Baxter JB 1991a).

Basal and GH-stimulated synthesis and secretion of IGF-I, has been reported in rodent leukocytes (Baxter JB 1991b), as well as in human immortalised T-cells (Geffner ME 1990), PBMN (Auernhammer CJ, unpublished observation) and thymocytes (Sabharwal P 1996). The IGF-I secretion of mouse bone marrow cultures can be increased by various cytokines as G-CSF and GM-CSF (Arkins S 1995).

IGF binding proteins (IGFBPs) are important regulators of IGF function by modulating the bioavailability of IGFs (Kelley, 1996; Jones, 1995). Expression of IGFBP mRNA or protein synthesis of different IGFBPs by immunocompetent cells has been reported in murine bone marrow stromal cells (Abboud 1991; Grellier P 1995), murine and human macrophages (Li YM 1996; Aston C 1995), human T and B lymphoblasts (Neely EK 1991, Elmlinger MW 1996) and human PBMN cells (Nyman T 1993; Gayle 1996; Auernhammer CJ 1997). In human B and T lymphoblast cell lines secretion of IGFBP-2 and IGFBP-4 has been demonstrated (Neely EK 1991; Elmlinger MW 1996). By RT-PCR the cDNA of IGFBP-2, -3, -4, -5 and -6 has been found in human PBMN cells (Nyman T 1993; Gayle CM 1996). T-cells were found to express higher levels of IGFBP-2, -3 and -5 mRNA than B-cells (Gayle CM 1997). By western ligand blotting of serum-free conditioned medium of unstimulated human PBMN cells, the most prominent band was observed at 24 KDa, followed by a band at 34 KDa and a doublet at 39/44 KDa. According to their molecular weight these bands were compatible with deglycosilated IGFBP-4, IGFBP-2, and IGFBP-3, respectively (Auernhammer CJ 1997). The synthesis and secretion of IGFBPs by PBMN cells was differentially stimulated by the lymphocyte mitogens PHA, Con A and PWM (Gayle CM 1996; Auernhammer CJ 1997). In contrast, GH had no effect on IGFBP synthesis or secretion (Gayle CM 1996; Auernhammer CJ 1997).

Lymphopoesis

Ames dwarf mice (df) and Snell-Bagg dwarf mice (dw) are both congenitally deficient in GH, prolactin and TSH. While Snell-Bagg (dw) mice have a mutation of the Pit-1 gene (Camper SA 1990; Cohen LE 1996), Ames dwarf mice have a mutation in the Prophet of Pit-1 gene (Prop-1), a homeodomain transcription factor which is required in pituitary ontogeny to initiate cell-line specific differentaiation and activation of Pit-1 (Andersen B 1995; Sornson MW 1996; Gage PJ 1996). Ames and Snell-Bagg mice, both have reduced spleen and thymus weights, and GH or IGF-I application is able to increase relative spleen and thymus weights and the number of lymphocytes in these organs (Villanua MA 1992; Murphy WJ 1993). In normal adult mice and bone marrow transplanted mice, IGF-I-induced an increase in bone marrow and splenic B-cell numbers (Jardieu P 1994). The increase in splenic B-cells was caused by a proliferative effect of IGF-I, increasing the percentage of B-cells

going into S-phase (Jardieu P 1994). An inhibitory action of IGF-I on apoptosis would be another possible mechanism to increase cellularity. In IL-3 dependent murine myeloid progenitor cells, IGF-I has been shown to inhibit apoptosis of IL-3-deprived cells by maintaining high levels of Bcl-2 (Minshall C 1997). A recent study compared hematopoesis in Snell-Bagg (dw/dw) mice, GH-deficient Little (lit/lit) mice, IGF-I knockout (IGF-I-/-) mice and primary hypothyroid (hyt/hyt) mice (Montecino-Rodriguez E 1997). A reduced absolute cellularity in bone marrow, thymus and spleen was observed in all four different strains, compared to normal control mice. However, when normalizing the absolute cellularity to the size of the mouse, the cellularity index (number of cells/weight of mouse) was found to be normal in every strain, with exception of the spleen cellularity index in Snell-Bagg (dw/dw) mice. Absolute numbers of bone marrow B lineage and myeloid lineage cells were reduced in all four strains, while the B-cellularity index was only reduced in (dw/dw) and (hyt/hyt) mice, but not in (lit/lit) and (IGF-I-/-) mice. In Snell-Bagg (dw/dw) mice, treatment with GH and IGF-I increased the number of B lineage cells in the bone marrow but did not normalize the frequency of bone marrow B lineage cells (Mentecino-Rodriguez E 1996). In contrast, thyroxine treatment increased the frequency of B lineage cells in (dw/dw) and (hyt/hyt) mice to normal (Mentecino-Rodriguez E 1996; Mentecino-Rodriguez E 1997). These data suggest, that, although GH and IGF-I significantly stimulate bone marrow cellularity in GH and IGF-I deficient mice, they do not appear to give a specific differentiation signal for B-cell lymphopoesis. In contrast, Gala et al. reported the splenocyte cellularity index to be significantly decreased in Snell-Bagg (dw/dw) mice. The decreased percentage of B220+ splenocytes was only partially restored by thyroxine treatment alone, but fully normalized by additional substitution with rbGH and rbPRL (Gala RR 1995).
In normal mice, IGF-I treatment increased lymphocyte numbers in all peripheral lymphoid organs, stimulated primary B lymphopoiesis and caused higher antibody titers after suboptimal immunization (Jardieu P 1994; Robbins K 1994). Also, IGF-I was found to stimulate immunoglobulin synthesis in splenocytes from antigen-primed mice in vitro (Robbins K 1994). Human B-lymphoblastoid cell lines and multiple myeloma cells both express IGF-I receptors (Freund GG 1994; Jelinek DF 1997). In vitro, the immunoglobulin synthesis and proliferation of several human lymphoblastoid B-cell lines (i.e. IM-9; CBL; GM-1056), as well as the immunoglobulin synthesis of human mononuclear cells, is stimulated by hGH (Yoshida A 1992; Kimata H 1994a; Kimata H 1994b) or IGF-I (Kimata H 1994a; Kimata H 1994b). The GH-induced stimulation of the B-cell lines was blocked by a specific GH antibody, but not by antibodies against IGF-I or the IGF-I-receptor. In contrast, the IGF-I-induced stimulation of the B-cell lines was blocked by either IGF-I or IGF-I-receptor antibodies, but not by a GH antibody (Yoshida A 1992; Kimata H 1994a; Kimata H 1994b).

IGF-I in vitro increases DNA synthesis in multiple myeloma cell lines and stimulates the cells to enter the S-phase (Jelinek DF 1997). In purified human tonsil B-cells, GH stimulated immunoglobulin synthesis and thymidine uptake in vitro (Yoshida A 1992). A clinical correlate might exist in GH and IGF-I substitution therapy, where development of tonsillar and adenoidal hypertrophy has been reported in some children during substitution therapy (Ranke MB 1995; Gerard JM

1997). However adenotonsillar hypertrophy in childhood is frequent and the most common cause of obstructive sleep apnea at this age (Marcus CL 1996).
Paracrine/autocrine effects of GH play an important role in lymphocyte proliferation (Weigent DA 1991c). While the in vitro proliferative effect of GH on immortalised human T lymphocytes has been reported to be indirect by paracrine / autocrine derived IGF-I (Geffner ME 1990), another study in murine T lymphocytes suggested a direct action of GH (Postel-Vinay MC 1997). Another report did not find an effect of IGF-I on the proliferation of bovine PBMN cells (Olsen SC 1993).

Two mechanisms of interaction between the local GH-IGF-I axis and the lymphoproliferative immune reaction have been described: Firstly, enhanced responsiveness of activated immune cells to the mitogenic effects of GH action was reported. In vitro activation of murine T splenocytes with ConA or anti-CD3 antibody increased the percentage of GH receptor positive cells (Gagnerault MC 1996) and subsequent GH-binding by these cells (Postel-Vinay MC 1997). Furthermore, the CD3 antibody-induced lymphoproliferative response could be further augmented by GH (Postel-Vinay MC 1997). Thus, lymphocyte activation can result in an increased mitogenic response to GH, mediated by an upregulation of GH receptors on the lymphocyte surface. Secondly, GH or IGF-I mediate enhanced lymphoproliferative response to other mitogenic signals. IL-2 can increase the PHA-stimulated proliferation of PBMN cells (Olsen SC 1993; Schillaci R 1995). Both, IGF-I and IL-2, were found to increase the expression of the IL-2 receptor alpha chain on human PBMN cells and to show a synergistic effect on their proliferation (Schillaci R 1995). Also in Snell-Bagg dwarf mice, the CD4+ T-cells beared a reduced number of IL-2 receptors which could be signifgicantly upregulated by GH treatment (Gala RR 1995). In human peripheral T-cells, IGF-I was found to increase the secretion of IL-2 (Kooijman R 1996). Thus, IGF-I might enhance IL-2 effects on T-cells by upregulating the expression of IL-2 and the IL-2 receptor.

In hGH-deficient children, several studies did not find any significant difference in basal and PHA-, Con A- or PWM-stimulated lymphocyte proliferation in comparison to controls (Rapaport R 1987; Petersen BH 1990; Spadoni GL 1991), but during hGH replacement therapy, an increase of basal or PHA-stimulated proliferation was reported (Rapaport R 1987; Petersen BH 1990). In fetal rhesus monkeys, the direct administration of IGF-I resulted in an increase of thymic and spleen weights, total lymphocyte numbers and red blood cell counts (Tarantal AF 1997).

Several studies reported lymphocyte subsets (T-cells, CD4+ and CD8+ T-cell subsets, B-cells and natural killer cells) in GH-deficient children to be within the normal range (Kiess W 1986; Kiess W 1988; Rapaport R 1987; Petersen BH 1990; Spadoni GL 1991). During hGH substitution therapy no significant changes in these lymphocyte subsets have been reported (Rapaport R 1987; Petersen BH 1990; Spadoni GL 1991; Rekers-Mombarg LT 1995), besides a few studies observing an isolated decrease of B-cells during GH therapy (Petersen BH 1990; Spadoni GL 1991). In GH-deficient adults, only limited data exist on the lymphocyte subset distribution. One study reported increased CD8+ T-cell and CD20+ B-cell numbers in GH-deficient adults, which both decreased during GH substitution therapy (Balducci R 1994). In contrast, we observed normal lymphocyte subsets (CD2+, CD4+, CD8+, CD4/CD8 ratio, and CD19+) in GH-deficient adults (Auernhammer

CJ 1998). During GH substitution therapy the CD4/CD8-ratio significantly increased, primarily caused by a decrease of the CD8+ T-cell number (Balducci R 1994; Auernhammer CJ 1998).

The immunglobulin serum levels (IgM, IgG and IgA) were found to be in the normal range in GH-deficient children (Spadoni GL 1986; Rekers-Mombarg LT 1995) and GH-deficient adults (Balducci R 1994; Auernhammer CJ 1998). However, discrepant data exist on the effects of GH substitution therapy on immunoglobulin serum levels. While some studies did not report any significant changes (Spadoni GL 1986; Balducci R 1994), others observed a modest decline in IgG, IgM and IgA serum levels (Auernhammer CJ 1998), or an isolated decrease of IgA serum levels (Rekers-Mombarg LT 1995), respectively, during GH therapy. In these studies (Rekers-Mombarg LT 1995; Auernhammer CJ 1998), although immunoglobulin serum levels in GH-deficient patients being in the normal reference range, the mean pre-treatment immunoglobulin serum levels were found to be significantly higher in the GH-deficient patients than in the control group (Rekers-Mombarg LT 1995; Auernhammer CJ 1998).

NK cell activity

hGH- (Badolato R 1994; Rapaport R 1995; Bidlingmaier M 1997) and IGF-I-receptors (Stuart CA 1991; Kooijman R 1992) have been reported to exist on human NK cells in peripheral blood. IGF-I in vitro can increase NK cell activity in PBMN cells derived from normal subjects, as well as from GH-deficient patients (Kooijman R 1992; Auernhammer CJ 1996). The exact mechanism of this phenomenon is not yet understood. IL-2 is an effective endogenous stimulus of NK cell activity (Miller JS 1997). GH and IGF-I have been reported to increase the secretion of IL-2 from human T-cells in vitro (Kooijman R 1996b; Hochberg Z 1997). Therefore, a possible mechanism of the GH and IGF-I mediated increase in NK cell activity might be the autocrine or paracrine secretion of IL-2. Further studies are needed to evaluate this possible interaction.

Normal NK cell numbers in GH-deficient children and adults have been reported in several studies (Kiess W 1986; Kiess W 1988; Petersen BH 1990; Spadoni GL 1991; Auernhammer CJ 1996), while one large study found a significantly lower percentage and absolute number of NK cells in 22 GH-deficient adults when compared to 100 control subjects (Span JPT 1996). Despite normal NK cell numbers, GH-deficient patients show a decreased NK cell activity in comparison to normal controls (Kiess W 1986; Kiess W 1988; Bozzola M 1989; Crist DM 1987; Auernhammer CJ 1996), while acromegalic patients have no difference of NK cell activity in comparison to controls (Kotzmann H 1994). In GH-deficient adults and normal controls we observed a positive correlation between serum IGF-I levels and NK cell activity (Auernhammer CJ 1996). GH substitution therapy in GH-deficient patients was found to have either no effect on NK cell activity (Kiess W 1986; Kiess W 1988), or increase NK cell activity (Crist DM 1987; Bozzola M 1989). Also in healthy adults with normal GH secretion, supplemental GH therapy increased NK cell activity (Crist DM 1990).

Thymic epithelial cells (TECs) and thymocytes

GH mRNA positive cells have been demonstrated in normal thymus tissue (Binder G 1994; Maggiano N 1994; Wu H 1996), as well as thymomas (Wu H 1996; Lauriola L 1997). Human thymocytes synthesize and secrete GH and IGF-I (Sabharwal P 1996). Functional IGF-I receptors have been characterized on human thymocytes, with an average density of 260+/-30 receptors per thymocyte (Kooijman R 1995b and c). In vitro, the IGF-I secretion of thymocytes and thymic epithelial cells can be stimulated by GH (Tsuji Y 1994; Sabharwal P 1996). The proliferation of thymocytes and thymic epithelial cells is stimulated by GH in vitro and can be blocked by specific antibodies against GH and IGF-I, respectively (Tsuji Y 1994; Sabharwal P 1996). In addition, it has beeen shown that conditioned medium from GH-treated TECs can stimulate thymocyte proliferation (Lin B 1997), while - the other way around - conditioned medium from GH-treated thymocytes can stimulate TECs proliferation (Sabharwal P 1996). These data suggest, that the proliferation of thymic epithelial cells and thymocytes, respectively, is effected by locally derived growth hormone, which acts indirectly via paracrine/autocrine derived IGF-I.

In old Wistar-Furth rats, implantation of GH3 cells reversed the accumulation of CD4-CD8- lymphocytes in thymus glands of aged rats (Li YM 1992). In contrast, in C57BL/6J mice, hGH and oGH did not show any effect on lymphocyte subpopulations in the thymus (Knyszynski A 1992), indicating a species specific phenomenon (Dubiski S 1989; Knyszynski A 1992; Goya RG 1992). Treatment of immunodeficient SCID mice with hGH promotes human lymphocyte engraftment in the thymus (Murphy WJ 1992b). This increased engraftment may not only be caused by a direct effect of GH on trafficing the donor cells to the thymus (Murphy WJ 1992b) but also by an effect on TECs, as specific interactions between thymocytes and thymic epithelial cells have also been shown to be modulated by GH (Taub DD 1994; de Mello-Coelho V 1997). In vitro preincubation of human T-cells with hGH increased the cell adhesion to purified ICAM-1, VCAM-1 and fibronectin (Taub DD 1994). In addition, in vitro pretreatment of human peripheral blood lymphocytes with hGH promoted subsequent T-cell engraftment in the thymus of SCID mice (Taub DD 1994). Both effects could be blocked by antibodies against human integrin beta 1 (CD29) and beta 2 (CD18) (Taub DD 1994), while the expression of other cell adhesion molecules (i.e. CD4, CD8, CD11a, CD11b, CD11c, LFA-1, VLA-2, VLA-4, VLA-5, VLA-6, CD18, CD29, ICAM-1, VCAM-2, ELAM-1) was not increased on the T-cells by GH, respectively (Taub DD 1994). Preincubation of murine thymic epithelial cells with GH caused an increase in thymocyte adhesion, which could be blocked by antibodies against fibronectin, laminin, VLA-5 and VLA-6 (de Mello-Coelho V 1997). In primary cultures of fetal thymic stromal cells, GH was also found to increase the transcription rate of IL-1beta mRNA (Tseng YH 1997). Although the exact mechanisms, by which GH is promoting the thymocyte - thymic epithelial cell interactions are not yet defined, the upregulation of specific receptors and ligands, as well as their "activation", e.g. increase in receptor affinity, are possible underlying mechanisms.

The production of the zinc-dependent nonapeptide thymulin (Dardenne M 1982) by thymic epithelial cells has been shown to be stimulated by hGH and IGF-I in vitro (Timsit J 1992). GH in this effect is acting through a paracrine / autocrine mechanism mediated via IGF-I (Timsit J 1992). In comparison to normal controls, thymulin plasma levels are increased in acromegalic patients (Timsit J 1990) and decreased in GH-deficient children and adults (Mocchegiani E 1991; Mocchegiani E 1996). In both studies, a significant positive correlation was reported between plasma levels of IGF-I and thymulin (Timsit J 1990; Mocchegiani E 1996). Although plasma zinc levels in GH-deficient subiects were in the normal range, there was also a positive correlation of IGF-I and zinc plasma levels (Mocchegiani E 1996). As zinc supplementation in old mice fully restores thymulin bioactivity (Mocchegiani E 1995), the observed positive correlation of IGF-I and thymulin plasma levels might be partially caused by the positive correlation between IGF-I and zinc (Mocchegiani 1996).

Granulopoesis and macrophage / neutrophil function

GH immunoreactivity is found in about 25% of human neutrophil granulocytes, and GH-N gene mRNA expression can be demonstrated by RT-PCR (Kooijman R 1997 b). GH and IGF-I receptors are expressed on human neutrophils (Bierknes R 1997; Kooijman R 1997 b). Granulopoesis is stimulated by GH and IGF-I, GH acting indirect by paracrine IGF-I (Merchav S 1988a; Merchav S 1993a).

In vitro, the oxidative burst of human macrophages and neutrophils has been reported to be primed by GH (Fu YK 1991; Fu Y-K 1992; Warwick-Davies J 1995a) and IGF-I (Fu YK 1991; Bjerknes R 1995), while others observed an enhancement by GH, but not IGF-I, respectively (Warwick-Davies J 1995a). GH- and IGF-I-induced priming of oxidative burst *in vitro* could be blocked by antibodies against GH or IGF-I, respectively, while the GH-induced priming could not be blocked by an IGF-I antibody (Fu Y-K 1992), thus indicating a direct effect of hGH on the oxuidative burst, not mediated by IGF-I. The hGH-induced priming of superoxide anion production has been found to be caused by the binding of hGH to PRL-receptors (Fu Y-K 1992), which is dependent on zinc (Fu Y-K 1992; Cunningham BC 1990).

Although incubation with GH increased the burst activity of human monocytes *in vitro*, phagocytosis and killing of Mycobacterium tuberculosis were not affected by GH (Warwick-Davies J 1995b). However, in another study, IGF-I increased burst activity, as well as phagocytosis of immunoglobulin G-opsonized Staphylococcus aureus and complement-opsonized Candida albicans in human neutrophils (Bjerknes 1995). The complement receptors 1 (CD35) and 3 (CD11b), but not the FcgammaRI (CD64), the FcgammaRII (Cdw32), and the FcgammaRIII (CD16) were found to be more expressed after IGF-I priming (Bjerknes 1995). Thus, the IGF-I induced priming of neutrophil phagocytosis seems to be mediated by several mechanisms, including up-regulation of specific receptors.

In hypophysectomized rats, *in vivo* substitution of GH caused a priming of superoxide anion production by peritoneal macrophages (Edwards III CK 1988). GH treatment of hypophysectomized rats, infected with salmonella typhimurium, partially restored the reduced killing capacity of their peritoneal macrophages *in vitro* and improved their decreased survival rate in comparison to control rats (Edwards III CK 1991). In contrast, *in vivo* administration of GH in piglets resulted in a decrease of neutrophil phagocytosis; after induction of E.coli sepsis no significant difference of phagocytosis was observed in GH pretreated vs. untreated piglets (Balteskard L 1997).

Neutrophils of patients with GH-deficiency have been reported to show a low oxidative burst capacity (Rovensky J 1982; Reinisch N 1996) and impaired phagocytic function (Manfredi R 1994). In contrast, one study observed a high oxidative burst capacity in neutrophils of acromegalic patients (Rovensky J 1982), while another study found an increase of phagocytosis, but no difference of oxidative burst activity in acromegalic patients in comparison to controls (Kotzmann H 1994). In GH-deficient patients phagocytic activity and oxidative burst capacity were significantly increased and restored to normal values after GH substitution therapy (Rovensky 1982; Manfredi R 1994; Reinisch N 1996). In patients with Laron syndrome, no difference of phagocytic capacity and oxidative burst of unprimed neutrophils was observed in comparison to normal controls (Bierknes R 1997). However, before the onset of IGF-I therapy, IGF-I receptors on neutrophils from patients with Laron syndrome were significantly up-regulated, and showed a significantly increased response to *in vitro* IGF-I priming of phagocytic capacity and oxidative burst (Bierknes R 1997). Therefore, the normal neutrophil function in untreated patients with Laron syndrome might be due to a compensatory upregulation of the IGF-I receptor.While neutrophils of elderly persons show a decreased oxidative metabolism compared to neutrophils from younger individuals, they can be primed by hGH *in vitro* to the same extent as neutrophils from younger persons (Wiedermann CJ 1991a; Wiedermann CJ 1991b).

The effects of GH on the chemotaxis of neutrophils and monocytes are not fully understood. GH has been found to either stimulate spontaneous migration of monocytes (Wiedermann CJ 1992 and 1993a), or to have no effect on the spontaneous migration of polymorphonuclear neutrophils (Wiedermann CJ 1991b; Fornari MC 1994a). In contrast, the chemoattractant-stimulated chemotaxis of polymorphonuclear neutrophils and monocytes has been found to be decreased by GH (Wiedermann R 1991b, 1992, 1993a and b; Fornari MC 1994a). This inhibition has been suggested to be caused by a stimulation of cell adhesiveness by GH (Wiedermann CJ 1991b). Also in acromegalic patients a decreased spontaneous and chemoattractant-stimulated migration of polymorphonuclear cells was observed in comparison to controls (Fornari MC 1994b).

Despite these experimental data suggesting a possible role of GH and IGF-I in neutrophil function and host defense (Saito H 1996), so far, no human studies exist, evaluating the effect of GH therapy in GH-deficient subjects, in elderly persons, or in patients with sepsis on important end-point parameters as incidence of morbidity and mortality caused by infectious diseases.

Erythropoiesis

Scatchard plot analysis of IGF-I binding to human erythrocytes has shown about 2 to 13 highly specific binding sites / erythrocyte with an affinity constant of 1.2 - 1.8 x 10^{-9}M (Izumi T 1986; Catanese VM 1986; Eshet R 1993a-c). Binding of IGF-I to erythrocytes could be inhibited by a monoclonal antibody against the IGF-I receptor (alpha IR-3) (Catanese VM 1986). The IGF-I receptor expression on erythrocytes has been shown to be negatively correlated to serum IGF-I levels in different states as GH deficiency, GH insensitivity, or acromegaly. In patients with GH deficiency (Eshet R 1993 b; Mandel S 1995; El-Andere W 1995) or GH insensitivity (Eshet R 1993c), the number of receptors for IGF-I on erythrocytes is increased in comparison to controls, while the affinity is unchanged. Substitution therapy with GH (Eshet R 1993b; Mandel S 1995) or IGF-I (Eshet 1993c) causes a decrease of IGF-I receptor number on erythrocytes, which is negatively correlated to the increase in serum IGF-I (Eshet R 1993b; Mandel S 1995). Also erythrocytes from acromegalic patients with elevated serum levels of IGF-I were found to have a decreased number of receptors for IGF-I with decreased affinity, in comparison to normal controls (El-Andere W 1995).

Experimental and clinical data show a specific role of GH and IGF-I in erythropoiesis. *In vitro* GH (50-100 ng/ml) and IGF-I (0.5-1.0 ng/ml) stimulate the growth of human erythroid progenitor cells (Claustres M 1987; Merchav S 1988b; Merchav S 1993b), with GH acting indirect through paracrine derived IGF-I (Merchav S 1988b; Merchav S 1993b). Hypophysectomized rats present with anemia (Nagy E 1989; Engstrom KG 1990). Treatment of hypophysectomized rats with bGH or IGF-I was reported to correct the anemia (Kurtz A 1988; Nagy E 1989), while another study observed no normalization of red blood cell counts and haemoglobin levels, despite an early increase in reticulocyte numbers (Engstrom KG 1990). Transgenic mice expressing either bovine growth hormone or human growth hormone releasing hormone showed splenic hyperplasia with increased absolute numbers of splenic erythroid and megakaryocytic progenitor cells, showing higher proliferative activity, than non-transgenic controls (Blazar BR 1995). In contrast, the bone marrow of transgenic animals did not show an increased number of progenitor cells of any lineage, in comparison to non-transgenic animals. Also in peripheral blood, leukocyte, erythrocyte and platelet number did not differ significantly in transgenic and non-transgenic animals (Blazar BR 1995).

In GH-deficient children the blood hemoglobin concentration has been reported to be positively correlated with serum IGF-I and IGFBP-3 levels (Vihervuori E 1996). Hemoglobin levels, hematocrit, and red blood cell count were found to be decreased in GH deficient children in comparison to normal controls (Valerio G 1997b), and to be increased after GH substitution therapy (Ardizzi A 1993; Vihervuori E 1996; Valerio G 1997b). Also in GH-deficient adults an increase of red cell mass after 3 months of GH therapy (Christ ER 1997) and haemoglobin concentrations after 9 months of GH treatment (Ten Have SMTH 1997) were observed. Another study did not observe effects of GH therapy on peripheral red blood cell and reticulocyute numbers, however, a significant increase of erythroid and myeloid progenitor precursor cells after 18 and 24 months of GH therapy was found (Kotzmann H 1996). Serum erythropoetin levels were unchanged during GH therapy of GH-deficient adults (Christ ER 1997; Kotzmann H 1996). Patients with GH-deficiency have a decreased number of erythroid precursor cells (Kotzmann H 1996), and also seem to have modestly reduced red blood cell parameters in comparison to matched normal controls (Ardizzi A 1993; Vihervuori E 1996; Valerio G 1997; Christ ER 1997), but, they do not present with anemia. Therefore, while stimulation of the GH-IGF-I axis can cause a steep increase in erythroid progenitor cells, the measurable peripheral changes are modest.

GH AND IGF-I - CLINICAL IMMUNOMODULATION ?

Acromegaly

Retrospective studies in acromegalic patients show an increased incidence of malignant neoplasias (Ezzat S 1991). The incidence of hyperplastic and adenomatous colonic polyps (Colao A 1997), as well as colonic adenocarcinomas (Jenkins PJ 1997) is increased in acromegalic patients. Analysis of the colonic lamina propria lymphocyte subsets in acromegalic patients vs. normal controls showed a significant decrease in CD20+ and CD19+ B-lymphocyte subsets, and CD16+ natural killer cells. CD3+ T lymphocytes were found to be increased, while TCR gamma/delta and CD4+/leu8- subsets were decreased (Colao A 1997). So far, however, data indicating a pathophysiological link between these observations and increased incidence of colonic polyps and neoplasms are lacking. Further studies should focus on a possible role of changed local immune surveillance and malignancy.

GH-deficiency/GH-insensitivity and GH/IGF-I therapy

Many animal and *in vitro* studies revealed an important role of GH, and IGF-I in the cellular and humoral immune system. In hGH-deficient patients single case-reports associated hGH-deficiency with different immune disorders, i.e. hypogammaglobulinemia (Fleisher TA 1980; Maghnie M 1992; Haraldsson A 1993; Ohzeki T 1993), or combined immunodeficiency (Tang ML 1993). A causative role of GH-deficiency in these cases was however not proven and random coincidence must be considered. Generally, patients with GH-deficiency or GH insensitivity syndrome do not present with a clinically significant cellular or humoral immunodeficiency, although there are some abnormalities in their immune functions in comparison to normal subjects. The lack of clinically overt immune system deficits in states of GH-deficiency might be explained by the para- or autocrine action of immune cell derived GH and IGF-I, probably compensating the systemic endocrine failure. In GH insensitivity syndrome, some authors raised the hypothesis of an increased mortality possibly caused by higher susceptibility to infectious diseases (Rosenfeld RG 1994; Clark R 1997). The GH insensitivity syndrome affects the paracrine / autocrine immune GH-IGF-I axis as well. Therefore, from a theoretical aspect, patients with GH insensitivity syndrome are more likely to have severe failure of the immune GH-IGF-I system, than pituitary GH-deficient patients. In this respect, the thourough study of patients with Pit-1 mutations appears to be of special interest (Melen L 1997).

GH therapy of GH-deficient children in Japan has been discussed to cause a higher incidence of leukemia than in the age-related normal population of children (Stahnke N 1992). However, recent studies do not show evidence of an increased incidence of leukemia during or after GH substitution therapy, in GH-deficient children without additional risk factors (Allen DB 1997). While an increased chromosome fragility was reported in short normal children treated with GH (Tedeschi B 1993), two recent studies observed no increased somatic cell mutation frequency (Lin YW 1996) and no increased incidence of chromosomal abnormalities (Bozzola M 1997) in peripheral blood leukocytes of GH-deficient patients receiving GH substitution therapy. However, an *in vitro* study found hGH and IGF-I to increase blast colony numbers of acute leukemia cells (Zadik Z 1993). A recent study described hGH receptor expression on 44 examined primary acute leukemias and 13 leukemic cell lines. While GH caused a proliferative response in several leukemic cell lines, only one of 19 primary leukemias showed a mitogenic response to GH (Giesbert S 1997). These experimental data support the theoretical possibility of an increased risk of tumorgenesis or tumor recurrence by GH therapy. As suggested in clinical studies on GH therapy of children after leukemia treatment (Ogilvy-Stuart AL 1995; Shalet 1997), the confidence interval is very large, and despite a statistical significant incidence of an increased tumor recurrence in GH-treated children is lacking, close monitoring and further databases are needed. First pilot studies reported a positive effect of GH treatment on growth velocity in 7 children after bone marrow transplantation (Brauner R 1997). Prospective studies evaluating GH treatment after bone marrow transplantation will also have to evaluate the various immunological aspects, e.g. graft survival and tumor recurrence.

Besides GH therapy, GH-deficiency itself has been suggested to be a possible risk factor for the development of leukemia (Rapaport R 1995; Kubota M 1995). Whether the decreased NK cell activity in GH-deficient patients (Kiess W 1986; Kiess W 1988; Bozzola M 1989; Crist DM 1987; Auernhammer CJ 1996) might play a causative role in tumorigenesis, or enhancement of NK cell activity by GH and IGF-I (Crist DM 1990; Auernhammer CJ 1996) might be of benefit in controlling tumorigenesis is hypothetical and needs further evaluation.

GH therapy in children with renal transplants

Treatment of growth-retarded children after renal allograft transplantation with hGH has been shown to improve growth rates (Broyer M 1996; Mentser M 1997). However, concerns exist about possible adverse immunological effects of GH treatment on graft survival and renal graft function. A European placebo-controlled study enrolling 203 children after renal graft transplantation found a tendency of an increased incidence of acute rejection during GH therapy (1.0 IU/kg/week), particularly in those patients who had more than one previous rejection episode (Broyer M 1996). However, the mean change in the glomerular filtration rate did not differ significantly in the GH treated vs. the placebo group (Broyer M 1996). Another multicentre study from Japan, enrolling 23 children after renal graft transplantation, reported acute rejection episodes in 20% of children treated with 0.5 IU rhGH/kg/week, in comparison to 38% in children treated with 1.0 IU rhGH/kg/week (Kawaguchi H 1996). In contrast, GH treatment (4 vs. 8 IU/m^2/day, respectively) in 18 adolescents with renal allografts did not reveal a significantly higher reduction in the glomerular filtration rate over 2 yrs, than in non-GH treated controls (Hokken-Koelega ACS 1994). A placebo-controlled trial in 11 prepubertal children did not show an increase of glomerular filtration rate during GH treatment (4 IU/m2/day) (Hokken-Koelega ACS 1996). In a recently published study from the USA, 59 children after renal graft transplantation were treated with rhGH for up to 5 years (Mentser M 1997). No significant deterioration of renal graft function or increase of acute rejection episodes was found in the GH treated children in comparison to the North American Pediatric Renal Transplant Cooperative Study cohort, envolving 2390 transplant recipients (Mentser M 1997).

As GH therapy in growth retarded children after renal graft transplantation is still controversial (Benfiled MR 1997; Friedman AL 1997), GH should not be used outside clinical studies in growth-retarded children with renal allograft transplants and renal graft function should be monitored closely.

GH and IGF-I therapy in catabolic states

In several diseases with catabolic states, e.g. burn injuries (Herndon DN 1995), criticall illness as trauma and sepsis (Ziegler TR 1994; Kudsk KA 1994; Petersen SR 1995; Jeevanandam M 1996; Yarwood GD 1997), AIDS wasting syndrome (Krentz AJ 1993; Mulligan K 1993; Lieberman SA 1994, Ellis KJ 1996; Lee PD 1996; Waters D 1996, Schambelan M 1996; Hirschfeld S 1996; McNurlan MA 1997), and cancer (Tayek JA 1995), clinical studies have been performed to evaluate the potential therapeutic benefits of the anabolic effects of GH and IGF-I. However, the theoretical positive and negative aspects of immunomodulatory effects caused by GH or IGF-I administration in these states have not been subject to clinical studies, yet.

Only one study examined the effects of GH treatment in humans with burn injuries (Herndon DN 1995). In a placebo-controlled trial 10 patients with burns covering > 40% of total body surface area received 0.2 mg/kg/day of rhGH for 7 days after operation for split-thickness skin grafts. Treatment with rhGH caused a significantly increased basal lamina coverage and decreased healing time of the donor sites in comparison to placebo (Herndon DN 1995). Expression of IGF-I receptors, laminin, type IV collagen, type VII collagen, and cytokeratin-14 was increased in the donor sites after GH therapy (Herndon DN 1995).

In animal models of burn injuries, GH and IGF-I therapy improved hepatic metabolism (Dong YL 1993) and nitrogen balance (Sugiura T 1997; Shimoda N 1997), increased lean muscle weight (Fang CH 1997; Meyer NA 1996), reduced gut atrophy and bacterial intestinal translocation (Shimoda N 1997; Sugiura T 1997; Huang KF 1993). No data are available to answer the question, whether the reduction of bacterial intestinal translocation after GH or IGF-I therapy is only caused by the prevention of gut atrophy during parenteral nutrition, or also by changes in the local and systemic immune system. Re-epitheliazation of burn wounds in rats was significantly more improved by combined systemic GH and IGF-I treatment, than GH or IGF-I alone (Meyer NA 1996), while local administration of a small dose of a IGF-I cDNA liposome preparation (0.9µg/kg/week) was as effective as systemic administration of a higher dose of IGF-I (5mg/kg/day) plus GH (Pierre EJ 1997). Whether these effects of GH and IGF-I are only caused by local anabolic and mitogenic effects on the scar tissue (Herndon DN 1995; Gahary A 1995), or also by changes in the local and systemic immune system remains unclear. In burninjured rats, delayed-type hypersensivity was enhanced in GH and IGF-I treated animals, respectively (Shimoda N 1997). In burninjured mice infected with herpes simplex virus type I, treatment with rhGH with a 4mg/kg dose every other day, restored the interferon-gamma production by splenic mononuclear cells, as well as the cytostatic function of peritoneal macrophages, and improved mortality as important end-point parameter (Takagi K 1997).

In criticall illness such as trauma and sepsis pilot studies with GH or IGF-I have shown beneficial anabolic effects (Ziegler TR 1994; Kudsk KA 1994; Petersen SR 1995; Jeevanandam 1996; Yarwood GD 1997). However, scarce immunological data are available. IGF-I treatment in patients with severe head injury restored the depressed CD4/CD8 ratio to normal values (Kudsk KA 1994). GH treatment in patients with abdominal surgery was reported to improve serum immunoglobulin concentrations and cell-mediated immune response (Ziegler TR 1994). No data are available on important end-point parameter such as days of ICU treatment and survival rates.

A few studies also evaluated the value of GH treatment in non GH-deficient animals with sepsis. In piglets with E.coli sepsis, no difference of phagocytosis and oxidative burst capacity was observed in GH-pretreated (16IU; 4hrs before E. coli infusion) vs. untreated animals (Balteskard L 1997). However, systemic TNF-α levels were reduced in the GH-pretreated vs. untreated group (Balteskard L 1997). In a murine model of E. coli sepsis, pretreatment for 6 days with high or low dose rhGH (4.8 or 0.48 mg/kg/day) or high or low-dose rhIGF-I (24 or 2.4 mg/kg/day) resulted in significantly prolonged survival in all groups in comparison to placebo (Inoue T 1995). GH and IGF-I treatment lead to lower systemic cytokine levels (e.g. TNF-α, IL-1, IL-6) during sepsis. However, GH and IGF-I treatment increased the number of peritoneal exsudative cells and the *in vitro* production of cytokines by these cells. In addition, the number of viable bacterial counts in the peritoneal fluid and in the liver was reduced (Inoue T 1995). In contrast, pretreatment of rats with GH for 3 days before injection of endotoxin caused deterioration of renal and liver function, and hypoglycemia and hyperlipidaemia, in comparison to non GH pretreated controls (Liao W 1996). In piglets, intravenous pretreatment with 16IU hGH deteriorated cardiac compensation after haemorrhagic shock and promoted metabolic acidosis during E.coli sepsis (Unneberg K 1996).

Thus, the *in vivo* data on the potential therapeutic effects of GH during sepsis are not congruent. Immune stimulation by GH and IGF-I may cause therapeutic benefits under certain conditions caused by specific pathogens. Further evaluation of the subsets of posttrauma and sepsic states, in which GH therapy might be useful, is necessary.

In AIDS waisting syndrome the anabolic effects of GH and/or IGF-I therapy were evaluated in several clinical studies (Krentz AJ 1993; Mulligan K 1993; Lieberman SA 1994, Ellis KJ 1996; Lee PD 1996; Waters D 1996, Schambelan M 1996; Hirschfeld S 1996; McNurlan MA 1997). Several animal studies indicated possible positive and negative immunomodulatory effects of a GH therapy in patients with HIV infection or AIDS. In C57BL/6 mice, the significant myelotoxic side effects of azidothymidine could be partially restored by rhGH administration (20 µg, i.p. every other day), causing an increase in bone marrow cellularity, haematocrit values, white blood cell counts and splenic hematopoietic progenitor cell content (Murphy WJ 1992a). *In vitro*, however, rhGH augmented viral replication in acutely HIV-1 infected peripheral blood mononuclear cells by 2- to 20-fold, while having no effect on HIV-1 viral replication in chronically infected promonocytic, monocytic and CD4+T-cell lines (Laurence J 1992). In mice with severe combined immune deficiency (SCID) rhGH administration (20 µg, i.p. every other day) caused increased human lymphocyte engraftment, but also increased the incidence of human Epstein-Barr virus-infected B-cell lymphoma, when the donor was EBV infected (Murphy WJ 1992b). These mechanisms appeared to be potential risk factors, when patients with AIDS wasting syndrome are treated with GH and/or IGF-I. However, several placebo-controlled studies observed no significant effects of GH and/or IGF-I therapy on CD4+ or CD8+ lymphocyte counts (SA Lieberman 1994; PDK Lee 1996; Schambelan 1996; D Waters 1996), serum levels of HIV p24 antigen (SA Lieberman 1994; PDK Lee; Schambelan 1996; D Waters 1996), and HIV RNA levels (SA Lieberman 1994; Schambelan 1996; D Waters 1996; MA McNurlan). Two placebo-controlled studies in 178 (Schambelan 1996) and 142 HIV-infected patients (PDK Lee 1996), respectively, treating patients with GH (0.1 mg/kg/day) or GH (0.34 mg twice daily) plus IGF-I (5.0 mg, twice daily) for 12 weeks, respectively, did not observe a significant difference in the incidence of Kaposi's sarcomas or opportunistic infections between drug- and placebo-treated patients. The latter study (PDK Lee 1996), however, reported a non-significant higher incidence of diarrhea (44 vs. 31%), retinitis (10 vs. 2%) and herpes simplex (5 vs. 1%) in the placebo group. These observations should be further evaluated, as in the animal model of herpes simplex infected mice GH treatment improves immune function and survival (Takagi K 1997).

Therefore, during GH and/or IGF-I treatment in patients with AIDS wasting syndrome, a careful monitoring of immunolgical parameters including HIV replication and replication of opportunistic viral infections should be performed.

GH and IGF-I therapy in aging

GH secretion and IGF-I serum levels decline with age and the "somatopause" has been suggested to be partially responsible for the aging process (Corpas E 1993). As one aspect, the age-related decline in immune function might be partially caused by the "somatopause" (Gelato MC 1995). Therapy of the somatopause with GHRH and synthetic orally administered GHRPs, as well as GH is currently under clinical evaluation (Thorner MO 1997; Lieberman SA 1997).

However, so far there are only very limited *in vivo* studies, evaluating the immunolgical aspects of these therapeutic concepts. The immunolgical effects of a therapy with rhGH (100 µg/kg/day), rhIGF-I (120 µg/kg/day), or a combination of both, were studied in aged female monkeys (LeRoith D 1996). When compared to untreated control animals, all treatment groups showed a higher *in vitro* lymphoproliferation after PHA-stimulation as well as higher serum antibody titers after tetanus toxoid immunization. Although these data did not reach significance due to the small number of animals and the high intersubject variability, they suggest a possible activation of the immune system in aged primates by GH and IGF-I (LeRoith D 1996).

A recent clinical study evaluated the immunological effects of GHRH 1-29-NH2 (10 µg/kg/day s.c.) therapy in 20 elderly men and women (Khorram O 1997). After 16 weeks there was a 30% increase of CD20+ B-cells, however, only a transient increase in IgG and IgA serum levels was observed at 4 weeks. The number of CD3+ T-cells, CD4+ T helper cells and CD8+ T suppressor cells as well as CD57+ natural killer cells was unchanged during the study period. In contrast, GHRH therapy induced T-cell activation by significantly increasing the expression of IL-2 receptors (CD25) and transferrin receptors (CD71), as well as the *in vitro* secretion of IL-2 and the lymphoproliferative response to PHA (Khorram O 1997). The immunolgical effects of GHRH therapy could be mediated by different factors. Firstly, the significant increase in integrated GH secretion and IGF-I serum levels during GHRH therapy (Khorram O 1997) could play an important role. Secondly, a direct effcet of GHRH on the function of immune cells and their paracrine / autocrine secretion of GH has to be considered. For example, *in vitro* low concentrations of GHRH 1-29 increased PHA-stimulated lymphoproliferation, while higher concentrations had an inhibitory effect (Valtorta A 1991).

Experimental perspectives for the future

IGF-I treatment has been reported to be of benefit in animal models of acute experimental autoimmune encephalomyelitis (Yao DL 1995; Liu X 1997), by reducing lesion number and size, as well as functional deficits. In a double-blind, placebo-controlled clinical study in 266 patients with amyotrophic lateral sclerosis treatment with recombinant human insulin-like growth factor (0.05 mg/kg/day or 0.10 mg/kg/day) resulted in a significantly slower clinical deterioration (Lange DJ 1996). In the animal model of acute non-demyelinative experimental autoimmune encephalomyelitis the numbers of infiltrating CD4-positive T-cells, alpha/beta TCR-positive cells, and ED-1-positive macrophages were significantly reduced by IGF-I treatment (Liu X 1997), thus suggesting an immunomodulatory effect of IGF-I. Future studies will be needed to discriminate the direct trophic actions of IGF-I on neurons and myelinization (Yao DL 1995), from prevention of neuronal cell death by IGF-I induced immunomodulation (Liu X 1997).

Furthermore, in a murine model of autoimmune diabetes, IGF-I treatment could lower the successful transfer of diabetes after injection of autoreactive Thyq-1.2+T-cells from diabetic NOD mice into congenic NOD-N Thy-1.1 mice (Bergerot I 1995 and 1996). IGF-I treatment lowered the percentage of islet-cell infiltration and the percentage of autoreactive Thyq-1.2+T- cells in the spleen, but not in the thymus (Bergerot I 1995 and 1996). Thus, IGF-I might cause a specific mechanism of T-cell trafficking. Whether this immunomodulatory mechanism represents the key effect of its protective effect against autoimmune diabetes, has to be further evaluated.

SUMMARY AND PERSPECTIVES

The multitude of data presented in this chapter can be summarized by the following statement: Most investigations on GH and immune function add to the existing evidence for a significant impact of the GH-IGF-I axis on the immune system. However, until now there is no conclusive interpretation of all these findings, and the clinical significance is much less clear than *in vitro* experiments or animal models might suggest. For future studies on the effect of GH and IGF-I on the immune system aiming at more specific information about mechanisms of endocrine-immune interactions, one has to take into account some lessons from former experiments:

Firstly, there are important differences between GH actions in animals and in humans. The observed lack of clearcut immunodeficiency in patients with Laron type dwarfism is in contrast what we would predict extrapolating from observations in rodents. Furthermore, human prolactine and GH both bind to the prolactine receptor in primates, but not in rodents. This fact might hint a more „redundant" network in humans in the sense of the absence of one hormone being substituted for by another. Therefore, a careful interpretation is necessary when extrapolating results from animal models to humans.

Secondly, peripheral blood contains only about 2% of whole body immunocompetenT-cells (Westermann J 1990) and changes in primary or secondary lymphoid tissues are not always reflected in the periphery (Sharp B 1993). Therefore, when *in vivo* studies on the effects of GH or IGF-I on human peripheral blood cells are carried out, one has to be aware of the fact, that information on a small diagnostic window of the immune system is being obtained. Besides these "surrogate markers" more reliable endpoints, e.g. survival rates in infectious states, should be evaluated in clinical studies.

For the interpretation of *in vivo* findings a thorough discrimination between endocrine systemically derived and para / autocrine immune cell derived GH and IGF-I has to be performed. The paracrine/autocrine GH-IGF-IGFBP system of immunocompetent cells seems to be partially independent from pituitary function, as recent studies suggest a regulation of GH production in leucocytes different from the classical pituitary mechanisms. This might explain the lack of major disturbances of the immune system in GH-deficient patients. More experiments on the impact of the auto- and paracrine GH-IGF-IGFBP system of immunocompetent cells are needed to elucidate its specific regulation. This understanding will also help to adress the usefulness of systemic GH or IGF-I therapy for clinical immunomodulation.

Furthermore, the historically derived division of protein messenger molecules into „hormones" on the one hand and „cytokines" on the other appears increasingly arbitrary. Hormones influence the production and effects of several cytokines in immunocompetent cells. The other way around, cytokines have a major impact on pituitary neuroendocrine function. Further studies will have to focus on these interactions to obtain more information on the cytokine-hormone network.

REFERENCES

Abboud SL, Bethel CR, Aron DC. Secretion of insulinlike growth factor I and insulinlike growth factor-binding proteins by murine bone marrow stromal cells. J Clin Invest 1991;88:470-475

Allen DB, Rundle AC, Graves Andersen B, Pearse RV 2nd, Jenne K, Sornson M, Lin SC, Bartke A, Rosenfeld MG. The Ames dwarf gene is required for Pit-1 gene activation. Dev Biol 1995;172:495-503

Allen DB, Rundle AC, Graves DA, Blethen SL. Risk of leukemia in children treated with human growth hormone: review and reanalysis. J Pediatr 1997;131:S32-S36

Andersen B, Pearse RV 2nd, Jenne K, Sornson M, Lin SC, Bartke A, Rosenfeld MG. The Ames dwarf gene is required for Pit-1 gene activation. Dev Biol 1995;172:495-503

Ardizzi A, Guzzaloni G, Grugni G, Moro D, Calo G, Mazzzilli G, Tonelli E, Morabito F. The effect of GH on erythropoiesis in vivo. Minerva Endocrinol 1993;18:83-85

Argetsinger LS, Carter-Su C. Mechanism of signaling by growth hormone receptor. Physiol Rev 1996;76:1089-1107

Arkins S, Rebeiz N, Brunke-Reese DL, Minshall C, Kelley KW. The colony-stimulating factors induce expression of insulin-like growth factor -I messenger ribonucleic acid during hematopoiesis. Endocrinology 1995;136:1153-1160

Aston C, Jagirdar J, Lee TC, Hur T, Hintz RL, Rom WN. Enhanced insulin-like growth factor molecules in idiopathic pulmonary fibrosis. Am J Respir Crit Care Med 1995;151:1597-1603

Auernhammer CJ, Strasburger CJ. Effects of growth hormone and insulin-like growth factor I on the immune system. Eur J Endocrinol 1995;133: 635-645

Auernhammer CJ, Feldmaier H, Nass R, Pachmann K, Strasburger CJ. Insulin-like growth factor I is an independent coregulatory modulator of natural killer (NK) cell activity. Endocrinology 1996;137:5332-5336

Auernhammer CJ, Weber MM, Fottner C, Engelhardt D. Human lymphocytes secrete IGF binding proteins (IGFBPs) in vitro. 79th Annual Meeting of the Endocrine Society 1997;P3-259

Auernhammer CJ, , Pachmann K, Nass R, Feldmaier H, Horn K, Belohradsky BH, Strasburger CJ. Impact of growth hormone replacement therapy on immune parameters in GH-deficient adults. submitted 1998

Badolato R, Bond HM, Valerio G, Petrella A, Morrone G, Waters MJ et al. Differential expression of surface membrane growth hormone receptor on human peripheral blood lymphocytes by dual fluorochrome flow cytometry. J Clin Endocrinol Metab 1994;79:984-90

Balducci R, Toscano V, Pasquino AM, Anselmi M, Mangiantini A, Municchi G, Armenise P, Di Rito A, Picone S, Finocchi A, Boscherini B. Lymphocyte subsets in adults with panhypopituitarism before and during GH treatment. 76th Annual Meeting of the Endocrine Society, Anaheim, CA, 1994:303

Balteskard L, Unneberg K, Halvorsen D, Ytrebo LM, Waage A, Sjursen H, Revhaug A. The influence of growth hormone on tumour necrosis factor and neutrophil leukocyte function in sepsis. Scand J Infect Dis 1997;29:393-399

Bamberger AM, Bamberger CM, Pu LP, Puy LA, Loh YP, Asa SL. Expression of pit-1 messenger ribonucleic acid and protein in the human placenta. J Clin Endocrinol Metab 1995;80:2021-2026

Baxter JB, Blalock JE, Weigent DA. Expression of immunoreactive growth hormone in leukocytes in vivo. J Neuroimmunol 1991a;33:43-54

Baxter JB, Blalock JE, Weigent DA. Characterization of immunoreactive insulin-like growth factor-I from leukocytes and its regulation by growth hormone. Endocrinology 1991b;129:1727-1734

Bazan JF. A novel family of growth factor receptors: a common binding domain in the growth hormone, prolactin, erythropoietin and IL-6 receptors, and the p75 IL-2 receptor beta-chain. Biochem Biophys Res Commun 1989;164:788-795

Benfield MR, Kohaut EC. Growth hormone is safe in children after renal transplantation. J Pediatr 1997;131:S28-S31

Ben-Jonathan N, Mershon JL, Allen DL, Steinmetz RW. Extrapituitary prolactin: distribution, regulation, functions, and clinical aspects. Endocr Rev 1996;17:639-669

Bergerot I, Fabien N, Maguer V, Thivolet C. Insulin-like growth factor-1 (IGF-1) protects NOD mice from insulitis and diabetes. Clin Exp Immunol 1995;102:335-340

Bergerot I, Fabien N, Thivolet C. Effects of insulin like growth factor-1 and insulin on effector T-cells generating autoimmune diabetes. Diabetes Metab 1996;22:235-239

Bidlingmaier M, Auernhammer CJ, Feldmeier H, Strasburger CJ. Effects of growth hormone and insulin-like growth factor I binding to natural killer cells. Acta Paediatr Suppl 1997;423:80-81

Binder G, Revskoy S, Gupta D. In vivo growth hormone gene expression in neonatal rat thymus and bone marrow. J Endocrinol 1994;140:137-143

Bjerknes R, Aarskog D. Priming of human polymorphonuclear neutrophilic leukocytes by insulin-like growth factor I: increased phagocytic capacity, complement receptor expression, degranulation and oxidative burst. J Clin Endocrinol Metab 1995;80:1948-1955

Bjerknes R, Vesterhus P, Aarskog D. Increased neutrophil insulin-like growth factor-I (IGF-I) receptor expression and IGF-I-induced functional capacity in patients with untreated Laron syndrome. Europ J Endocrinol 1997;136:92-95

Blazar BR, Brennan CA, Broxmeyer HE, Shultz LD, Vallera DA. Transgenic mice expressing either bovine growth hormone (bGH) or human GH releasing hormone (hGRH) have increased splenic progenitor cell colony formation and DNA synthesis in vitro and in vivo. Exp Hematol 1995;23:1397-1406

Bozzola M, Cisternino M, Larizza D, Maghnie M, Moretta A, Valtorta A, et al. Correspondence: To the editor. Metabolism 1989;38:193
Bozzola M, Tettoni K, Severi F, Cpara E, Danesino C, Scappaticci S. Does growth hormone treatment increase chromosomal abnormalities? Clin Endocrinol 1997;47:363-366
Brauner R, Adan L, Souberbielle JC, Esperou H, Michon J, Devergie A, Gluckman E, Zucker JM. Contribution of growth hormone deficiency to the growth failure that follows bone marrow transplantation. J Pediatr 1997;785-792
Broyer M, on behalf of the Pharmacia&Upjohn Study Group. Results and side-effects of treating children with growth hormone after kidney transplantation - a preliminary report. Acta Pediatr Suppl 1996;417:76-79
Camper SA, Saunders TL, Katz RW, Reeves RH. The Pit-1 transcription factor gene is a candidate for the murine Snell dwarf mutation. Genomics 1990;8:586-590
Catanese VM, Grigorescu F, King GL, Kahn CR. The human erythrocyte insulin-like growth factor I receptor: characterization and demonstration of ligand-stimulated autophosphorylation. J Clin Endocrinol Metab 1986;62:692-699
Christ ER, Cummings MH, Westwood NB, Sawyer BM, Pearson TC, Sonksen PH, Russell-Jones DL. The importance of growth hormone in the regulation of erythropoiesis, red cell mass, and plasma volume in adults with growth hormone deficiency. J Clin Endocrinol Metab 1997;82:2985-2990
Claustres M, Chatelain P, Sultan C. Insulin-like growth factor I stimulates human erythroid colony formation in vitro. J Clin Endocrinol Metab 1987;65:78-82
Clayton PE, Day RN, Silva CM, Hellmann P, Day KH, Thorner MO. Growth hormone induces tyrosine phosphorylation but does not alter insulin-like growth factor-I gene expression in human IM9 lymphocytes. J Mol Endocrinol 1994;13:127-136
Clark R. The somatogenic hormones and insulin-like growth factor-1: stimulation of lymphopoiesis and immune function. Endocr Rev 1997;18:157-179
Cohen LE, Wondisford FE, Radovick S. Role of Pit-1 in the gene expression of growth hormone, prolactin and thyrotropin. Endocrinol Metab Clin North Am 1996;25:523-540
Colao A, Balzano A, Ferone D, Panza N, Grande G, Marzullo P, Bove A, Iodice G, Merola B, Lombardi G. Increased prevalence of colonic polyps and altered lymphocyte subset pattern in the colonic lamina propria in acromegaly. Clin Endocrinol 1997;47:23-28
Corpas E, Harman SM, Blackman MR. Human growth hormone and human aging. Endocr Rev 1993;14:20-39
Crist DM, Peake GT, Mackinnon LT, Sibbitt Jr WL, Kraner JC. Exogenous growth hormone treatment alters body composition and increases natural killer cell activity in women with impaired endogenous growth hormone secretion. Metabolism 1987;36:1115-7
Crist DM, Kraner JC. Supplemental growth hormone increases the tumor cytotoxic activity of natural killer cells in healthy adults with normal growth hormone secretion. Metabolism 1990;39:1320-4
Cross RJ, Elliott LH, Morford LA, Roszman TL, McGillis JP. Functional characterization of the insulin-like growth factor I receptor on Jurkat T-cells. Cell Immunol 1995;160:205-210

Cunningham BC, Bass G, Fuh G, Wells JA. Zinc mediation of the binding of human growth hormone to the human prolactin receptor. Science 1990;250:1709-12

Dardenne M, Pleau JM, Nabarra B, Lefrancier P, Derrien M, Choay J, et al. Contribution of zinc and other metals to the biological activity of the serum thymic factor (FTS). Proc Natl Acad Sci USA 1982;79:5370-3

Delhase M, Vergani P, Malur A, Hooghe R, Hooghe-Peters EL. The transcription factor Pit-1/GHF-1 is expressed in hematopoietic and lymphoid tissues. Eur J Immunol 1993;23:951-955

Dong YL, Huang KF, Xia ZF, Chung DH, Yan TZ, Herndon DN, Waymack JP. Impact of exogenous insulinlike growth factor 1 on hepatic energy metabolism in burn injury. Arch Surg 1993;128:703-708

Dubiski S, Ponnappan U, Cinader B. Strain polymorphism in progression of aging: changes in CD4, CD8 bearing subpopulations. Immunol Lett 1989;23:1-8

Edwards III CK, Ghiasuddin SM, Schepper JM, Yunger L, Kelley K. A newly defined property of somatotropin: priming of macrophages for production of superoxide anion. Science 1988;239:769-71

Edwards III CK, Yunger LM, Lorence RM, Dantzer R, Kelley KW. The pituitary gland is required for protection against lethal effects of Salmonella typhimurium. Immunology 1991;88:2274-7

El-Andere W, Lerario AC, Netto DG, Wajchenberg BL. Erythrocyte insulin-like growth factor-I receptor evaluation in normal subjects, acromegalics, and growth hormone-deficient and insulin-dependent diabetic children. Metab 1995;44:923-928

Ellis KJ, Lee PD, Pivarnik JM, Bukar JG, Gesundheit N. Changes in body composition of human immunodeficiency virus-infected males receiving insulin-like growth factor I and growth hormone. J Clin Endocrinol Metab 1996;81:3033-3038

Elmlinger MW, Wimmer K, Biemer E, Blum WF, Ranke MB, Dannecker GE. Insulin-like growth factor binding protein 2 is differentially expressed in leukaemic B- and T-cell lines. Growth Regul 1996;6:152-157.

Engstrom KG, Ohlsson L, Oscarsson J. Effect of hypophysectomy and growth hormone substitution on red blood cell morphology and filterability in rats. J Lab Clin Med 1990;116:196-205

Eshet R, Werner H, Klinger B, Silbergeld A, Laron Z, Le Roith D, et al. Up-regulation of insulin-like growth factor-I (IGF-I) receptor gene expression in patients with reduced serum IGF-I levels. J Mol Endocrinol 1993a;10:115-20

Eshet R, Klinger B, Silbergeld A, Laron Z. Modulation of IGF-I-receptors by exogenous hGH treatment in constitutionally short children. Clin Endocrinol 1993b;39:687-93

Eshet R, Klinger B, Silbergeld A, Laron Z. Modulation of insulin-like growth factor I (IGF-I) binding sites on erythrocytes by IGF-I treatment in patients with Laron syndrome (LS). Regul Pept 1993c;48:233-239

Ezzat S, Melmed S. Clinical review 18: Are patients with acromegaly at increased risk for neoplasia? J Clin Endocrinol Metab 1991;72:245-249

Fabris N, Pierpaoli W, Sorkin E. Hormones and the immunological capacity III. The immunodeficiency disease of the hypopituitary Snell-Bagg dwarf mouse. Clin Exp Immunol 1971;9:209-25

Fang CH, Li BG, Wang JJ, Fischer JE, Hasselgren PO. Insulin-like growth factor 1 stimulates protein synthesis and inhibits protein breakdown in muscle from burned rats. J Parenter Enteral Nutr 1997;21:245-251

Fleisher TA, White RM, Broder S, Nissley SP, Blaese RM, Mulvihill JJ, Olive G, Waldmann TA. X-linked hypogammaglobulinemia and isolated growth hormone deficiency. N Engl J Med 1980;302:1429-1434

Fornari MC, Scolnik MP, Palacios MF, Intebi AD, Diez RA. Growth hormone inhibits normal B-cell differentiation and neutrophil chemotaxis in vitro. Int J Immunopharmac 1994a;16:667-673

Fornari MC, Placios MF, Diez RA, Intebi AD, Decreased chemotaxis of neutrophils in acromegaly and hyperprolactinemia. Europ J Endocrinol 1994b;130:463-468

Frank SJ, Gilliland G, Van Epps C. Treatment of IM-9 cells with human growth hormone (GH) promotes rapid disulfide linkage of the GH receptor. Endocrinology 1994a;135:148-156

Frank SJ, Gilliland G, Kraft AS, Arnold CS. Interaction of the growth hormone receptor cytoplasmic domain with the JAK2 tyrosine kinase. Endocrinology 1994b;135:2228-2239

Freund GG, Kulas DT, Way BA, Mooney RA. Functional insulin and insulin-like growth factor-1 receptors are preferentially expressed in multiple myeloma cell lines as compared to B-lymphoblastoid cell lines. Cancer Res 1994;54:3179-85

Friedman AL. Growth hormone is not safe for children with renal transplants. J Pediatr 1997;131:S25-S27

Fu YK, Arkins S, Wang BS, Kelley KW. A novel role of growth hormone and insulin-like growth factor-I. Priming neutrophils for superoxide anion secretion. J Immunol 1991;146:1602-1608

Fu Y-K, Arkins S, G. F, Cunningham C, Wells JA, Fong S, et al. Growth hormone augments superoxide anion secretion of human neutrophils by binding to the prolactin receptor. J Clin Invest 1992;89:451-7

Gage PJ, Brinkmeier ML, Scarlett LM, Knapp LT, Camper SA, Mahon KA. The Ames dwarf gene, df, is required early in pituitary ontogeny for the extinction of Rpx transcription and initiation of lineage-specific cell proliferation. Mol Endocrinol 1996;10:1570-1581

Gagnerault MC, Postel-Vinay MC, Dardenne M. Expression of growth hormone receptors in murine lymphoid cells analyzed by flow cytofluorometry. Endocrinology 1996;137:1719-1726

Gala RR. Influence of thyroxine and thyroxine with growth hormone and prolactin on splenocyte subsets and on the expression of interleukin-2 and prolactin receptors on splenocyte subsets of Snell dwarf mice. Proc Soc Exp Biol Med 1995;210:117-125

Gayle CM, Miel JP, McGregor AM, Crossey PA. IGF binding proteins (IGFBPs) in normal human lymphocytes: endogenous gene expression and changes with lymphocyte stimulation. 10th International Congress of Endocrinology; San Francisco 1996;OR33-2

Gayle CM, Miel JP, McGregor AM, Crossey PA. Endogenous gene expression of IGF binding proteins (IGFBPs) and growth hormone receptor (GHR) in normal human lymphocyte B and T-cell subpopulations. 79th Annual Meeting of the Endocrine Society 1997;P3-258

Geffner ME, Bersch N, Lippe BM, Rosenfeld RG, Hintz RL, Golde DW. Growth hormone mediates the growth of T-lymphoblast cell lines via locally generated insulin-like growth factor-I. J Clin Endocrinol Metab 1990;71:464-9

Gelato MC. Aging and immune function: a possible role for growth hormone. Horm Res 1996;45:46-49

Gerard JM, Garibaldi L, Myers SE, Aceto T Jr, Kotagal S, Gibbons VP, Stith J, Weber C. Sleep apnea in patients receiving growth hormone. Clin Pediatr (Phila) 1997;36:321-326

Giesbert S, Panzer S, Kovar H, Fischer S, Printz D, Gadner H, Panzer-Grumayer ER. Acute leukemias express a functional receptor for the human growth hormone. Ann Hematol 1997;74:253-257

Goya RG, Gagnerault M-C, Leite De Moraes MC, Savino W, Dardenne M. In vivo effects of growth hormone on thymus function in aging mice. Brain Behav Immun 1992;6:341-354

Grellier P, Yee D, Gonzalez M, Abboud SL. Characterization of insulin-like growth factor binding proteins (IGFBP) and regulation of IGFBP-4 in bone marrow stromal cells. Br J Haematol 1995;90:249-257

Guarcello V, Weigent DA, Blalock JE. Growth hormone releasing hormone receptors on thymocytes and splenocytes from rats. Cell Immunol 1991;136:291-302

Hansen B, Hjorth S, Welinder B, Skriver L, De Meyts P. The growth hormone (hGH)-binding protein cloned from human IM-9 lymphocytes modulates the down-regulation of hGH receptors by 22- and 20-kilodalton human hGH in IM-9 lymphocytes and the biological effects of the hormone in Nb2 lymphoma cells. Endocrinology 1993;133:2809-2817

Haraldsson A, van der Burgt CJ, Weemaes CM, Otten B, Bakkeren JA, Stoelinga GB. Antibody deficiency and isolated growth hormone deficiency in a girl with Mulibrey nanism. Eur J Pediatr 1993;152:509-512

Hattori N , Shimomura K, Ishihara T, Moridera K, Hino M, Ikekubo K, Kurahachi H. Growth hormone (GH) secretion from human lymphocytes is up-regulated by GH, but not affected by insulin-like growth factor. J Clin Endocrinol Metab 1993; 76:937-939

Hattori N, Ikekubo K, Ishihara T, Moridera K, Hino M, Kurahachi H. Spontaneous growth hormone (GH) secretion by unstimulated human lymphocytes and the effects of GH-releasing hormone and somatostatin. J Clin Endocrinol Metab 1994;79:1678-1680

Hattori Y, Vera JC, Rivas CI, Bersch N, Bailey RC, Geffner ME, Golde DW. Decreased insulin-like growth factor I receptor expression and function in immortalized African Pygmy T-cells. J Clin Endocrinol Metab 1996;81:2257-2263

Herndon DN, Hawkins HK, Nguyen TT, Pierre E, Cox R, Barrow RE Characterization of growth hormone enhanced donor site healing in patients with large cutaneous burns. Ann Surg 1995;221:649-656

Hirschfeld S. Use of human recombinant growth hormone and human recombinant insulin-like growth factor-I in patients with human immunodeficiency virus infection. Horm Res 1996;46:215-221

Hochberg Z, Aviram M, Rubin D, Pollack S. Decreased sensitivity to insulin-like growth factor I inTurner's syndrome: a study of monocytes and T lymphocytes. Eur J Clin Invest 1997;27:543-547

Hokken-Koelega AC, Stijnen T, de Ridder MA, de Muinck Keizer-Schrama SM, Wolff ED, de Jong MC, Donckerwolcke RA, Groothoff JW, Blum WF, Drop SL, et al. Growth hormone treatment in growth-retarded adolescents after renal transplant. Lancet 1994;343(8909):1313-1317

Hokken-Koelega AC, Stijnen T, de Jong RC, Donckerwolcke RA,Groothoff JW, Wolff ED, Blum WF, de Muinck Keizer-Schrama SM, Drop SL. A placebo-controlled, double-blind trial of growth hormone treatment in prepubertal children after renal transplant. Kidney Int Suppl 1996;53:S128-S134

Huang KF, Chung DH, Herndon DN. Insulinlike growth factor 1 (IGF-1) reduces gut atrophy and bacterial translocation after severe burn injury. Arch Surg 1993;128:47-53

Ilondo MM, Damholt AB, Cunningham BA, Wells JA, De Meyts P, Shymko RM. Receptor dimerization determines the effects of growth hormone in primary rat adipocytes and cultured human IM-9 lymphocytes. Endocrinology 1994;134:2397-2403

Inoue T, Saito H, Fukushima R, Inaba T, Lin MT, Fukatsu K, Muto T. Growth hormone and insulinlike growth factor I enhance host defense in a murine sepsis model. Arch Surg 1995 Oct;130(10):1115-1122

Izumi T, Kasuga M, Kadowaki T, Hizuka N, Takaku F, Akanuma Y. Characteristics of human erythrocyte insulin-like growth factor I receptors. J Clin Endocinol Metab 1986;62:1206-1212

Jardieu P, Clark R, Mortensen D, Dorshkind K. In vivo administration of insulin-like growth factor-I stimulates primary B lymphopoiesis and enhances lymphocyte recovery after bone marrow transplantation. J Immunol 1994;152:4320-4327

Jelinek DF, Witzig TE, Arendt BK. A role for insulin-like growth factor in the regulation of IL-6-responsive human myeloma cell line growth. J Immunol 1997;159:487-496

Jenkins PJ, Fairclough PD, Richards T, Lowe DG, Monson J, Grossman A, Wass JAH, Besser M. Acromegaly, colonic polyps and carcinoma. Clin Endocrinol 1997;47:12-22

Jeevanandam M, Holaday NJ, Petersen SR. Integrated nutritional, hormonal, and metabolic effects of recombinant human growth hormone (rhGH) supplementation in trauma patients. Nutrition 1996;12:777-787

Jones JI, Clemmons DR. Insulin-like growth factors and their binding proteins: biological actions. Endocr Rev 1995;16:3-34

Kao TL, Harbour DV, Meyer III WJ. Immunoreactive growth hormone production by cultured lymphocytes. Ann NY Acad Sci 1992;650:179-81

Kawaguchi H, Ito K. rhGH use in children with CRI and undergoing dialysis post-transplant in Japan: a multicentre study. MultiCenter Study Group Japan. Br J Clin Pract Symp Suppl 1996;85:26-31

Kelley KM, Oh Y, Gargosky SE, Gucev Z, Matsumoto T, Hwa V, Ng L, Simpson DM, Rosenfeld RG. Insulin-like growth factor-binding proteins (IGFBPs) and their regulatory dynamics. Int J Biochem Cell Biol 1996;28:619-637

Khorram O, Yeung M, Vu L, Yen SS. Effects of (Norleucine 27) growth hormone-releasing hormone (GHRH) (1-29)-NH2 administration on the immune system of aging men and women. J Clin Endocrinol Metab 1997;82:3590-3596

Khosraviani M, Davis SL. Hormonal regulation of peripheral blood mononuclear cells in sheep. Domest Anim Endocrinol 1996;13:139-150

Kiess W, Doerr H, Eisl E, Butenandt O, Belhoradsky BH. Lymphocyte subsets and natural-killer activity in growth hormone deficiency. Letter. N Engl J Med 1986;314:321

Kiess W, Malozowski S, Gelato M, Butenandt M, Doerr H, Crisp B, et al. Lymphocyte subset distribution and natural killer activity in growth hormone deficiency before and during short-term treatment with growth hormone releasing hormone. Clin Immunol Immunopathol 1988;48:85-94

Kiess W, Butenandt O. Specific growth hormone receptors on human peripheral mononuclear cells: reexpression, identification and characterization. J Clin Endocrinol Metab 1985;60:740-746

Kimata H, Yoshida A. Effect of growth hormone and insulin-like growth factor-I on immunoglobulin production by and growth of human B-cells. J Clin Endocrinol Metab 1994;78:635-41

Knyszynski A, Adler-Kunin S, Globerson A. Effects of growth hormone on thymocyte development from progenitor cells in the bone marrow. Brain Behav Immun 1992;6:327-40

Kooijman R, Willems M, De Haas CJ, Rijkers GT, Schuurmans AL, Van Buul-Offers SC, et al. Expression of type I insulin-like growth factor receptors on human peripheral blood mononuclear cells. Endocrinology 1992;131:2244-2250

Kooijman R, Scholtens LE, Rijkers GT, Zegers BJM. Differential expression of type-I insulin-like growth factor receptors in different stages of human T-cells. Eur J Immunol 1995a;25:931-935

Kooijman R, Scholtens LE, Rijkers GT, Zegers BJ. Type I insulin-like growth factor receptor expression in different developmental stages of human thymocytes. J Endocrinol 1995b;147:203-209

Kooijman R, Lauf JJ, Kappers AC, Rijkers GT. Insulin-like growth factor induces phosphorylation of immunoreactive insulin receptor substrate and its association with phosphatidylinositol-3 kinase in human thymocytes. J Exp Med 1995c;182:593-597

Kooijman R, Hooghe-Peters EL, Hooghe R. Prolactin, growth hormone, and insulin-like growth factor-I in the immune system. Adv Immunol 1996a;63:377-454

Kooijman R, Rijkers GT, Zegers BJ. IGF-I potentiates interleukin-2 production in human peripheral T-cells. J Endocrinol 1996b;149:351-356

Kooijman R, Malur A, Van Buul-Offers SC, Hooghe-Peters EL. Growth hormone expression in murine bone marrow cells is independent of the pituitary transcription factor Pit-1. Endocrinology 1997a;138:3949-3955

Kooijman R, Berus D, Malur A, Delhase M, Hooghe-Peters EL. Human neutrophils express GH-N gene transcripts and the pituitary transcription factor Pit-1b. Endocrinology 1997b;138:4481-4484

Kotzmann H, Köller M, Czernin S, Clodi M, Svoboda T, Riedl M, Boltz-Nitulescu G, Zielinski CC, Luger A. Effect of elevated growth hormone concentrations on the phenotype and functions of human lymphocytes and natural killer cells. Neuroendocrinology 1994;60:618-625

Kotzmann H, Riedl M, Clodi M, Barnas U, Kaider A, Hoecker P, Luger A. The influence of growth hormone substitution therapy on erythroid and myeloid progenitor cells and on peripheral blood cells in adult patients with growth hormone deficiency. Eur J Clin Invest 1996;26:1175-1181

Kozak RW, Haskell JF, Greenstein LA, Rechler MM, Waldmann TA, Nissley SP. Type I and II insulin-like growth factor receptors on human phytohemagglutinin-activated T-lymphocytes. Cell Immunity 1987;109:318-31

Krentz AJ, Koster FT, Crist DM, Finn K, Johnson LZ, Boyle PJ, Schade DS. Anthropometric, metabolic, and immunological effects of recombinant human growth hormone in AIDS and AIDS-related complex. Acquir Immune Defic Syndr 1993;6:245-251

Kubota M, Fujii K, Yamanaka C, Akiyama Y, Momoi T, Hori C, Watanabe S. Leukemia in children with growth hormone deficiency not treated with growth hormone. Eur J Pediatr 1995;154:418-419

Kudsk KA, Mowatt-Larssen C, Bukar J, Fabian T, Oellerich S, Dent DL, Brown R. Effect of recombinant human insulin-like growth factor I and early total parenteral nutrition on immune depression following severe head injury. Arch Surg 1994;129:66-70

Kurtz A, Zapf J, Eckardt KU, Clemons G, Froesch ER, Bauer C. Insulin-like growth factor I stimulates erythropoiesis in hypophysectomized rats. Proc Natl Acad Sci USA 1988;85:7825-7829

Lange DJ, Felice KJ, Festoff BW, Gawel MJ, Gelinas DF, Kratz R, Lai EC, Murphy MF, Natter HM, Norris FH, Rudnicki S. Recombinant human insulin-like growth factor-I in ALS: description of a double-blind, placebo-controlled study. North American ALS/IGF-I Study Group. Neurology 1996;47(4 Suppl 2):S93-S94

Laurence J, Grimison B, Gonenne A. Effect of recombinant human growth hormone on acute and chronic human immunodeficiency virus infection in vitro. Blood 1992;79:467-472

Lauriola L, Maggiano N, Serra FG, Nori S, Tardio M, Capelli A, Piantelli M, Ranelletti FO. Immunohistochemical and in situ hybridization detection of growth-hormone-producing cells in human thymoma. Am J Pathol 1997;151:55-61

Lee PD, Pivarnik JM, Bukar JG, Muurahainen N, Berry PS, Skolnik PR, Nerad JL, Kudsk KA, Jackson L, Ellis KJ, Gesundheit N. A randomized, placebo-controlled trial of combined insulin-like growth factor I and low dose growth hormone therapy for wasting associated with human immunodeficiency virus infection. J Clin Endocrinol Metab 1996;81:2968-2975

LeRoith D, Yanowski J, Kalgjian EP, Jaffe ES, LeRoith T, Purdue K, Cooper BD, Pyle R, Adler W. The effects of growth hormone and insulin-like growth factor I on the immune system of aged female monkeys. Endocrinol 1996;137:1071-1079

Lesniak MA, Gorden P, Roth J, Gavin III JR. Binding of 125I-human growth hormone to specific receptors in human cultured lymphocytes. J Biol Chem 1974;249:1661-1667

Li YM, Arkins S, McCusker RH Jr, Donovan SM, Liu Q, Jayaraman S, Dantzer R, Kelley KW. Macrophages synthesize and secrete a 25-kilodalton protein that binds insulin-like growth factor-I. J Immunol 1996;156:64-72

Li YM, Schacher DH, Liu Q, Arkins S, Rebeiz N, McCusker RH Jr, Dantzer R Kelley KW. Regulation of myeloid growth and differentiation by the insulin-like growth factor I receptor. Endocrinology 1997;138:362-368

Liao W, Rudling M, Angelin B. Growth hormone potentiates the in vivo biological activities of endotoxin in the rat. Eur J Clin Invest 1996;26:254-258

Lieberman SA, Butterfield GE, Harrison D, Hoffman AR. Anabolic effects of recombinant insulin-like growth factor-I in cachectic patients with the acquired immunodeficiency syndrome. J Clin Endocrinol Metab 1994;78:404-410

Lieberman SA, Hoffman AR. The somatopause: should growth hormone deficiency in older people be treated? Clin Geriatr Med 1997; 13:671-684

Lin B, Kinoshita Y, Hato F, Tsuji Y. Enhancement of thymic lymphocyte proliferation by the culture supernatant of thymus epithelial cells stimulated by prolactin. Cell Mol Biol (Noisy-le-grand) 1997a;43:361-367

Lin B, Kinoshita Y, Hato F, Tsuij Y. Enhancement of DNA synthetic activity of thymic lymphocytes by the culture supernatant of thymus epithelial cells stimulated by growth hormone. Cell Mol Biol (Noisy-le-grand) 1997b;43:351-359

Lin YW, Kubota M, Wakazono Y, Hirota H, Okuda A, Bessho R, Usami I, Kataoka A, Yamanaka C, Akiyama Y, Furusho K. Normal mutation frequencies of somatic cells in patients receiving growth hormone therapy. Mutat Res 1996;362:97-103

Liu X, Linnington C, Webster HD, Lassmann S, Yao DL, Hudson LD, Wekerle H, Kreutzberg GW. Insulin-like growth factor-I treatment reduces immune cell responses in acute non-demyelinative experimental autoimmune encephalomyelitis. J Neurosci Res 1997;47:531-538

Lytras A, Quan N, Vrontakis ME, Shaw JE, Cattini PA, Friesen HG. Growth hormone expression in human burkitt lymphoma serum-free ramos cell line. Endocrinology 1993;132:620-8

Maggiano N, Piantelli M, Ricci R, Larocca LM, Capelli A, Ranelletti FO. Detection of growth hormone-producing cells in human thymus by immunohistochemistry and non-radioactive in situ hybridization. J Histochem Cytochem 1994;42:1349-1354

Maghnie M, Monafo V, Marseglia GL, Valtorta A, Avanzini A, Moretta A, Balottin U, Touraine JL, Severi F. Immunodeficiency, growth hormone deficiency and central nervous system involvement in a girl. Thymus 1992;20:69-76

Mandel S, Moreland E, Nichols V, Hanna C, Lafranchi S. Changes in insulin-like growth factor-I (IGF-I), IGF-binding protein-3, growth hormone (GH)-binding protein, erythrocyte IGF-I receptors, amd growth rate during GH treatment. J Clin Endocrinol Metab 1995;80:190-194

Manfredi R, Tumietto F, Azzaroli L, Chiodo F, Manfredi G. Growth hormone (GH) and the immune system: impaired phagocytic function in children with idiopathic GH deficiency is corrected by treatment with biosynthetic GH. J Pediatr Endocrinol 1994;7:245-251

Marcus CL, Loughlin GM. Obstructive sleep apnea in children. Semin Pediatr Neurol 1996;3:23-28

McNurlan MA, Garlick PJ, Steigbigel RT, DeCristofaro KA, Frost RA, Lang CH, Johnson RW, Santasier AM, Cabahug CJ, Fuhrer J, Gelato MC. Responsiveness of muscle protein synthesis to growth hormone administration in HIV-infected individuals declines with severity of disease. J Clin Invest 1997;100:2125-2132

Melen L, Hennen G, Dullaart RP, Heinen E, Igout A. Both pituitary and placental growth hormone transcripts are expressed in human peripheral blood mononuclear cells. Clin Exp Immunol 1997;110:336-340

Mellado M, Rodriguez-Frade JM, Kremer L, von Kobbe C, de Ana AM, Merida I, Martinez-A C. Conformational changes required in the human growth hormone receptor for growth hormone signaling. J Biol Chem 1997;272:9189-9196

de Mello-Coelho V, Villa-Verde DM, Dardenne M, Savino W. Pituitary hormones modulate cell-cell interactions between thymocytes and thymic epithelial cells. J Neuroimmunol 1997;76:39-49

Mentser M, Breen TJ, Sullivan EK, Fine RN. Growth-hormone treatment of renal transplant recipients: the National Cooperative Growth Study experience - a report of the National Cooperative Growth Study and the North American Pediatric Renal Transplant Cooperative Study. J Pediatr 1997;131(1 Pt 2):S20-S24

Merchav S, Tatarsky I, Hochberg Z. Enhancement of human granulopoiesis in vitro by biosynthetic insulin-like growth factor I / somatomedin C and human growth hormone. J Clin Invest 1988a;81:791-7

Merchav S, Tatarsky I, Hochberg Z. Enhancement of erythropoiesis in vitro by human growth hormone is mediated by insulin-like growth factor I. Br J Haematol 1988b;70:267-271

Merchav S, Lake M, Skottner A. Comparative studies of the granulopoietic enhancing effects of biosynthetic human insulin-like growth factors I and II. J Cell Physiol 1993a;157:178-83

Merchav S, Grauf Z, Skottner A. In-vitro response of erythroid progenitors from children with thalassaemia major to human growth hormone and insulin-like growth factor I. Clin Endocrinol 1993b;39:207-211

Meyer NA, Barrow RE, Herndon DN. Combined insulin-like growth factor-1 and growth hormone improves weight loss and wound healing in burned rats. J Trauma 1996;41:1008-1012

Miller JS, Tessmer-Tuck J, Blake N, Lund J, Scott A, Blazar BR, Orchard PJ. Endogenous IL-2 production by natural killer cells maintains cytotoxic and proliferative capacity following retroviral-mediated gene transfer. Exp Hematol 1997;25:1140-1148

Minshall C, Arkins S, Straza J, Conners J, Dantzer R, Freund GG, Kelley KW. IL-4 and insulin-like growth factor-I inhibit the decline in Bcl-2 and promote the survival of IL-3-deprived myeloid progenitors. J Immunol 1997;159:1225-1232

Mocchegiani E, Fabris N, Travaglini P, Sartorio A, De Min C, Paolucci P. Thymic endocrine activity in children with idiopathic growth-hormone deficiency. Int J Neurosci 1991;59:151-157

Mocchegiani E, Santarelli L, Muzzioli M, Fabris N. Reversibility of the thymic involution and of age-related peripheral immune dysfunction by zinc supplementation in old mice. Int J Immunopharmacol 1995;17:703-718

Mocchegiani E, Sartorio A, Santarelli L, Ferrero S, Fabris N. Thymulin, zinc and insulin-like growth factor-I (IGF-I) activity before and during recombinant growth hormone (rec-GH) therapy in children and adults with GH deficiency. J Endocrinol Invest 1996;19:630-637

Montecino-Rodriguez E, Clark R, Johnson A, Collins L, Dorshkind K. Defective B-cell development in Snell dwarf (dw/dw) mice can be corrected by thyroxine treatment. J Immunol 1996;157:3334-3340

Montecino-Rodriguez E, Clark R, Powell-Braxton L, Dorshkind K. Primary B-cell development is impaired in mice with defects of the pituitary/thyroid axis. J Immunol 1997;159:2712-2719

Mulligan K, Grunfeld C, Hellerstein MK, Neese RA, Schambelan M. Anabolic effects of recombinant human growth hormone in patients with wasting associated with human immunodeficiency virus infection. J Clin Endocrinol Metab 1993;77:956-962

Murphy WJ, Tsarfaty G, Longo DL. Growth hormone exerts hematopoietic growth-promoting effects in vivo and partially counteracts the myelosuppressive effects of azidothymidine. Blood 1992a;80:1443-1447

Murphy WJ, Durum SK, Anver M, Frazier M, Longo DL. Recombinant human growth hormone promotes human lymphocyte engraftment in immunodeficient mice and results in an increased incidence of human epstein barr virus-induced B-cell lymphoma. Brain Behav Immun 1992b;6:355-3 64

Murphy WJ, Durum SK, Longo DL. Differential effects of growth hormone and prolactin on murine T-cell development and function. J Exp Med 1993;178:231-6

Nagy E, Berczi I. Pituitay dependence of bone marrow function. Br J Haematol 1989;71:457-462

Neely EK, Smith SD, Rosenfeld RG. Human leukemic T and B lymphoblasts produce insulin-like growth factor binding proteins 2 and 4. Acta Endocrinol (Copenh) 1991;124:707-714

Nyman T, Pekonen F. The expression of insulin-like growth factors and their binding proteins in normal human lymphocytes. Acta Endocrinol (Copenh) 1993;128:168-172

Ogilvy-Stuart A. Safety of growth hormone after treatment of a childhood malignancy. Horm Res 1995;44:73-79

Ohzeki T, Hanaki K, Motozumi H, Ohtahara H, Hayashibara H, Harada Y, Okamoto M, Shiraki K, Tsuij Y, Emura H. Immunodeficiency with increased immunoglobulin M associated with growth hormone insufficiency. Acta Paediatr 1993;82:620-623

Olsen SC, Stevens MG. Effects of recombinant human cytokines on mitogen-induced bovine peripheral blood mononuclear cell proliferation. Cytokine 1993;5:498-505

Petersen BH, Rapaport R, Henry DP, Huseman C, Moore WV. Effect of treatment with biosynthetic human growth hormone (GH) on peripheral blood lymphocyte

populations and function in growth hormone-deficient children. J Clin Endocrinol Metab 1990;70:1756-60
Petersen SR, Jeevanandam M, Holaday NJ. Adjuvant recombinant human growth hormone stimulates insulin-like growth factor binding protein-3 secretion in critically ill trauma patients. J Trauma 1995;39:295-300
Pierre EJ, Perez-Polo JR, Mitchell AT, Matin S, Foyt HL, Herndon DN. Insulin-like growth factor-I liposomal gene transfer and systemic growth hormone stimulate wound healing. J Burn Care Rehabil 1997;18:287-291
Postel-Vinay MC, Kelly PA. Growth hormone receptor signalling. Baillieres Clin Endocrinol Metab 1996;10:323-336
Postel-Vinay MC, De Mello Coelho V, Gagnerault MC, Dardenne M. Growth hormone stimulates the proliferation of activated mouse T lymphocytes. Endocrinology 1997;138:1816-1820
Ranke MB, Savage MO, Chatelain PG, Preece MA, Rosenfeld RG, Blum WF, Wilton P. Insulin-like growth factor improves height in growth hormone insensitivity: two years, results. Horm Res 1995;44:253-264
Rapaport R, Oleske J, Ahdieh H, Skuza K, Holland BK, Passannante MR, et al. Effects of human growth hormone on immune functions: in vitro studies on cells of normal and growth hormone-deficient children. Life Sci 1987;41:2319-24
Rapaport R, Sillis IN, Green L, Barrett P, Labus J, Skuza KA, Chartoff A, Goode L, Stene M, Petersen BH. Detection of human growth hormone receptors on IM-9 cells and peripheral blood mononuclear subsets by flow cytometry: correlation with growth hormone-binding protein levels. J Clin Endocrinol Metab 1995a;80:2612-2619
Rapaport R, Oberfield SE, Robinson L, Salisbury S, David R, Rao J, Redmond GP. Relationship of growth hormone deficiency and leukemia. J Pediatr 1995b;126:759-761
Rapaport R, Stein D, Louzao A, Goode L, Garibaldi L, Stene M. Growth hormone secretion by lymphocytes of growth hormone deficient children. 10th International Congress of Endocrinology, San Francisco 1996;P1-661
Reinisch N, Schratzenberger P, Finkenstedt G, Kahler CM, Wiedermann CJ. Superoxide anion release from neutrophils in growth hormone deficient adults before and after replacement therapy with recombinant human growth hormone. Naunyn Schmiedebergs Arch Pharmacol 1996;354:369-373
Rekers-Mombarg LT, Rijkers GT, Massa GG, Wit JM. Immunologic studies in children with idiopathic short stature before and during growth hormone therapy. Dutch Growth Hormone Working Group. Horm Res 1995;44:203-207
Robbins K, McCabe S, Scheiner T, Strasser J, Clark R, Jardieu P. Immunological effects of insulin-like growth factor-I-enhancement of immunoglobulin synthesis. Clin Exp Immunol 1994;95:337-342
Rohn WM; Weigent DA. Cloning and nucleotide sequencing of rat lymphocyte growth hormone cDNA. Neuroimmunomodulation 1995;2:108-114
Rosenfeld RG, Rosenbloom AL, Guevara-Aguirre J. Growth hormone (GH) insensitivity due to primary GH receptor deficiency. Endocrine Rev 1994;15:369-90

Rosenow F, Reuter S, Deuss U, Szelies B, Hilgers RD, Winkelmann W, Heiss WD. Sleep apnoea in treated acromegaly: relative frequency and predisposing factors. Clin Endocrinol (Oxf) 1996;45:563-569

Rovensky J, Vigas M, Lokaj J, Cuncik P, Lukac P, Takac A. Effect of growth hormone on the metabolic activity of phagocytes of peripheral blood in pituitary dwarfs and acromegaly. Endocrinol Exp (Bratisl) 1982;16:129-34

Sabharwal P, Varma S. Growth hormone synthesized and secreted by human thymocytes acts via insulin-like growth factor I as an autocrine and paracrine growth factor. J Clin Endocrinol Metab 1996;81:2663-2669

Saito Y, Ikebuchi H, Yamazaki T, Sawada J. Release of a soluble form of growth hormone receptors (growth hormone-binding proteins) from human IM-9 cells by proteolytic cleavage. J Biochem (Tokyo) 1995;118:521-525

Saito H, Inoue T, Fukatsu K, Ming-Tsan L, Inaba T, Fukushima R, Muto T. Growth hormone and the immune response to bacterial infection. Horm Res 1996;45:50-54

Schambelan M, Mulligan K, Grunfeld C, Daar ES, LaMarca A, Kotler DP, Wang J, Bozzette SA, Breitmeyer JB Recombinant human growth hormone in patients with HIV-associated wasting. A randomized, placebo-controlled trial. Serostim Study Group. Ann Intern Med 1996;125:873-882

Schillaci R, Brocardo MG, Roldan A. Co-operative effect between insulin-like growth factor-1 and interleukin-2 on DNA synthesis and interleukin-2 receptor-alpha chain expression in human lymphocytes. Immunol Cell Biol 1995;73:340-345

Shalet SM, Brennan BM, Reddingius RE. Growth hormone therapy and malignancy. Horm Res 1997;48:29-32

Shimoda N, Tashiro T, Yamamori H, Takagi K, Nakajima N, Ito I. Effects of growth hormone and insulin-like growth factor-1 on protein metabolism, gut morphology, and cell-mediated immunity in burned rats. Nutrition 1997;13:540-546

Silva CM, Weber MJ, Thorner MO. Stimulation of tyrosine phosphorylation in human cells by activation of the growth hormone receptor. Endocrinology 1993a;132:101-108

Silva CM, Day RN, Weber MJ, Thorner MO. Human growth hormone (GH) receptor is characterized as the 134-kilodalton tyrosine-phosphorylated protein activated by GH treatment in IM-9 cells. Endocrinology 1993b;133:2307-2312

Silva CM, Lu H, Weber MJ, Thorner MO. Differential tyrosine phosphorylation of JAK1, JAK2, and STAT1 by growth hormone and interferon-gamma in IM-9 cells. J Biol Chem 1994;269:27532-27539

Silva CM, Lu H, Day RN. Characterization and cloning of STAT5 from IM-9 cells and its activation by growth hormone. Mol Endocrinol 1996;10:508-518

Slabaugh MB, Liebermann ME, Rutledge JJ, Gorski J. Growth hormone and prolactin synthesis in normal and homozygous Snell and Ames dwarf mice. Endocrinology 1981;109:1040-6

Smal J, Closset J, Hennen G, De Meyts P. Receptor-binding and down-regulatory properties of 22000-M human growth homone and its natural 20000-M variant on IM-9 human lymphocytes. Biochem J 1985;225:283-289

Sornson MW, Wu W, Dasen JS, Flynn SE, Norman DJ, O,Connell SM, Gukovsky I, Carriere C, Ryan AK, Miller AP, Zuo L, Gleiberman AS, Andersen B, Beamer WG, Rosenfeld MG. Pituitary lineage determination by the Prophet of Pit-1 homeodomain factor defective in Ames dwarfism. Nature 1996;384:327-333

Spadoni GL, Rossi P, Ragno W, Galli E, Cianfarani S, Galasso C, et al. Immune funtion in growth hormone-deficient children treated with biosynthetic growth hormone. Acta Paediatr Scand 1991;80:75-9

Span JPT, Pieters GFFM, Smals AGH, Koopmans PP, Kloppenborg PWC. Number and percentage of NK cells are decreased in growth hormone-deficient adults. Clin Immunol Immunopathol 1996;78:90-92

Stahnke N. Leukemia in growth-hormone-treated patients: an update 1992. Horm Res 1992;38(Suppl 1):56-62

Stephanou A, Knight RA, Lightman SL. Production of a growth hormone-releasing hormone-like peptide and its mRNA by human lymphocytes. Neuroendocrinology 1991;53:628-33

Stuart CA, Meehan RT, Neale LS, Cintron NM, Furlanetto RW. Insulin-like growth factor-I binds selectively to human peripheral blood monocytes and B-lymphocytes. J Clin Endocrinol Metab 1991;72:1117-1122

Sugiura T, Tashiro T, Yamamori H, Morishima Y, Otsubo Y, Hayashi N, Furukawa K, Nitta H, Nakajima N, Ishizuka T, Tatibana M, Ino H, Ito I. Effects of insulin-like growth factor-1 on endotoxin translocation in burned rats receiving total parenteral nutrition. Nutrition 1997;13:783-787

Takagi K, Suzuki F, Barrow RE, Wolf SE, Kobayashi M, Herndon DN. Growth hormone improves immune function and survival in burned mice infected with herpes simplex virus type 1. J Surg Res 1997;69:166-170

Tang ML, Kemp AS. Growth hormone deficiency and combined immunodeficiency. Arch Dis Child 1993;68:231-232

Tapson VF, Boni-Schnetzler M, Pilch PF, Center DM, Berman JS. Structural and functional characterization of the human T-lymphocyte receptor for insulin-like growth factor I in vitro. J Clin Invest 1988;82:950-7

Tarantal AF, Hunter MK, Gargosky SE. Direct administration of insulin-like growth factor to fetal rhesus monkeys (Macaca mulatta). Endocrinology 1997;138:3349-3358

Taub DD, Tsarfaty G, Lloyd AR, Durum SK, Longo DL, Murphy WJ. Growth hormone promotes human T-cell adhesion and migration to both human and murine matrix proteins in vitro and directly promotes xenogeneic engraftment. J Clin Invest 1994;94:293-300

Tayek JA, Brasel JA. Failure of anabolism in malnourished cancer patients receiving growth hormone: a clinical research center study. J Clin Endocrinol Metab 1995;80:2082-2087

Tedeschi B, Spadoni GL, Sanna ML, Vernole P, Caporossi D, Cianfarani S, Nicoletti B, Boscherini B. Increased chromosome fragility in lymphocytes of short normal children treated with recombinant human growth hormone. Hum Genet 1993;91:459-463

Ten Have SMTH, van der Lely AJ, Lamberts SWJ. Increase in haemoglobin concentrations in growth hormone deficient adults during human recombinant growth hormone replacement therapy. Clin Endocrinol 1997;47:565-570

Thorner MO, Chapman IM, Gaylinn BD, Pezzoli SS, Hartman ML. Growth hormone-releasing hormone and growth hormone-releasing peptide as therapeutic agents to enhance growth hormone secretion in disease and aging. Recent Prog Horm Res 1997;52:215-244
Timsit J, Safieh B, Gagnerault M-C, , Savino W, Lubetzki J, Bach J-F, Dardenne M. Increase of circulating levels of thymulin in hyperprolactinemia and acromegaly. C R Acad Sci III 1990;310:7-13
Timsit J, Savino W, Safieh B, Chanson P, Gagnerault M-C, Bach J-F, et al. Growth hormone and insulin-like growth factor-I stimulate hormonal function and proliferation of thymic epithelial cells. J Clin Endocrinol Metab 1992;75:183-8
Tseng YH, Kessler MA, Schuler LA. Regulation of interleukin (IL)-1alpha, IL-1beta, and IL-6 expresion by growth hormone and prolactin in bovine thymic stromal cells. Mol Cell Endocrinol 1997;128:117-127
Tsuij Y, Kinoshita Y, Hato F, Tominaga K, Yoshida K. The in vitro proliferation of thymus epithelial cells stimulated with growth hormone and insulin-like growth factor-I. Cell Mol Biol (Noisy-le-grand) 1994;40:1135-1142
Unneberg K, Balteskard L, Mjaaland M, Revhaug A. Growth hormone impaired compensation of hemorrhagic shock after trauma and sepsis in swine. J Trauma 1996;41:775-780
Valerio G, Bond HM, Badolato R, Petrella A, Di Maio S, Slaerno M, Waters MJ, Venuta S, Tenore A. Expression of growth hormone receptor by peripherical blood lymphocytes in children: evaluation in clinical conditions of impaired growth. Clin Endocrinol 1997a;47:329-335
Valerio G, DiMaio S, Salerno M, Argenziano A, Badolato R, Tenore A. Assessment of red blood cell indices in growth-hormone-treated children. Horm Res 1997b;47:62-66
Valtora A, Moretta A, Maccario R, Bozzola M, Severi F. Influence of growth hormone-releasing hormone (GHRH) on phytohemagglutinin-induced lymphocyte activation: comparison of two synthetic forms. Thymus 1991;18:51-9
Varma S, Sabharwal P, Sheridan JF, Malarkey WB. Growth hormone secretion by human peripheral blood mononuclear cells detected by an enzyme-linked immunoplaque assay. J Clin Endocrinol Metab 1993;76:49-53
Vihervuori E, Virtanen M, Koistinen H, Koistinen R, Seppaelae M, Siimes MA. Hemoglobin level is linked to growth hormone-dependent proteins in short children. Blood 1996;87:2075-2081
Villanua MA, Szary A, Bartke A, Esquifino AL. Changes in lymphoid organs of Ames dwarf mice after treatment with growth hormone, prolactin or ectopic pituitary transplants. J Endocrinol Invest 1992;15:587-95
Warwick-Davies J, Lowrie DB, Cole PJ. Growth hormone is a human macrophage activating factor. Priming of human monocytes for enhanced release of H_2O_2. J Immunol 1995a;154:1909-1918
Warwick-Davies J, Lowrie DB, Cole PJ. Growth hormone activation of human monocytes for superoxide production but not tumor necrosis factor production, cell adherence, or action against mycobacterium tuberculosis. Infection Immunity 1995b;63:4312-4316
Waters D, Danska J, Hardy K, Koster F, Qualls C, Nickell D, Nightingale S, Gesundheit N, Watson D, Schade D. Recombinant human growth hormone,

insulin-like growth factor 1, and combination therapy in AIDS-associated wasting. A randomized, double-blind, placebo-controlled trial. Ann Intern Med 1996;125:865-872
Weigent DA, Baxter JB, Wear WE, Smith LR, Bost KL, Blalock JE. Production of immunoreactive growth hormone by mononuclear leukocytes. FASEB J 1988;2:2812-8
Weigent DA, Blalock JE. Immunoreactive growth hormone-releasing hormone in rat leukocytes. J Neuroimmunol 1990;29:1-13
Weigent DA, Blalock JE. The production of growth hormone by subpopulations of rat mononuclear leukocytes. Cell Immunol 1991a;135:55-65
Weigent DA, Riley JE, Galin FS, LeBoeuf RD, Blalock JE. Detection of growth hormone and growth hormone-releasing hormone-related messenger RNA in rat leucocytes by the polymerase chain reaction. Proc Soc Exp Biol Med 1991b;198:643-8
Weigent DA, Blalock JE, LeBoeuf RD. An antisense oligonucleotide to growth hormone messenger ribonucleic acid inhibits lymphocyte proliferation. Endocrinology 1991 c;128:2053-7
Weigent DA, Baxter JB, Blalock JE. The production of growth hormone and insulin-like growth factor-I by the same subpopulation of rat mononuclear leukocytes. Brain Behav Immun 1992;6:365-76
Weigent DA, Blalock JE. Effect of administration of growth-hormone-producing lymphocytes on weight gain and immune function in dwarf mice. Neuroimmunomodulation 1994;1:50-58
Weigent DA. Immunoregulatory properties of growth hormone and prolactin. Pharmacol Ther 1996;69:237-257
Wiedermann CJ, Niedermühlbichler M, Beimpold H, Braunsteiner H. In vitro activation of neutrophils of the aged by recombinant human growth hormone. J Infect Dis 1991a;164:1017-20
Wiedermann CJ, Niedermühlbichler M, Geissler D, Beimpold H, Braunsteiner H. Priming of human neutrophils by recombinant human growth hormone. Br J Haematol 1991b;78:19-22
Wiedermann CJ, Reinisch N, Kahler C, Geisen F, Zilian U, Herold M, Braunsteiner H. In vivo activation of circulating monocytes by exogenous growth hormone in man. Brain Behav Immun 1992;6:387-393
Wiedermann CJ, Reinisch N, Braunsteiner H. Stimulation of monocyte chemotaxis by growth hormone and its deactivation by somatostatin. Blood 1993a;82:954-960
Wiedermann CJ, Reinisch N, Niedermühlbichler M, Braunsteiner H. Inhibition of recombinant human growth hormone-induced and prolactin-induced activation of neutrophils by octreotide. Naunyn-Schmiedeberg's Arch Pharmacol 1993b;347:336-41
Wu H, Devi R, Malarkey WB. Localization of growth hormone messenger ribonucleic acid in the human immune system - a clinical research center study. J Clin Endocrinol Metab 1996;81:1278-1282
Xu X, Mardell C, Xian CJ, Zola H, Read LC. Expression of functional insulin-like growth factor-I receptor on lymphoid cell subsets of rats. Immunol 1995;85:394-399

Yang Y, Hoeflich A, Kessler U, Barenton B, Blum W, Schwarz HP, Kiess W. Human IM-9 lymphoblasts as a model of the growth hormone-insulin-like growth factor axis: gene expression, and interactions of ligands with receptors and binding proteins. Regul Peptides 1993;48:41-53

Yao DL, Liu X, Hudson LD, Webster HD Insulin-like growth factor I treatment reduces demyelination and up-regulates gene expression of myelin-related proteins in experimental autoimmune encephalomyelitis. Proc Natl Acad Sci U S A 1995;92:6190-6194

Yarwood GD, Ross RJ, Medbak S, Coakley J, Hinds CJ. Administration of human recombinant insulin-like growth factor-I in critically ill patients. Crit Care Med 1997;25:1352-1361

Yoshida A, Ishioka C, Kimata H, Mikawa H. Recombinant human growth hormone stimulates B-cell immunoglobulin synthesis and proliferation in serum-free medium. Acta Endocrinol (Copenh) 1992;126:524-529

Yu-Lee LY. Molecular actions of prolactin in the immune system. Proc Soc Exp Biol Med. 1997;215:35-52.

Zadik Z, Estrov Z, Karov Y, Hahn T, Barak Y. The effect of growth hormone and IGF-I on clonogenic growth of hematopoietic cells in leukemic patients during active disease and during remission - a preliminary report. J Pediatr Endocrinol 1993;6:79-83

Ziegler TR, Leader I . Adjunctive human growth hormone therapy innutrition support: potential to limit septiccomplications in intensive care unit patients. Semin Respir Infect 1994;9:240-247

14 EFFECTS OF GROWTH HORMONE ON MUSCLE MASS AND FUNCTION

Jens Otto Lunde Jørgensen, Rolf Dall, Nina Vahl, Helene Nørrelund, Jens Sandahl Christiansen

Medical Department M (Endocrinology and diabetes), Aarhus University Hospital, Aarhus, Denmark

The anabolic nature of growth hormone (GH) is dramatically exhibited by patients with acromegaly, and indirectly by patients with GH-deficiency. A large number of *in vitro* and animal studies for several decades has documented stimulating effects of GH on skeletal muscle growth (1). The methods employed to document *in vivo* effects of GH on muscle mass in humans have been exhaustive including whole body retention of nitrogen and potassium, total and regional muscle protein metabolism using labeled amino acids, estimation of lean body mass by total body potassium or dual x-ray absorptiometry (DEXA), and direct calcualtion of muscle area or volume by computerised tomography (CT) and magnetic resonance imaging (MRI).

The introduction of biosynthetic human GH a little more than 10 years ago has resulted in renewed interest in this field, which to some extent has been fueled by the pharmaceutical industry. This new condition did, however, provide the means to conduct both large, long-term clinical studies, and a number of more experimental studies.

The present review will focus on muscle mass and strength in GH-deficient adults before and after GH substitution, but attention will also be paid to recent studies on the short-term effects of GH on muscle protein metabolism in normal subjects.

Effects of GH on skeletal muscle metabolism *in vitro* and *in vitro*.

The clinical picture of acromegaly and gigantism includes increased lean body mass of which skeletal muscle mass accounts for approximately 50%. Moreover, retention of nitrogen was one of the earliest observed and most reproducible effects of GH administration in humans (2). Thoroughly conducted studies with GH administration in GH-deficient children using a variety of classic anthropometric techniques strongly suggested that skeletal muscle mass increased significantly during treatment (1, 3). Indirect evidence of an increase in muscle cell number following GH treatment was also presented (1).

These early clinical studies were paralleled by equally impressive experimental studies in rodent models. Growth hormone administration in hypohysectomised rats increased not only muscle mass, but also muscle cell number (i.e. muscle DNA content) (1). Interestingly, the same series of experiments revealed that work-induced muscle hypertrophy could occur in the absence of GH. The ability of GH to stimulate RNA synthesis and amino acid incorporation into protein of isolated rat diaphragm suggested direct mechanisms of actions, whereas direct effects of GH on protein synthesis could not be induced in liver cell cultures (4). Another important observation of that period was made by Goldberg, who studied protein turnover in skeletal muscle of hypophysectomised rats with ^{3}H-leucine tracer techniques. He convincingly demonstrated that GH directly increased the synthesis of both sarcoplasmic and myofibrillar protein without affecting proteolysis (5).

The most substantial recent contributions within the field derive from human *in vivo* studies on the effects of systemic and local GH and IGF-I administration on total and regional protein metabolism by means of amino acid isotope dilution techniques. Horber and Haymond demonstrated that systemic GH administration for 7 days in normal adults increased whole body protein synthesis without affecting proteolysis (6), and similar data were subsequently obtained in GH-deficient adults (7). Fryburg and Barret (8) infused GH systemically for 8 hours in normal adults and reported an acute stimulation of forearm (muscle) protein synthesis without any effects on whole body protein turnover. By contrast Copeland and Nair (9) observed an acute stimulatory effect of GH on whole body protein synthesis, but no stimulatory effect on leg protein synthesis, in a design which also included co-administration of somatostatin to suppres insulin. Finally, Fryburg et al. (10) infused GH into the intrabracheal artery, which was accompanied by a local increase in forearm muscle protein synthesis.

Based on these recent studies it is clear that the nitrogen retaining properties of GH predominantly involve stimulation of protein synthesis without affecting (lowering) proteolysis, and clues are also provided about the underlying mechanisms. Theoretically, the protein anabolic effects of GH could be either direct, or mediated through IGF-I, insulin or lipid intermediates. Growth hormone receptors are present in skeletal muscle (11), which combined with Fryburgs intraarterial GH studies, makes a direct GH effect possible. An alternative interpretation of Fryburgs data could be that GH stimulates local muscle IGF-I release, which subsequnetly acts in a autocrine/paracrine manner. The effects of systemic IGF-I administration on whole body protein metabolism seem to depend on ambient amino acid levels in the sense that IGF-I administered alone suppresses proteolysis (12), whereas IGF-I in combination with an amino acid infusion increase protein synthesis (13). Moreover, intraarterial IGF-I in combination with systemic amino acid infusion increased protein synthesis (14). It is therefore likely that the muscle anabolic effects of GH at least to some extent is mediated via IGF-I. By contrast, it is repeatedly shown that insulin solely acts through suppression of proteolysis, and this effects appears to be blunted by co-administration of GH (15). The degree to which mobilisation of lipids contributes to the muscle anabolic actions of GH has so far not been specifically investigated.

In conclusion, several experimental lines of evidence strongly suggests that GH stimulates muscle protein synthesis. This effect is presumably mediated through binding of GH to GH receptors in skeletal muscle. This does not rule out a significant role of IGF-I being produced either systemically or locally.

A very interesting recent discovery has been that infusion of GH and IGF-I into the brachial artery increase forearm blood flow several fold (10, 16). This effect appears to be mediated through stimulation of endothelial nitric oxid release leading to local vasodilatation (17, 18). Moreover, co-infusion of a nitric oxide inhibitor with IGF-I appeared to blunt the stimulatory effect of IGF-I on forearm protein synthesis (17). It thus appears that an IGF-I mediated increase in muscle nitric oxide release accounts for some of the effects of GH on skeletal muscle protein synthesis. These intriguing observations may have many other implications. It is, for instance, tempting to speculate that this increase in skeletal muscle blood flow contributes to the GH-induced increase in resting energy expenditure, since skeletal muscle metabolism is a major determinant of REE (19). Moreover, it is plausible that the reduction in total peripheral resistance seen after GH administration in GHDA are mediated by nitric oxide (18).

Muscle mass and function in GH-deficient adults

Muscle mass. Since 1989 several studies have evaluated body composition in GH-deficient adults before and after GH substitution. A number of methods for body composition measurements have been employed including conventional anthropometrics, bioimpedance (BIA), total body potassium (TBK), DEXA scan and CT scanning (20, 21). It is obvious that each of these methods has its limitations some of which may be of special relevance in GH-deficient adults. As an example, calculation of LBM from BIA assumes a normal and constant hydration of lean tissues. There is substantial evidence to suggest that GHDA is associated with reduced extracellular water (ECW), whereas GH substitution leads to a subtantial increase in ECW. As a consequence, LBM derived from BIA may be underestimated in the untreated state and overestimated after GH replacement (21). Moreover, it appears that TBK content of LBM is reduced in GHDA, which may underestimate LBM in the untreated state (22). Nevertheless, all studies comparing LBM in untreated patients with normal subjects have shown signifcantly reduced LBM in untrated GHDA of 5-20 % (20, 23-25). More specific measures of muscle mass calculated from anthrompometry and CT scans of the thighs showed muscle mass to be reduced by 5-15 % (26-28).

Following GH substitution an increase in LBM of 5-11 % has been documented in several placebo-controlled trials with a duration between 4 to 12 months (20, 29). Five placebo-controlled trials (30-34) have included CT measurements of thigh muscle content reporting an increase of 5-11 % after GH treatment with the largest increase in the study of longest duration (12 months)(fig.1)(33). Moreover, GH substitution for 3 years in an open design was associated with a continued increase in muscle mass to a level no longer different from the background population (27).

As previously alluded to it is known that GH susbtitution also increases lean tissue hydration, wherefore part of the observed increments in LBM and muscle mass could be attributed to fluid retention. Indeed, magnetic resonance imaging of the thigh region in GH-deficient adults did suggest increased water content of the muscles after 12 weeks of GH substitution (35). It should, however, be recalled that untreated patients are dehydrated. It is also noteworthy that clinically recordable fluid retention usually subsides after 3-6 months, and at these time points the pixel density of the muscle compartments by CT has been shown to be unchanged (30, 31).

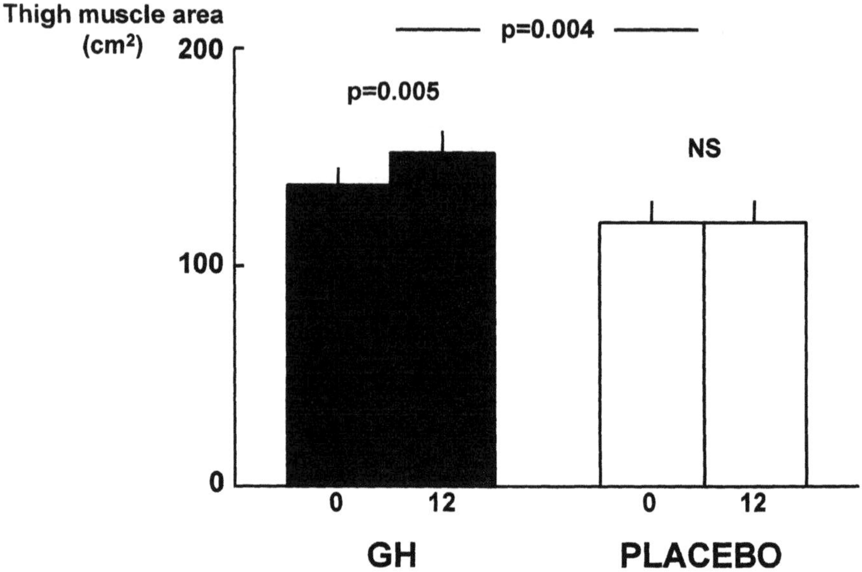

Figure 1.
Thigh muscle volume in GHDA at baseline (0) and after 12 months (12) of treatment with either GH (left panel) or placebo (right panel). Data are derived from ref.33

Only one published study has evaluated muscle composition in GH-deficient patients by means of muscle biposies (36). Baseline biposies were obtained in 21 patients of whom 15 also underwent a repeat biposy after 6 months of treatment with GH (n=8) or placebo (n=7). The amount and distribution of type I and II muscle fibers did not differ from a control group, nor did it change after treatment. Likewise, glycogen stores and the content of fat and connective tissue remained comparable. It was emphasised in the study that the lack of detectable difference could be attributed to the small sample size, since *post hoc* power calculations estimated that 250 would be required in each group to detect a 5% increase in muscle area.

Muscle strength. Maximal and sub-maximal exercise performance as assessed with bicycle ergometry and/or treadmills are distinctly subnormal in untreated GHDA and have been shown to increase in some (30, 32, 33, 37), but not all (38-40) placebo-controlled trials. Open GH treatment up to 3 years has shown a normalisation of exercise capacity (27).

There is also convincing evidence to suggest that skeletal muscle strength is significantly reduced in GHDA. This is particularly well documented regarding the ismoteric strength of the quadriceps muscle (20, 26, 27). By contrast, placebo-controlled trials lasting for 4-12 months have reported only marginally improvement in muscle strength (30-32, 38). Uncontrolled GH substitution for up to 3 years has, on the other hand, been associated with dramatic improvements in muscle strength.

Uncontrolled data must however be interpreted with great caution owing to a substantial positive placebo effect as evidenced from figure 2. What is the explanation for the discrepancy between the changes in muscle mass versus performance? It appears that the composition of the increased muscle tissue is unchanged with GH therapy, and the observed significant increase in exercise performance does suggest that 'general' physical fitness increases after GH. There is little doubt that strenous muscle training is important or even essential for obtaining an increase in strength. In all studies the patients were encouraged not to change their level of physical activity, and pedometer recordings have not revealed differences in physical activity during GH substitution (31). Another important aspect is the large inter- and intraindividual variation in muscle strength measurements. In fact, Cuneo et al. (31) estimated that at least 26 subjects in each group would have been necessary to demonstrate a significant increase in quadriceps force.

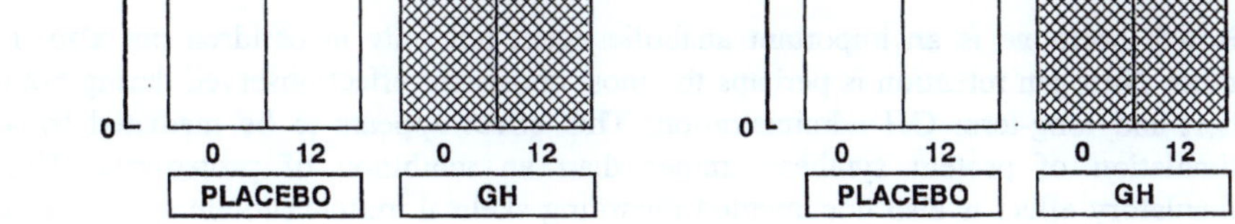

Figure 2.
Quadriceps strength (left panel) and exercise capacity (right panel) in GHDA at baseline (0) and after 12 months of treatment (12) with either placebo or GH. Data are derived from ref.33

Effects of GH administration on muscle mass and function in adults without GH- deficiency

As previously mentioned the ability of acute and more prolonged GH administration to retain nitrogen in normal adults has been known for decades, and more recent studies have documented a stimulatory effect on whole body and forearm protein synthesis.

Rudman et al. was the first to hypothesise that the senescent changes in body composition were causally linked to the concomitant decline in circulating GH and IGF-I levels (41). This paradigm, which by some is known as the somatopause, has recently been reviewed (42), and a number of studies with GH and other anabolic agents for treating the sarcopenia of aging are currently in progress.

Placebo-controlled GH administration in young healthy adults (21-34 years) undergoing a resistance exercise programme for 12 weeks showed a GH-induced increase in LBM, whole body protein balance, and whole body protein synthesis, whereas quadriceps muscle protein synthesis rate and muscle strength increased to the same degree in both groups during training (43). In a similar study in older men (67 years) GH also increased LBM and whole body protein synthesis, without significantly amplifying the effects of exercise on muscle protein synthesis or muscle strength (44). An increase in LBM but unaltered muscle strength following 10 weeks of GH administration plus resistance exercise training was also recorded by Taafe et al. (45). A more recent study of 52 older men (70-85 years) treated with either GH or placebo for 6 months, without concomitant exercise, observed a significant increase (4.4%) in LBM with GH, but no significant effects on muscle strength (46).

Numerous studies have evaluated the effects of GH administration in chronic and acute catabolic illness. A comphrehensive survey of the proliferating litterature within this field is beyond the scope of this review, but it is noteworthy, that HIV-associated wasting is now an officially approved indication for GH treatment in USA. In this patient category GH treatment for 12 weeks has been associated with significant increments in LBM and physical fitness (47, 48).

Conclusion

Growth hormone is an important anabolic agent not only in children but also in adults. Nitrogen retention is perhaps the most consistent effect observed during both short and long-term GH administration. This effect appears to be mediated by a stimulation of protein synthesis rather than an inhibition of proteolysis. This stimulatory effect is also documented regarding skeletal muscle protein synthesis in most studies. It does, however,appear evident that muscle training in adults is far more important for maintaining or increasing muscle mass, and this effect of exercise is not amplified by GH administration. Growth hormone apparently exerts direct effects on muscle protein synthesis, but this direct or local effect may well involve *in situ* production and action of IGF-I, and it is also likely that circulating IGF-I contributes to the overall effect of GH. Whether or not the associated increase in circulating insulin and lipid intermediates contribute to the muscle anabolic actions of GH remains to be investigated. The ability of GH and IGF-I to increase skeletal muscle blood flow is a fascinating new aspect, which may explain some of the actions of GH, not only on muscle function, but also regarding cardiovascular function and REE.

From a clinical point of view the beneficial effects of GH substitution in GHDA is thoroughly documented also regarding the effects on lean body mass and exercise capacity. It does, however, remain to be demonstrated that GH administration increases muscle strength, which probably will require a large number of patients. Growth hormone administration in elderly people without evidence of pituitary disease induces moderate increments in LBM, but does not improve physical fitness or quality of life. It is therfore premature to ascertain whether GH or GH secretagogues can be used in the treatment of senescent sarcopenia. We will undoubtedly know more about that in the near future.

References

1. Cheek DB. Effect of growth hormone on cell and somatic growth. In: Handbook of physiology (Eds Knobil and Sawyer) Washington DC 1974, pp 159-186
2. Raben M. Growth hormone. N Engl J Med 1962;266:31-35
3. Tanner JM, Hughes PCR, Whitehouse RH. Comparative rapidity of response of height, limb muscle and limb fat to treatment with human growth hormone in patients with and without growth hormone deficiency. Acta Endocrinol 1977;84:681-696
4. Korner A. Growth hormone control of biosynthesis of protein and ribonucleic acid. Recent Progr Horm Res 1965;21:205-240
5. Goldberg AL. Protein turnover in skeletal muscle. The Journal of Biological Chemistry 1969;244:3217-3222
6. Horber FF, Haymond MW. Human growth hormone prevents the protein catabolic side effects of prednisone in humans. J Clin Invest 1990;86:265-272
7. Russell-Jones Dl, Weissberger Aj, Bowes Sb, Kelly JM, Thomason M, Umpleby AM, Jones RH, Sönksen PH. The effects of growth hormone on protein metabolism in adult growth hormone deficient patients. Clin Endocrinol 1993;38:427-431
8. Fryburg D, Barret EJ. Growth hormone acutely stimulates skeletal muscle but not whole-body protein synthesis in humans. Metabolism 1993;42:1223-1227
9. Copeland KC, Nair KS. Acute growth hormone effects on amino acid and lipid metabolism. J Clin Endocrinol Metab 1994;78:1040-1047
10. Fryburg DA, Gelfand RA, Barrett EJ. Growth hormone acutely stimulates foreearm muscle protein synthesis in normal humans. Am J Physiol 1991;260:E499-E504
11. Kelly PA, Dijane J, Postel-Vinay M-C, Edery M. The prolactin/growth hormone receptor family. Endocr Rev 1991;12:235-251
12. Turkalj I, Keller U, Ninnis R, Vosmeer S, Stauffacher W. Effect of increasing doses of recombinant human insulin-like growth factor-I on

glucose, lipid, and leucine metabolism in man. J Clin Endocrinol Metab 1992;75:1186-1191

13. Russell-Jones DL, Umpleby AM, Hennesy TR, Bowes SB, Hopkins KD, Jackson NC, Kelly JM, Jones RH, Sonksen PH. Use of a leucine clamp to demonstrate that IGF-I actively stimulates protein synthesis in normal humans. Am J Physiol 1994;267:E591-E598
14. Fryburg DA, Jahn LA, Hill SA, Oliveras DM, Barrett EJ. Insulin and insulin-like growth factor-I enhance human skeletal muscle protein anabolism during hyperaminoacidemia by different mechanisms. J Clin Invest 1995;96:1722-1729
15. Fryburg DA, Louard RJ, Gerow KE, Gelfano RA, Barrett EJ. Growth hormone stimulates skeletal muscle protein synthesis and antagonizes insulin's antiproteolytic action in humans. Diabetes 1992;41:424-429
16. Copeland KC, Nair KS. Recombinant human insulin-like growth factor-I increases forearm blood flow. J Clin Endocrinol Metab 1994;79:230-232
17. Fryburg D. N^G-monomethyl-L-arginine inhibits the blood flow but not the insulin-like response of forearm muscle to IGF-I. J Clin Invest 1996;97:1319-1328
18. Böger RH, Skamira C, Bode-Böger SM, Brabant G, von zur Mühlen A, Fröhlich JC. Nitric oxide may mediate the hemodynamic effects of recombinant growth hormone in patients with acquired growth hormone deficiency. J Clin Invest 1996;98:2706-2713
19. Zurio F, Larson K, Bogardus C, Ravussin E. Skeletal muscle metabolism is a major determinant of resting energy expenditure. J Clin Invest 1990;86:1423-1427
20. Cuneo RC, Salomon F, Sönksen PH. The syndrome of growth hormone deficiency in adults. In: Growth hormone in adults (Eds. Juul and Jørgensen), Cambridge University Press 1996.pp 145-168
21. De Boer H, Blok G-J, Van Der Veen EA. Clinical aspects of growth hormone deficiency in adults. Endocr Rev 1995;16:63-86
22. Davies JS, Bell W, Evans W, Villis RJ, Scanlon MF. Body composition derived from whole body counting of potassium in growth hormone-deficient adults: a possible low intracellular potasssium concentration. J Clin Endocrinol Metab 1996;81:1720-1723
23. Binnerts A, Deurenberg P, Swart GR, Wilson JHP, Lamberts SWJ. Body composition in growth hormone-deficient adults. Am J Clin Nutr 1992;55:918-923
24. Hoffman DM, O'Sullivan AJ, Freund J, Ho KKY. Adults with growth hormone deficiency have abnormal body composition but normal energy metabolism. J Clin Endocrinol Metab 1995;80:72-77
25. Beshyah SA, Freeemantle C, Thomas E, Rutherford O, Page B, Murphy M, Johnston DG. Abnormal body composition and reduced bone mass in growth hormone deficient hypopituitary adults. Clin Endocrinol 1995;42:179-189
26. Jørgensen JOL, Pedersen SA, Thuesen L, Jørgensen J, Møller J, Müller J, Skakkebæk NE, Christiansen JS. Long-term growth hormone treatment in growth hormone deficient adults. Acta Endocrinol 1991;125:449-453

27. Jørgensen JOL, Thuesen L, Müller J, Ovesen P, Skakkebæk NE, Christiansen JS. Three years growth hormone treatment in growth hormone deficient adults European Journal of Endocrinology 1994;130:224-228

28. De Boer H, Blok GJ, Voerman HJ, De Vriess PMJM, Van Der Veen EA. Body composition in adult growth hormone-deficient men, assessed by anthropometry and bioimpedance analysis. J Clin Endocrinol Metab 1992;75:833-837

29. Hansen TB, Vahl N, Jørgensen JOL, Christiansen JS, Hagen C. Whole body and regional soft tissue changes in growth hormone deficient adults after one year of growth hormone treatment: a double blind, placebo-controlled study. Clin Endocrinol 1995;43:689-696

30. Jørgensen JOL, Pedersen SA, Thuesen L, Jørgensen J, Ingemann-Hansen T, Skakkebæk NE, Christiansen JSC. Beneficial Effects of growth hormone treatment in GH deficient adults. Lancet 1989;i:1221-25.

31. Cuneo RC, Salomon F, WIles CM, Hesp R, Sönksen PH. Growth hormone treatment in growth hormone-deficient adults.I. Effects on muscle mass and strength. J Appl Physiol 1991;70:688-94

32. Whitehead HM, Boreham C, Mcllrath EM, Sheridan B, Kennedy L, Atkinson AB, Haden DR. Growth hormone treatment of adults with growth hormone deficiency:results of a 13-month placebo controlled cross-over study. Clin Endocrinol 1992;36:45-52

33. Jørgensen JOL, Vahl N, Hansen TB, Thuesen L, Hagen C, Christiansen JS. Growth hormone versus placebo treatment for one year in growth hormone deficient adults:increase in exercise capacity and normalization of body composition. Clin Endocrinol 1996;45:681-688

34. Bengtsson B-Å, Edén S, Lönn L et al. Treatment of adults with growth hormone (GH) deficiency with recombinant human GH. J Clin Endocrinol Metab 1993;76:309-17

35. Møller J, Jørgensen JOL, Møller N, Stødkilde-Jørgensen H, Christiansen JS. Assessment of fat and muscle volume and water content by magnetic resonance imaging (MRI) and proton spectroscopy in growth hormone (GH) deficient adults before and after GH treatment. Endocrinology and Metabolism 1994;1:131-135

36. Cuneo RC, Salomon F, Wiles CM, Round JM, Jones D, Hesp R, Sönksen RH. Histology of skeletal muscle in adults with GH deficiency:comparison with normal muscle and response to GH treatment. Horm Res 1992;37:23-28

37. Cuneo RC, Salomon F, WIles CM, Hesp R, Sönksen PH. Growth hormone treatment in growth hormone-deficient adults.II. Effects on exercise performance. J Appl Physiol 1991;70:695-700

38. Caidahl K, Edén S, Bengtsson B-Å. Cardiovascular and renal effects of growth hormone. Clin Endocrinol 1994;40:393-400

39. Beshyah SA, Freeemantle C, Shahi M, Anyaoku V, Merson S, Lynch S, Skinner E, Sharp P, Foale R, Johnston D. Replacement treatment with biosynthetic human growth hormone in growth hormone-deficient hypopituitary adults. Clin Endocrinol 1995;42:73-84

40. Degerblad M, Almkvist O, Grunditz Hall K, Kaijser L, Knutsson E, Ringertz H, Thorén M. Physical and psychological capabilities during substitution therpay with recombinant growth hormone in adults with growth hormone deficiency. Acta Endocrinol 1990;123:185-193
41. Rudman D, Kutner MH, Rogers CM, Lubin MF, Fleming A, Bain R. Impaired growth hormone secretion in the adult population. J Clin Invest 1981;67:1361-1369
42. Corpas E, Harman SM, Blackman MR. Human growth hormone and human aging. Endocr Rev 1993;14:20-39
43. Yarasheski KE, Campbell JA, Smith K, Rennie MJ, Holloszy JO, Bier DM. Effects of growth hormone and resistance exercise on muscle growth in young men. Am J Physiol 1992;262:E261-E267
44. Yarasheski KE, Zachwieja JJ, Campbell JA, Bier DM. Effect of growth hormone and resistance exercise on muscle growth and strength in older men. Am J Physiol 1995;268:E268-E276
45. Taafe DR, Pruitt L, Reim J, Hintz RL, Butterfield G, Hoffman AR, Marcus R. Effect of recombinant human growth hormone on the muscle stregth response to resistance exercise in elderly men. J Clin Endocrinol Metab 1994;79:1361-1366
46. Papadakis MA, Grady D, Black D, Tierny MJ, Gooding GAW, Schamblean M, Grunfeld C. Growth hormone replacement in healthy older men improves body composition but not functional ability. Ann Intern Med 1996;124:708-716
47. Waters D, Danska J, Hardy K, Koster F, Qualls C, Nickell D, Nightingale S, Gesundheit N, Watson D, Schade D. Recombinant human growth hormone, insuin-like growth factor I, and combination therapy in AIDS-associated wasting. Ann Intern Med 1996;125:865-872
48. Schamblean M, Mulligan K, Grunfeld C, Daar ES, LaMarca A, Kotler DP, Wang J, Bozzette SA, Breitmeyer JB. Recombinant human growth hormone in patients with HIV-associated wasting. Ann Intern Med 1996;125:873-882

15 EFFECTS OF GH ON BONE METABOLISM AND BONE MASS

Claes Ohlsson

Research Centre for Endocrinology and Metabolism
Sahlgrenska University Hospital, Göteborg, Sweden
and Diabetes Bransch, NIDDK, NIH, Bethesda, MD, USA

I. INTRODUCTION

It is well known that growth hormone (GH) is important in the regulation of longitudinal bone growth (1; 2). Its role in the regulation of bone metabolism in man has not been understood until recently (1). Bone remodeling is regulated by a balance between bone resorption and bone formation. Recent studies, which demonstrate that GH exerts potent effects on bone remodeling, will be discussed in this chapter. Finally a hypothetical model for the mechanism of action of GH in the regulation of bone remodeling and bone mass will be presented.

II. IN VITRO EFFECTS OF GH ON OSTEOBLASTS AND OSTEOCLASTS

Effects of GH on osteoblasts: Osteoblasts express GH-receptors (GHR) (3; 4). Serum decreases the number of GH receptors (5). In a search for the factors in serum responsible for this reduction in GHR expression, it was found that IGF I and II decrease the number of GHR (6). Conversely, it was found that IGFBPs upregulated the GH receptor number and activity, possibly through the inhibition of IGF activity (6; 7). These findings suggest a local feedback of the GH/IGF axis at the tissue level. Thus, hypothetically, IGF decreases the GHR number and activity, fine-tuned by the presence of IGF binding proteims (IGFBPs) (6; 8). Estrogen is known to exert important effects on bone tissue and a recent study in vitro indicates that estrogen interacts with GH action at the cellular level. In rat and human osteoblasts, 17β-estradiol promoted GH-stimulated proliferation and increased GHR expression (9). In conclusion, a regulation of GHR expression in osteoblasts may be important for a local autocrine feedback loop in the GH/IGF-I axis and for estrogen and other factors modulating the effect of GH in osteoblasts. GH induces proliferation and differentiated functions of rodent, as well as human osteoblasts (3; 10-12). It is difficult to find a good model system for the identification of the actual target cell of GH action in osteoblasts. However, bone marrow derived precursors of human bone cells are responsive to GH (13; 14), suggesting that GH interacts with progenitor cells. It has been suggested that at least some of the GH effect on osteoblasts is mediated via a local stimulation of IGF production. IGFs are expressed in osteoblasts and exert anabolic effects on these cells (15; 16). GH induces IGF-I expression in rodent osteoblasts, while the induction of IGF-I by GH in human osteoblasts is uncertain (10; 17-19).

The bioactivity of IGFs in bone tissue is modulated by several IGFBPs, mainly IGFBP 3, 4 and 5 (20). GH increases the IGFBP-3 production in rat osteoblasts (21; 22) while no effect of GH is seen on the IGFBP-3 expression in human cells (10; 19; 23). In rat, as well as human osteoblasts, it was found that GH decreases IGFBP-4 as determined by ligand blotting (23; 24). In contrast, the IGFBP-5 mRNA levels in rat osteoblasts were increased twofold after GH-treatment (22). In conclusion, there are some indications of a GH induced regulation of IGFBPs that may potentially have a regulatory role in bone metabolism.

Effects of GH on Osteoclasts: GH may also influence bone metabolism via a regulation of the recruitment and/or the activity of osteoclasts. GH increases the number of osteoclasts in the metaphysial bone of the proximal tibia of hypophysectomized rats (25). In a recent study by Nishiyama et al, using mouse stromal cells and hemopoetic blast cells, it was found that GH stimulates osteoclastic bone resorption through both direct and indirect actions on osteoclast differentiation and indirect activation of mature osteoclasts (26). In another study by Ransjö et al (27), using mouse marrow cultures, GH caused an inhibition of osteoclast formation by means of an IGF-I independent mechanism. In summary, available in vitro data suggest that GH regulates osteoclast formation but both stimulatory and inhibitory mechanisms have been presented, probably due to differences in culture conditions.

III. EFFECTS OF GH ON BONE METABOLISM IN ANIMALS

Effects of GH in Rats: Hypophysectomy of rats with subsequent replacement with thyroxine and glucocorticoid is followed by a rapid and pronounced decrease in the amount of bone which is reversed by GH treatment (25; 28). Furthermore, GH administration increases cortical bone mass in old rats with normal GH-secretion. Tetracycline labeling of the mineralisation front in these rats demonstrates that GH induces subperiosteal bone formation without influencing the endosteal bone surface (29). There is also an increase in mechanical strength of the whole bone, corresponding to the increased bone mass, and the mechanical quality of the bone itself is almost the same in GH-injected animals as in the controls (29). In summary, these results from GH treated rats with a normal GH secretion, clearly demonstrate that GH increases cortical bone mass by inducing subperiosteal bone formation while no major effect on cancellous bone mass is seen.

Ovariectomized rats have been accepted as an animal model of postmenopausal bone loss. As GH is a potential anabolic agent, there have recently been a number of studies, in which GH was systemically given to ovariectomized rats (30-34). GH treatment results in increased bone mass in the OVX model. In addition. the mechanical strength of the vertebral body is increased and correlates well with the increase in bone mass (31; 35). The results of GH treatment on cancellous bone in ovariectomized rats seems rather promising, but it must be emphasized that GH also stimulated linear bone growth which makes it difficult to interpret the results.

Effects of GH in transgenic mice: The effect of GH on bone growth and bone metabolism has also been studied in transgenic mice. In most bone metabolic studies, the metallothionein promoter (MT) fused to the GH gene has been used, resulting in very high serum levels of GH (36-40). The femora of MT-GH-transgenic mice demonstrate increased bone growth, increased BMC, and increased mechanical strength (40). The increase in mechanical strength was due to increased cortical width and not due to any improved quality of the bone. Rather, one of the parameters measuring the quality of the cortical bone, the E-module, was decreased in the GH transgenic mice (40). It should be emphasized that these mice have been exposed to supra physiological serum levels of GH (more than 10 times higher) from late prenatal life (40).

A local effect of GH in bone is supported by two new GH-transgenic lines with a bone specific GH-expression, resulting in high local concentrations of GH without affecting the serum concentrations of GH (41; 42). These mice demonstrate that the local expression of GH stimulates bone formation without regulating the systemic levels of GH.

Effects of GH in monkeys: As discussed above, GH exerts potent effects in rodents, resulting in increased bone formation. However, it is possible that some of these effects are due to bone growth and bone modeling, as rodents close the epiphyseal plate late in life. The monkey is a primate and the bone metabolism in this animal is more similar to humans. A recent study, using old female monkeys, demonstrated that GH but not IGF-I, administered for 7 weeks increased bone formation (43). The effect was seen both in the tibia and in the femur while no significant effect was seen in the vertebrae. This study of primates is promising for future human clinical studies. However, further long-term studies of GH treatment of primates are needed to elucidate whether or not the increased bone formation results in increased bone mass and mechanical strength.

IV. EFFECTS OF GH ON BONE METABOLISM IN HUMANS

Acromegaly: Many years ago, acromegaly was seen as a cause of osteoporosis, whereas later studies have revealed normal or more often increased bone mass in patients with acromegaly without gonadal insufficiency (1). Thus, most studies suggest that cortical bone mass is increased in acromegaly whereas trabecular bone seems largely unaffected (1).

Growth hormone deficiency (GHD): Several studies have shown low bone mass in adults with childhood onset GHD. Similar results have been observed in patients with multiple pituitary deficiencies and isolated GHD suggesting that the lack of GH is the most important factor behind the observed low bone mass in childhood onset GHD (44; 45). At present, there are no studies showing that GH replacement during childhood results in a normalization of bone mineral density (BMD) when peak bone

mass is reached, suggesting that GH is also important for the additional increase of bone mass that occurs after the completion of linear growth. It has been suggested that GH treatment should be continued until the attainment of peak bone mass, irrespective of the height achieved (46).
An increase in the prevalence of osteoporosis has also been found in several recent studies of patients with adult onset GHD (47-53). In a study by Rosén et al (50), of 95 patients (55 males and 40 women) with adult onset GHD with a mean age of 54 years, BMC was assessed in the third lumbar vertebra with dual-photon absorptiometry. BMC was found to be reduced in all males and in females with untreated, as well as treated gonadal deficiency. BMC was reduced in patients under 55 years of age and normal in patients over 55 years of age. The lack of effect on the older patients is supported by a recent study on elderly patients (over 60 years old) with adult onset GHD, demonstrating normal BMD in the hip and the lumbar spine (54). In summary, most studies demonstrate that patients under 55 years but not over 55 years of age with adult onset GHD have decreased bone mass.

Few studies have investigated whether or not GHD patients have an increased fracture rate. The reasons for this are probably that a huge number of GHD patients are required for a meaningful study and/or that additional pituitary hormone deficits may confound the results. However, an increased risk of osteoporotic vertebral fractures has been suggested in hypopituitary patients (47). The consequences of low BMD in GHD adults have only recently been delineated by Rosén et al (55), demonstrating a higher fracture rate in patients with adult-onset GHD compared with that of healthy controls.

Treatment of GHD patients with GH: GH treatment of GHD adults has consistently been shown to have marked effects on markers for bone formation and bone resorption and serum IGF-I levels (1). Similarly, increased bone turnover has been observed in patients with long standing acromegaly (56), indicating that bone turnover can be elevated for many years as a result of high plasma levels of GH. Administration of GH to healthy volunteers for 7 days displayed no discernible effect on serum calcium concentration, but urinary calcium excretion increased. Serum PTH concentrations also increased, as did the concentrations of phosphate and 1,25-dihydroxyvitamin D (57). In contrast, in 6 month double blind, placebo-controlled trials of adult GHD patients, no effect on 1,25-dihydroxyvitamin D concentrations (58) and no effect (58) or a decrease (59) in intact PTH concentrations was found. These changes were accompanied by a concomitant increase in total serum calcium concentrations (58; 59). A similar increase in serum calcium concentrations has been observed by others (60-62). Still, after two years of GH treatment, serum calcium concentration was elevated compared with baseline (63). GH may enhance 1 a-hydroxylase activity (64), thus increasing the concentration (65) or availability (66) of vitamin D3, which is conceivably the mechanism behind the sustained increase in serum calcium concentration. An alternative hypothesis is the enhanced mobilization of skeletal calcium due to increased bone turnover.

Trials involving adults with childhood onset GHD have yielded conflicting results regarding the effect of GH on bone mass. Several short-term, placebo-controlled (67-69) and short-term, open trials (61; 70-72) have failed to show any increase in BMD or bone mineral concentration (BMC) during GH treatment. In fact, some of these studies (61; 68; 69) reported a slight decrease of BMD or BMC after 3-6 months of treatment. After more prolonged treatment periods (12-30 months) several studies have disclosed more encouraging results (69; 73-75). Degerblad (73) showed an increase in distal and proximal forearm BMD in six patients by 12 and 3.8%, respectively after 24 months of treatment. Similarly, Vandeweghe et al (69), reported a significant and progressive increase in BMC above pretreatment values, reaching 7.8% of the total BMC in the lumbar spine and 9.9% of the total BMC in the forearm, after 30 months of GH administration.

Short-term trials (53; 60; 62; 76; 77), 6-18 months in adults with adult onset GHD, have failed to show any increase in BMC or BMD. In analogy with the findings observed in adults, with childhood onset GHD, several studies have shown a decrease in BMD and/or BMC after 6-12 months of treatment. In contrast, in the longest placebo-controlled trial reported so far (18 months), Baum et al (78), reported a significant increase in BMD in lumbar spine and femoral neck of 5.1 and 2.4%, respectively. Since bone absorptiometry only detects the mineralized component of the bone, the reduction of BMD observed after short periods of GH treatment is best explained by increased remodeling activity, with an increased in remodeling space and an increase in the proportion of new unmineralized bone. Johannsson et al (63), recently demonstrated that 2 years of GH treatment in 24 men and 20 women with adult-onset GHD induced a sustained increase in overall bone remodeling activity and a net gain in BMD in several weight bearing skeletal locations (Fig. 1). A significant increase in BMD first became apparent after 18 months, which may explain why previous trials of shorter duration were unable to demonstrate an increase in BMD. The study also demonstrates the importance of an adequate duration of treatment in order to include a sufficient number of remodeling cycles and sufficient time for mineralization to occur before a net gain in BMD can be detected with bone absorptiometry. Interestingly, patients with a z-score of less than -1SD demonstrated the most pronounced increase in BMD (Fig. 1).

Effects of the GH on bone metabolism and bone mass in patients with normal GH secretion: Several studies have demonstrated that GH increases markers for bone resorption as well as bone formation in subjects with normal GH secretions (79; 80). The responsiveness of osteoporotic patients to GH is similar to healthy subjects in the regulation of bone markers for bone formation and bone resorption (81). In a small study, in which 3 patients with primary and secondary osteoporosis were treated with GH, an increase in periosteal new bone formation was seen, as determined by bone histomorphometry (82). This increase in periosteal bone formation is interesting as it is similar to what has been found in experimental studies (29). However, larger clinical studies using bone histomorphometric analyses are needed to confirm that GH induces periosteal bone formation in patients with osteoporosis.

The present treatment modalities of osteoporosis rely almost exclusively on agents aimed at reducing bone resorption. In contrast, GH has a stimulatory effect on bone formation, as well as bone resorption, as measured with biochemical parameters, and the gain in BMC/BMD that was observed after long-term GH treatment in adult GHD, occurred after the first remodeling cycle (63). There are as yet no treatment trials with GH of adequate duration (> 18 months) in patients with postmenopausal osteoporosis. Aloia et al (83), demonstrated no net gain in total body calcium in osteoporotic women after 12 months of GH treatment. In a two year study of postmenopausal women, the addition of GH to continuous, combined or sequential calcitonin treatment had no additional effects on total body calcium

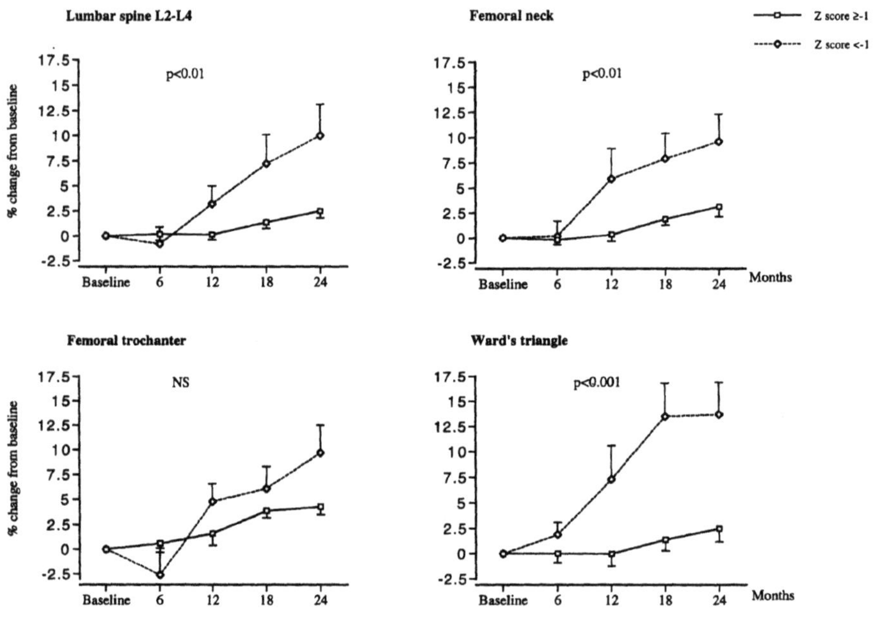

Fig. 1. BMD (BMC/area) in response to 2 yr of GH treatment in two subgroups of patients with adult-onset GH deficiency. One group is comprised of 13 patients with a baseline z-score of less than - 1 SD (broken line), and the second group is comprised of 31 patients with a baseline z-score of - 1 SD or more (solid line). Values are given as means ± SEM. P values denote the difference between the percent changes from baseline in the two groups of patients by two -way ANOVA. (Reproduced with permission from Johannsson et al (63))

(84; 85). No additional effect of the GH treatment was found on bone mineral mass in post-menopausal women treated with pamidronate for 12 months and with the addition of GH for 6 months. Holloway et al, has recently presented a study in which postmenopausal women were given a cyclic GH treatment for 7 days every 56th day. This cyclic treatment was repeated 12 times and resulted in a minor but statistical significant increase in BMD of the lumbar spine and of the hip (1-2%) (86). A cyclic block of bone resorption with calcitonin did not alter the effect of GH in this study. In summary, it is not yet shown if GH increases bone mass in patients with osteoporosis. However, most of the presented studies have been short-term studies and future long term studies are needed to investigate whether or not prolonged GH treatment increases bone mass.

Effects of GH in elderly: Studies in healthy elderly subjects have not shown any impressive effect of GH on bone mass. Rudman et al (87), studied elderly men above 60 years of age and demonstrated a slight (1.6%) increase in lumbar BMD after 6 months of treatment. A positive effect of GH on bone mass in GHD patients is not seen until after 18 months of treatment. Thus, long-term studies (>18 months), to explore whether or not GH increases bone mass in elderly patients, are needed before further conclusions regarding the effect of GH on bone mass in these subjects can be made.

Effect of IGF-I treatment on bone metabolism: Because of its potent mitogenic propensity on osteoblasts, IGF-I has been thought to have potential as a formation-stimulating drug in the treatment of osteoporosis. The first clinical study, in which IGF-I (160 μg/kgday) was given for 7 days to one male with idiopathic osteoporosis, indicated that IGF-I increases biochemical markers of both bone formation and bone resorption (88). Furthermore, normal women were treated for 6 days with different doses of IGF-I; 30, 60, 120 and 180 μg/kg body weight each day. Numerous side effects were observed with the two highest treatment doses but none with the lowest dose (89). A dose-dependent increase in biochemical markers for bone formation and bone resorption was seen in this clinical study. In contrast, Ghiron et al (90) observed that lower doses of IGF-I (15μg/kg/day) exerted no effect on resorption markers while osteocalcin and PICP increased progressively in elderly women. The authors claimed that a low dose of IGF-I is independently capable of stimulating bone formation without inducing bone resorption. In a study (91) of men with idiopathic osteoporosis, the effects of GH and IGF-I on bone metabolism were compared after 7 days of treatment. Both treatment regimes gave a similar increase in the osteocalcin concentration and an increase in urinary excretion of deoxypyridinoline. However, the IGF-I treatment increased PICP more than GH did. Urinary excretion of calcium increased during the GH treatment, whereas no changes occurred during the IGF-I treatment. The authors concluded that the differences between GH and IGF-I may be dose dependent, but could also indicate separate mechanisms of actions of the two peptides at the cellular level. In summary, both GH and IGF-I treatment increases biochemical markers for both bone formation and bone resorption. The GH treatment increases bone mass in GHD patients but it is not yet clear if IGF-I also has the capacity to increase bone mass in humans.

VI SUMMARY AND CONCLUSIONS

GH increases bone formation both via a direct interaction with GH receptors on osteoblasts and via an induction of endocrine and autocrine/paracrine IGF-I. GH deficiency results in decreased bone mass in both man and experimental animals. Long term treatment (>18 months) of GHD patients with GH results in increased bone mass. GH treatment also increases bone mass and the total mechanical strength of bones in rats with normal GH secretion. Because of the short duration of GH-treatment in man with normal GH secretion the effect on bone mass is still inconclusive.

The action of GH on bone metabolism in GHD adults is twofold: It stimulates both bone resorption and bone formation. "The Biphasic model" of GH action in bone remodeling has recently been proposed (Fig. 2). According to this model GH initially increases bone resorption with a concomitant bone loss that is followed by a phase of increased bone formation. At the time point when bone formation proceeds at a higher rate than bone resorption (transition point), bone mass increases. However, a net gain in bone mass caused by GH may take some time as the initial decrease in bone mass must first be replaced (Fig. 2). Taking all clinical studies of GH-treatment of GHD adults into account, it appears that the "transition point" occurs after approximately six months and that a net increase in bone mass will be seen after 12-18 months of GH treatment. It should be emphasized that "the Biphasic model" of GH action in bone remodeling is proposed based on findings in GHD adults. It remains to be clarified whether or not it is valid for subjects with normal GH-secretion. In conclusion, GH treatment increases bone mass in GHD adults and future clinical studies will determine if some patients with decreased bone mass for other reasons will benefit from treatment with GH alone or in combination with other treatments.

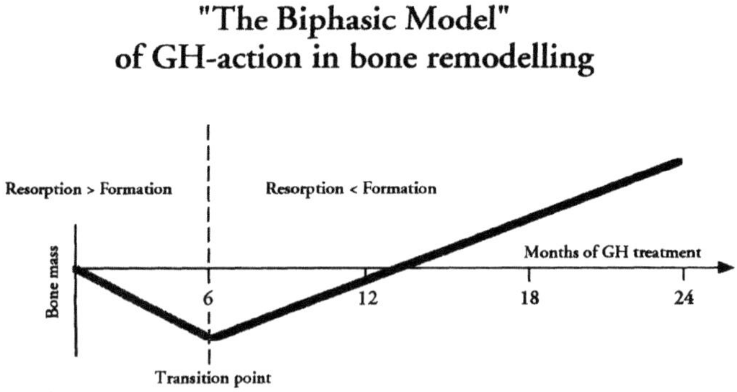

Fig. 2. "The Biphasic model" of GH action in bone remodeling". According to this model; GH results initially in an increased bone resorption with a concomitant bone loss followed by a later increased bone formation. After the moment when bon formation is stimulated more than bone resorption (transition point), bone mass is increased. However, a net gain of bone mass of GH may take some time as the initial decrease in bone mass first must be replaced.
(Reproduced with permission from Ohlsson et al (1))

REFERENCES

1 Ohlsson C., Bengtsson B., Isaksson O., Andreassen T. and Slootweg M. Growth hormone and bone. Endocr Rev 1998, in press.

2 Ohlsson C, Lindahl A, Isgaard J. and Isaksson O. The dual effector theory. In Handbook of Physiology. Kostyo J (Ed) 1997.

3 Nilsson A., Swolin D., Enerback S. and Ohlsson C. Expression of functional growth hormone receptors in cultured human osteoblast-like cells. J Clin Endocrinol Metab 1995, 80:3483-3488.

4 Barnard R., Ng K.W., Martin T.J. and Waters M.J. Growth hormone (GH) receptors in clonal osteoblast-like cells mediate a mitogenic response to GH. Endocrinology 1991, 128:1459-1464.

5 Salles J.P., De Vries C.P., Netelenbos J.C. and Slootweg M.C. Dexamethasone increases and serum decreases growth hormone receptor binding to UMR-106.01 rat osteosarcoma cells. Endocrinology 1994, 134:1455-1459.

6 Slootweg M.C., Ohlsson C., Salles J.P., de Vries C.P. and Netelenbos J.C. Insulin-like growth factor binding proteins-2 and -3 stimulate growth hormone receptor binding and mitogenesis in rat osteosarcoma cells. Endocrinology 1995, 136:4210-4217.

7 Slootweg M., Ohlsson C., van Elk E., Netelenbos J. and Andress D. Growth hormone receptor activity is stimulated by insulin-like growth factor binding protein 5 in rat osteosarcoma cells. Growth Regul 1996, 6:238-246.

8 Leung K., Rajkovic I.A., Peters E., Markus I., Van Wyk J.J. and Ho K.K. Insulin-like growth factor I and insulin down-regulate growth hormone (GH) receptors in rat osteoblasts: evidence for a peripheral feedback loop regulating GH action. Endocrinology 1996, 137:2694-2702.

9 Slootweg M., Swolin D., Netelenbos J., Isaksson O. and Ohlsson C. Estrogen enhances growth hormone receptor expression and growth hormone action in rat osteosarcoma cells and human osteoblast-like cells. J Endocrinol 1997, 151:159-164.

10 Kassem M., Blum W., Ristelli J., Mosekilde L. and Eriksen E.F. Growth hormone stimulates proliferation and differentiation of normal human osteoblast-like cells in vitro. Calcif Tissue Int 1993, 52:222-226.

11 Slootweg M.C., van Buul-Offers S.C., Herrmann-Erlee M.P., van der Meer J.M. and Duursma S.A. Growth hormone is mitogenic for fetal mouse osteoblasts but not for undifferentiated bone cells. J Endocrinol 1988, 116:R11-R13.

12 Swolin D. and Ohlsson C. Growth hormone increases interleukin-6 produced by human osteoblast-like cells. J Clin Endocrinol Metab 1996, 81:4329-4333.

13 Kassem M., Brixen K., Mosekilde L. and Eriksen E.F. Human marrow stromal osteoblast-like cells do not show reduced responsiveness to in vitro stimulation with growth hormone in patients with postmenopausal osteoporosis. Calcif Tissue Int 1994, 54:1-6.

14 Kassem M., Mosekilde L. and Eriksen E.F. Growth hormone stimulates proliferation of normal human bone marrow stromal osteoblast precursor cells in vitro. Growth Regul 1994, 4:131-135.

15 Okazaki R., Conover C.A., Harris S.A., Spelsberg T.C. and Riggs B.L. Normal human osteoblast-like cells consistently express genes for insulin-like growth factors I and II but transformed human osteoblast cell lines do not. Journal of Bone and Mineral Research 1995, 10:788-795.
16 Swolin D., Brantsing C., Matejka G. and Ohlsson C. Cortisol decreases IGF-I mRNA levels in human osteoblast-like cells. J Endocrinol 1996, 149:397-403.
17 Chenu C., Valentin-Opran A., Chavassieux P., Saez S., Meunier P.J. and Delmas P.D. Insulin like growth factor I hormonal regulation by growth hormone and by 1,25(OH)2D3 and activity on human osteoblast-like cells in short-term cultures. Bone 1990, 11:81-86.
18 Morel G., Chavassieux P., Barenton B., Dubois P.M., Meunier P.J. and Boivin G. Evidence for a direct effect of growth hormone on osteoblasts. Cell Tissue Res 1993, 273:279-286.
19 Kanzaki S., Baxter R.C., Knutsen R., Baylink D.J. and Mohan S. Evidence that human bone cells in culture secrete insulin-like growth factor (IGF)-II and IGF binding protein-3 but not acid-labile subunit both under basal and regulated conditions. Journal of Bone and Mineral Research 1995, 10:854-858.
20 Hayden J.M., Mohan S. and Baylink D.J. The insulin-like growth factor system and the coupling of formation to resorption. Bone 1995, 17:93S-98S.
21 Schmid C., Ernst M., Zapf J. and Froesch E.R. Release of insulin-like growth factor carrier proteins by osteoblasts: stimulation by estradiol and growth hormone. Biochem Biophys Res Commun 1989, 160:788-794.
22 McCarthy T.L., Casinghino S., Centrella M. and Canalis E. Complex pattern of insulin-like growth factor binding protein expression in primary rat osteoblast enriched cultures: regulation by prostaglandin E2, growth hormone, and the insulin-like growth factors. J Cell Physiol 1994, 160:163-175.
23 Mohan S., Strong D.D., Lempert U.G., Tremollieres F., Wergedal J.E. and Baylink D.J. Studies on regulation of insulin-like growth factor binding protein (IGFBP)-3 and IGFBP-4 production in human bone cells. Acta Endocrinol (Copenh) 1992, 127:555-564.
24 Chen T.L., Liu F., Bates R.L. and Hintz R.L. Further characterization of insulin-like-growth factor binding proteins in rat osteoblast-like cell cultures: modulation by 17 beta-estradiol and human growth hormone. Endocrinology 1991, 128:2489-2496.
25 Lewinson D., Shenzer P. and Hochberg Z. Growth hormone involvement in the regulation of tartrate-resistant acid phosphatase-positive cells that are active in cartilage and bone resorption. Calcif Tissue Int 1993, 52:216-221.
26 Nishiyama K., Sugimoto T., Kaji H., Kanatani M., Kobayashi T. and Chihara K. Stimulatory effect of growth hormone on bone resorption and osteoclast differentiation. Endocrinology 1996, 137:35-41.
27 Ransjo M., Lerner U. and Ohlsson C. Growth hormone inhibits formation of osteoclast-like cells in mouse bone marrow cultures. J Bone Miner Res 1996, 11:suppl T394.
28 Schmidt I.U., Dobnig H. and Turner R.T. Intermittent parathyroid hormone treatment increases osteoblast number, steady state messenger ribonucleic

acid levels for osteocalcin, and bone formation in tibial metaphysis of hypophysectomized female rats. Endocrinology 1995, 136:5127-5134.

29 Andreassen T.T., Jorgensen P.H., Flyvbjerg A., Orskov H. and Oxlund H. Growth hormone stimulates bone formation and strength of cortical bone in aged rats. Journal of Bone and Mineral Research 1995, 10:1057-1067.

30 Nowak J., Nygaard H. and Eschen C. Growth hormone and/or estradiol and gestagen as a treatment of osteoporosis in the aged ovariectomized rat. J Bone Miner Res 1995, suppl 1:S252.

31 Andreassen T. and Oxlund H. Additive anabolic effects of growth hormone and parathyroid hormone on vertebral body cortical and cancellous bone in old ovariectomized rats. J Bone Miner Res 1996, suppl 1:S457.

32 Noland K., Orhii P., Rutstein J. and Kalu D. Effect of growth hormone therapy on cortical bone in aged ovariectomized rats. J Bone Miner Res 1996, suppl 1:S458.

33 Orhii P., Noland K., Rutstein J. and Kalu D. Effect of growth hormone therapy on cancellous bone in aged ovariectomized rats. J Bone Miner Res 1996, suppl 1:S458.

34 Verhaeghe J., van Bree R., Van Herck E., Thomas H., Skottner A., Dequeker J Mosekilde L., Einhorn T.A. and Bouillon R. Effects of recombinant human growth hormone and insulin-like growth factor-I, with or without 17 beta-estradiol, on bone and mineral homeostasis of aged ovariectomized rats. Journal of Bone and Mineral Research 1996, 11:1723-1735.

35 Eschen C. and Andreassen T.T. Growth hormone normalizes vertebral strength in ovariectomized rats. Calcif Tissue Int 1995, 57:392-396.

36 Wolf E., Rapp K. and Brem G. Expression of metallothionein-human growth hormone fusion genes in transgenic mice results in disproportionate skeletal gigantism. Growth Dev Aging 1991, 55:117-127.

37 Sandstedt J., Ohlsson C., Norjavaara E., Nilsson J. and Tornell J. Disproportional bone growth and reduced weight gain in gonadectomized male bovine growth hormone transgenic and normal mice. Endocrinology 1994, 135:2574-2580.

38 Sandstedt J., Törnell J., Norjavaara E., Isaksson O. and Ohlsson C. Intact testis function is a prerequisite for an optimal effect of elevated GH levels on vertebral bone weight in mice. Endocrinology and Metabolism 1996, suppl 1A:10.

39 Sandstedt J., Tornell J., Norjavaara E., Isaksson O.G. and Ohlsson C. Elevated levels of growth hormone increase bone mineral content in normal young mice, but not in ovariectomized mice. Endocrinology 1996, 137:3368-3374.

40 Ohlsson C., Tornell J., Sandstedt J. and Andreassen T. Effect of increased growth hormone on mechanical strength of cortical bone in growth hormone-transgenic mice. J Bone Miner Res 1996, suppl 1:T634.

41 Baker A.R., Hollingshead P.G., Pitts-Meek S., Hansen S., Taylor R. and Stewart T.A. Osteoblast-specific expression of growth hormone stimulates bone growth in transgenic mice. Mol Cell Biol 1992, 12:5541-5547.

42 Saban J., Schneider G.B., Bolt D. and King D. Erythroid-specific expression of human growth hormone affects bone morphology in transgenic mice. Bone 1996, 18:47-52.
43 Sass D., Jerome C., Bowman A., Bennett-Cain A., LeRoith D. and Epstein S. Short term effects of GH and IGF-1 on cancellous bone formation in rhesus macaque monkeys. J Clin Endocrinol Metab 1997, 82:1202-1209.
44 de Boer H., Blok G.J., van Lingen A., Teule G.J., Lips P. and van der Veen E.A. Consequences of childhood-onset growth hormone deficiency for adult bone mass. Journal of Bone and Mineral Research 1994, 9:1319-1326.
45 Kaufman J.M., Taelman P., Vermeulen A. and Vandeweghe M. Bone mineral status in growth hormone-deficient males with isolated and multiple pituitary deficiencies of childhood onset. J Clin Endocrinol Metab 1992, 74:118-123.
46 Saggese G., Baroncelli G.I., Bertelloni S. and Barsanti S. The effect of long-term growth hormone (GH) treatment on bone mineral density in children with GH deficiency. Role of GH in the attainment of peak bone mass. J Clin Endocrinol Metab 1996, 81:3077-3083.
47 Wuster C., Slenczka E. and Ziegler R. Increased prevalence of osteoporosis and arteriosclerosis in conventionally substituted anterior pituitary insufficiency: need for additional growth hormone substitution? Klin Wochenschr 1991, 69:769-773.
48 Johansson A.G., Burman P., Westermark K. and Ljunghall S. The bone mineral density in acquired growth hormone deficiency correlates with circulating levels of insulin-like growth factor I. J Intern Med 1992, 232:447-452.
49 Bing-You R.G., Denis M.C. and Rosen C.J. Low bone mineral density in adults with previous hypothalamic-pituitary tumors: correlations with serum growth hormone responses to GH-releasing hormone, insulin-like growth factor I, and IGF binding protein 3. Calcif Tissue Int 1993, 52:183-187.
50 Rosen T., Hansson T., Granhed H., Szucs J. and Bengtsson B.A. Reduced bone mineral content in adult patients with growth hormone deficiency. Acta Endocrinol (Copenh) 1993, 129:201-206.
51 Holmes S.J., Economou G., Whitehouse R.W., Adams J.E. and Shalet S.M. Reduced bone mineral density in patients with adult onset growth hormone deficiency. J Clin Endocrinol Metab 1994, 78:669-674.
52 Beshyah S.A., Freemantle C., Thomas E., Rutherford O., Page B., Murphy M. and Johnston D.G. Abnormal body composition and reduced bone mass in growth hormone deficient hypopituitary adults. Clin Endocrinol (Oxf) 1995, 42:179-189.
53 Degerblad M., Bengtsson B.A., Bramnert M., Johnell O., Manhem P., Rosen T. and Thoren M. Reduced bone mineral density in adults with growth hormone (GH) deficiency: increased bone turnover during 12 months of GH substitution therapy. European Journal of Endocrinology 1995, 133:180-188.
54 Toogood A.A., Adams J.E., O'Neill P.A. and Shalet S.M. Elderly patients with adult-onset growth hormone deficiency are not osteopenic. J Clin Endocrinol Metab 1997, 82:1462-1466.
55 Rosen T., Wilhelmsen L., Landin-Wilhelmsen K., Lappas G. and Bengtsson B. Increased fracture rate in adults with hipopituitarism and growth

hormone (GH) deficiency. European Journal of Endocrinology 1997, In Press:

56 Kotzmann H., Bernecker P., Hubsch P., Pietschmann P., Woloszczuk W., Svoboda T., Geyer G. and Luger A. Bone mineral density and parameters of bone metabolism in patients with acromegaly. Journal of Bone and Mineral Research 1993, 8:459-465.

57 Marcus R., Butterfield G., Holloway L., Gilliland L., Baylink D.J., Hintz R.L. and Sherman B.M. Effects of short term administration of recombinant human growth hormone to elderly people. J Clin Endocrinol Metab 1990, 70:519-527.

58 Holmes S., Mawer E. and Shalet S. Effect of Growth Hormone Replacement on Calcium, Phosphate, Vitamin D and Parathyroid Hormone Metabolism in Adults with Growth Hormone Deficiency. Endocrinology and Metabolism 1997, 4:49-54.

59 Bengtsson B.A., Eden S., Lonn L., Kvist H., Stokland A., Lindstedt G., Bosaeus I., Tolli J., Sjostrom L. and Isaksson O.G. Treatment of adults with growth hormone (GH) deficiency with recombinant human GH. J Clin Endocrinol Metab 1993, 76:309-317.

60 Beshyah S.A., Thomas E., Kyd P., Sharp P., Fairney A. and Johnston D.G. The effect of growth hormone replacement therapy in hypopituitary adults on calcium and bone metabolism. Clin Endocrinol (Oxf) 1994, 40:383-391.

61 Binnerts A., Swart G.R., Wilson J.H., Hoogerbrugge N., Pols H.A., Birkenhager J.C. and Lamberts S.W. The effect of growth hormone administration in growth hormone deficient adults on bone, protein, carbohydrate and lipid homeostasis, as well as on body composition. Clin Endocrinol (Oxf) 1992, 37:79-87.

62 Hansen T.B., Brixen K., Vahl N., Jorgensen J.O., Christiansen J.S., Mosekilde L. and Hagen C. Effects of 12 months of growth hormone (GH) treatment on calciotropic hormones, calcium homeostasis, and bone metabolism in adults with acquired GH deficiency: a double blind, randomized, placebo-controlled study. J Clin Endocrinol Metab 1996, 81:3352-3359.

63 Johannsson G., Rosen T., Bosaeus I., Sjostrom L. and Bengtsson B.A. Two years of growth hormone (GH) treatment increases bone mineral content and density in hypopituitary patients with adult-onset GH deficiency. J Clin Endocrinol Metab 1996, 81:2865-2873.

64 Spencer E.M. and Tobiassen O. The mechanism of the action of growth hormone on vitamin D metabolism in the rat. Endocrinology 1981, 108:1064-1070.

65 Spanos E., Barrett D., MacIntyre I., Pike J.W., Safilian E.F. and Haussler M.R. Effect of growth hormone on vitamin D metabolism. Nature 1978, 273:246-247.

66 Lieberman S.A., Holloway L., Marcus R. and Hoffman A.R. Interactions of growth hormone and parathyroid hormone in renal phosphate, calcium, and calcitriol metabolism and bone remodeling in postmenopausal women. Journal of Bone and Mineral Research 1994, 9:1723-1728.

67 Whitehead H.M., Boreham C., McIlrath E.M., Sheridan B., Kennedy L., Atkinson A.B. and Hadden D.R. Growth hormone treatment of adults with

growth hormone deficiency: results of a 13-month placebo controlled cross-over study. Clin Endocrinol (Oxf) 1992, 36:45-52.

68 Thoren M., Soop M., Degerblad M. and Saaf M. Preliminary study of the effects of growth hormone substitution therapy on bone mineral density and serum osteocalcin levels in adults with growth hormone deficiency. Acta Endocrinol (Copenh) 1993, 128:41-43.

69 Vandeweghe M., Taelman P. and Kaufman J.M. Short and long-term effects of growth hormone treatment on bone turnover and bone mineral content in adult growth hormone-deficient males. Clin Endocrinol (Oxf) 1993, 39:409-415.

70 Amato G., Carella C., Fazio S., La Montagna G., Cittadini A., Sabatini D., Marciano-Mone C., Sacca L. and Bellastella A. Body composition, bone metabolism, and heart structure and function in growth hormone (GH)-deficient adults before and after GH replacement therapy at low doses. J Clin Endocrinol Metab 1993, 77:1671-1676.

71 Balducci R., Toscano V., Pasquino A.M., Mangiantini A., Municchi G., Armenise P., Terracina S., Prossomariti G. and Boscherini B. Bone turnover and bone mineral density in young adult patients with panhypopituitarism before and after long-term growth hormone therapy. European Journal of Endocrinology 1995, 132:42-46.

72 Sartorio A., Ortolani S., Conti A., Cherubini R., Galbiati E. and Faglia G. Effects of recombinant growth hormone (GH) treatment on bone mineral density and body composition in adults with childhood onset growth hormone deficiency. J Endocrinol Invest 1996, 19:524-529.

73 Degerblad M., Elgindy N., Hall K., Sjoberg H.E. and Thoren M. Potent effect of recombinant growth hormone on bone mineral density and body composition in adults with panhypopituitarism. Acta Endocrinol (Copenh) 1992, 126:387-393.

74 Juul A., Pedersen S.A., Sorensen S., Winkler K., Jorgensen J.O., Christiansen J.S. and Skakkebaek N.E. Growth hormone (GH) treatment increases serum insulin-like growth factor binding protein-3, bone isoenzyme alkaline phosphatase and forearm bone mineral content in young adults with GH deficiency of childhood onset. European Journal of Endocrinology 1994, 131:41-49.

75 O'Halloran D.J., Tsatsoulis A., Whitehouse R.W., Holmes S.J., Adams J.E. and Shalet S.M. Increased bone density after recombinant human growth hormone (GH) therapy in adults with isolated GH deficiency. J Clin Endocrinol Metab 1993, 76:1344-1348.

76 Holmes S.J., Whitehouse R.W., Swindell R., Economou G., Adams J.E. and Shalet S.M. Effect of growth hormone replacement on bone mass in adults with adult onset growth hormone deficiency. Clin Endocrinol (Oxf) 1995, 42:627-633.

77 Beshyah S.A., Kyd P., Thomas E., Fairney A. and Johnston D.G. The effects of prolonged growth hormone replacement on bone metabolism and bone mineral density in hypopituitary adults. Clin Endocrinol (Oxf) 1995, 42:249-254.

78 Baum H.B., Biller B.M., Finkelstein J.S., Cannistraro K.B., Oppenhein D.S., Schoenfeld D.A., Michel T.H., Wittink H. and Klibanski A. Effects of

physiologic growth hormone therapy on bone density and body composition in patients with adult-onset growth hormone deficiency. A randomized, placebo-controlled trial. Ann Intern Med 1996, 125:883-890.
79 Brixen K., Nielsen H.K., Mosekilde L. and Flyvbjerg A. A short course of recombinant human growth hormone treatment stimulates osteoblasts and activates bone remodeling in normal human volunteers. Journal of Bone and Mineral Research 1990, 5:609-618.
80 Brixen K., Kassem M., Nielsen H.K., Loft A.G., Flyvbjerg A. and Mosekilde L. Short-term treatment with growth hormone stimulates osteoblastic and osteoclastic activity in osteopenic postmenopausal women: a dose response study. Journal of Bone and Mineral Research 1995, 10:1865-1874.
81 Kassem M., Brixen K., Blum W.F., Mosekilde L. and Eriksen E.F. Normal osteoclastic and osteoblastic responses to exogenous growth hormone in patients with postmenopausal spinal osteoporosis. Journal of Bone and Mineral Research 1994, 9:1365-1370.
82 Kruse H.P. and Kuhlencordt F. On an attempt to treat primary and secondary osteoporosis with human growth hormone. Horm Metab Res 1975, 7:488-491.
83 Aloia J.F., Zanzi I., Ellis K., Jowsey J., Roginsky M., Wallach S. and Cohn S.H. Effects of growth hormone in osteoporosis. J Clin Endocrinol Metab 1976, 43:992-999.
84 Aloia J.F., Vaswani A., Kapoor A., Yeh J.K. and Cohn S.H. Treatment of osteoporosis with calcitonin, with and without growth hormone. Metabolism 1985, 34:124-129.
85 Aloia J.F., Vaswani A., Meunier P.J., Edouard C.M., Arlot M.E., Yeh J.K. and Cohn S.H. Coherence treatment of postmenopausal osteoporosis with growth hormone and calcitonin. Calcif Tissue Int 1987, 40:253-259.
86 Holloway L., Kohlmeier L., Kent K. and Marcus R. Skeletal effects of cyclic recombinant human growth hormone and salmon calcitonin in osteopenic postmenopausal women. J Clin Endocrinol Metab 1997, 82:1111-1117.
87 Rudman D., Feller A.G., Nagraj H.S., Gergans G.A., Lalitha P.Y., Goldberg A.F., Schlenker R.A., Cohn L., Rudman I.W. and Mattson D.E. Effects of human growth hormone in men over 60 years old. N Engl J Med 1990, 323:1-6.
88 Johansson A.G., Lindh E. and Ljunghall S. Insulin-like growth factor I stimulates bone turnover in osteoporosis. Lancet 1992, 339:1619.
89 Ebeling P.R., Jones J.D., O'Fallon W.M., Janes C.H. and Riggs B.L. Short-term effects of recombinant human insulin-like growth factor I on bone turnover in normal women. J Clin Endocrinol Metab 1993, 77:1384-1387.
90 Ghiron L.J., Thompson J.L., Holloway L., Hintz R.L., Butterfield G.E., Hoffman A.R. and Marcus R. Effects of recombinant insulin-like growth factor-I and growth hormone on bone turnover in elderly women. Journal of Bone and Mineral Research 1995, 10:1844-1852.
91 Johansson A.G., Lindh E., Blum W.F., Kollerup G., Sorensen O.H. and Ljunghall S. Effects of growth hormone and insulin-like growth factor I in men with idiopathic osteoporosis. J Clin Endocrinol Metab 1996, 81:44-48.
92 Hedner E., Linde A. and Nilsson A. Systemically and locally administered growth hormone stimulates bone healing in combination with

osteopromotive membranes: an experimental study in rats. Journal of Bone and Mineral Research 1996, 11:1952-1960.

93 Tseng K.F., Bonadio J.F., Stewart T.A., Baker A.R. and Goldstein S.A. Local expression of human growth hormone in bone results in impaired mechanical integrity in the skeletal tissue of transgenic mice. J Orthop Res 1996, 14:598-604.

94 Ejersted C., Andreassen T.T., Hauge E.M., Melsen F. and Oxlund H. Parathyroid hormone (1-34) increases vertebral bone mass, compressive strength, and quality in old rats. Bone 1995, 17:507-511.

95 Ejersted C., Andreassen T.T., Nilsson M.H. and Oxlund H. Human parathyroid hormone(1-34) increases bone formation and strength of cortical bone in aged rats. European Journal of Endocrinology 1994, 130:201-207.

16 GH EFFECTS ON CARDIAC FUNCTION

Jörgen Isgaard

Research Center for Endocrinology and Metabolism, Department of Internal Medicine, Sahlgrenska University Hospital, Göteborg, Sweden

INTRODUCTION

Although traditionally associated with regulation of somatic growth, body composition and metabolism, it has become increasingly evident that the GH/IGF-I axis is important for cardiac growth and function. Early observations in acromegalic patients included manifestations such as cardiomegaly, impaired cardiac performance and higher morbidity and mortality in cardiovascular diseases [1,2]. These findings together with more recent data of compromised cardiac function in adult patients with GH deficiency [3-5] have focused the attention to GH as a potentially important regulator of the cardiovascular system. Moreover, a recent study showing beneficial effects of GH treatment of patients with dilated cardiomyopathy has also led to the suggestion of GH as a possible future agent in the treatment of congestive heart failure [6].

In this review, I will focus mainly on the regulation of IGF-I, IGF-I receptor (IGF-I-R) and GH receptor (GH-R) expression in the heart and their relation to cardiac growth processes. Moreover, hemodynamic effects of GH and IGF-I in experimental models of impaired cardiac function and in humans with congestive heart failure will be discussed, as well as some aspects of possible mechanisms of action of GH and IGF-I on the heart.

GH/IGF-I AXIS AND CARDIAC GROWTH

In adult life, cardiomyocytes are generally considered to be terminally differentiated cells and cardiac growth at this stage is predominantly a result of hypertrophy. An hypertrophic growth response of the heart is generally a physiological adaptation to increased hemodynamic load. A concomitant increase in the wall thickness of the left ventricle (LV) and resistance vessels occurs to maintain a normal wall tension per unit muscle layer (wall stress)[7]. In general, the underlying mechanisms behind cardiac hypertrophy triggered by increased wall stress are largely unknown, although it has been suggested that a rapid and short lasting induction of proto-oncogenes such as *c-fos*, *c-jun*, and *c-myc* that occurs within hours after an increase in LV systolic wall stress [8] may be of importance as an initial triggering signal for protein synthesis and cell growth [9]. A subsequent slower induction of local growth factors including IGF-I and components of the renin-angiotensin system and changes in the composition of contractile proteins have been reported in various experimental models [10-12]

IGF-I

IGF-I is synthesized in most tissues generally under GH regulation, although nutritional status and a variety of conditions associated with tissue repair and/or compensatory visceral growth also appear to influence IGF-I production. GH-dependent IGF-I expression has been demonstrated in the rat heart, both at the mRNA level [13,14] and protein level [15]. Moreover, the human fetus express IGF-I mRNA in the heart predominantly in the epicardium and coronary vessel wall [16].

Multiple studies have demonstrated an increased LV expression of IGF-I in different rat models of hypertension, including suprarenal aortic constriction, uninephrectomized spontaneously hypertensive rats, uninephrectomized, deoxycorticosterone-treated, saline-fed rats (DOCA salt)[17] and in 2-kidney, 1-clip (2K1C) rats [18,19]. In the former two studies, IGF-I mRNA induction was shown to be accompanied by increased IGF-I protein immunoreactivity, which was most abundant in the inner layers of the LV [18], where both tension and wall stress are high with a gradual decline towards the epicardial surface [20]. No elevation of IGF-I was observed in the right ventricle (RV). The onset of IGF-I expression preceded or occurred in parallel with development of LV hypertrophy [17,18]. Taken together, these results indicate that hypertension and increased LV pressure load constitutes a mechanical stimulus for a local growth response in the challenged ventricle where locally synthesized IGF-I may play a role through autocrine/paracrine mechanisms. Results supporting the hypothesis that an increased wall stress may trigger a local induction of IGF-I expression have also been reported in experimental models where the RV is predominantly challenged either through a surgically induced fistula between the abdominal aorta and the caval vein (ACF) [21] or through an induction of pulmonary hypertension [22].

Although the exact role and importance of IGF-I for development of cardiac hypertrophy is difficult to evaluate from the above mentioned studies it should be noted that when rats subjected to ACF were treated with the ornithine decarboxylase inhibitor di-fluoro-methyl-ornithine (DFMO), the induction of RV IGF-I expression was abolished and the development of RV hypertrophy transiently blunted [12]. This would suggest that IGF-I, at least during certain conditions, has a central role in the hypertrophic growth response.

IGF-I receptor

Although data regarding upregulation of IGF-I during cardiac hypertrophy in different rat models of pressure and volume overload are reasonably consistent, studies regarding the regulation of the IGF-R are more conflicting. During experimental 2K1C hypertension, a rapid and transient induction of IGF-I-R was found already 2 days after surgery [19]. In contrast, levels of IGF-I-R mRNA were reported to be stable over time in alternative rat models of experimental hypertension [17]. It should be noted, however that in the latter study the first measurements of IGF-I-R mRNA expression took place as late as 2 weeks after induction of hypertension in this study. It was recently reported that both IGF-I-R and IGF-I expression, measured by reverse transcriptase polymerase chain reaction, was increased after two days in viable myocytes in rats subjected to experimentally induced myocardial infarction [23]. Hence, it appears that a rapid induction of IGF-I-R can occur during increased hemodynamic load and/or a regenerative process in the myocardium. It is reasonable to speculate that an initial elevation of the IGF-I-R expression would enhance the response to locally available IGF-I and contribute to an hypertrophic response. However, if the upregulation of the IGF-I-R gene expression during these conditions results in an increased number of functional receptors remains to be established.

Hitherto six binding proteins for IGF-I and IGF-II (IGF-BP) have been isolated and characterized [for review see: 24]. There is sparse information regarding the physiological role for IGF-BP:s in cardiac growth and development, or whether they are actually synthesized within cardiac tissue. It was recently demonstrated in an experimental model with urethral ligation and subsequent induction of IGF-I mRNA expression in the bladder, consistent with hemodynamic regulation of IGF-I in the heart, that gene expression of IGF-BP-2 and IGF-BP-4 was induced [25]. However, if a similar regulatory mechanism for these IGF-BP:s exists in the heart remains to be investigated.

GH receptor

GH appears to be the primary regulator of cardiac IGF-I [13-15]. The GH-R gene is expressed in the rat [26] and rabbit [27] heart at low levels around birth and then rising to adult levels in the following months. Although basal levels of cardiac GH-R are lower than in the liver, the GH-R expression in the heart appears to be relatively high compared to other tissues [26].

GH-R gene expression was shown to be elevated in the acutely overloaded RV of ACF-operated rats [21] and in the LV of 2K1C rats [19], suggesting a role for the GH-R in compensatory growth of a hemodynamically challenged heart. However, since induction of GH-R expression preceded IGF-I mRNA in the volume overload model [21] but not in the 2K1C rats [19], it is unclear whether increased GH-R expression is really required for the upregulation of IGF-I message. Alternatively, it may be speculated that an increased expression of GH-R sustains the local synthesis of IGF-I in the heart.

CARDIAC EFFECTS OF GH AND IGF-I

Accumulating evidence indicates the importance of GH for normal cardiac function and structure. GH deficient adults receiving GH substitution therapy showed improvement of systolic function and normalization of LV mass [4,5]. A number of experimental studies both in normal rats [28] and in rats with surgically induced myocardial infarction [29-31] show that GH and IGF-I, alone or in combination, improve cardiac function. Moreover, GH and IGF-I was still effective in improving cardiac function when added as a treatment to post-infarction rats receiving ACE inhibition [32]. However, the absence of positive effects of GH on cardiac function after experimental myocardial infarction has also been reported [33]. Interestingly, it was recently demonstrated that GH, apart from improving cardiac function, also attenuates left ventricular remodeling in rats after myocardial infarction, apparently mostly due to reduction of LV dilation and additional hypertrophy of the noninfarcted myocardium [34]. Until recently, studies regarding GH treatment in heart failure was limited to case reports [35-37] where GH administration dramatically improved cardiac function. In a small open study of seven patients with idiopathic dilated cardiomyopathy and CHF without GH deficiency who received GH treatment for 3 months, considerable improvement of cardiac function was reported [6]. More recent studies have demonstrated beneficial effects in patients with CHF due to both ischemic and idiopathic dilated cardiomyopathy with improvements in hemodynamics when GH was added both as a maintenance therapy and as short term infusion [38,39] However, concern has also been raised regarding increased levels of circulating IGF-I [40] and a possible risk for of arrhythmia's during prolonged GH treatment [41]. In a placebo controlled, 3 months study, GH treatment of patients with CHF of various etiologies was reported to be safe and well tolerated without serious adverse effects, although no significant improvement in cardiac function was seen [42].

The precise mechanisms of actions of GH and IGF-I on the heart are less clear and there are few *in vitro* studies concerning with GH/IGF-I effects on cardiomyocytes. In papillary muscle from rats exposed to very high GH plasma levels due to a GH secreting tumor, it was found that the maximum Ca^{2+} activated force per cross-sectional area was increased, suggesting a GH induced elevated Ca^{2+} responsiveness of the myofilaments [43]. Moreover, it was observed in the same model that the duration of the action potential was prolonged, implying an increased Ca^{2+} influx as a contribution to increased contractility. In a model with isolated buffer-perfused hearts from rats treated with high doses of GH or IGF-I for four weeks, systolic function was improved and findings supported the notion of an increased maximal response to Ca^{2+} [44].

In vitro studies, with addition of IGF-I to cultured neonatal rat cardiomyocytes, have demonstrated changes associated with hypertrophy including increments in cell size, protein synthesis and induced expression of myosin light chain-2 and troponin I [45]. In adult rat cardiomyocytes, it has been reported that IGF-I stimulates myofibril development [46].

Apoptosis, or programmed cell death, has been reported to occur in the myocardium both after ischemic injury [47] and in patients with dilated cardiomyopathy [48]. Some of the genes involved in the apoptotic process have been identified including the Fas-receptor which has been shown to mediate signals for apoptosis [49,50]. Upregulation of the Fas-receptor has been shown to be associated with apoptosis after myocardial infarction in rats [51]. It has been reported that administration of IGF-I prior to cardiac ischemia attenuates the increased rate of myocyte apoptosis [52] and that both GH and IGF-I can prevent an upregulation of the Fas-receptor expression in the heart after experimental myocardial infarction in rats [53].

SUMMARY

Both the GH-R and IGF-I-R are expressed in cardiac tissue. Moreover, IGF-I is also locally produced in the heart and in part regulated by GH. Hence, the IGF-I-R receptor can theoretically be activated through locally produced IGF-I acting via autocrine/paracrine mechanisms, or via circulating IGF-I exerting its effects as an endocrine agent. During conditions of pressure and volume overload, an increased systolic wall stress triggers an induction of gene expression of IGF-I, GH-R and possibly IGF-I-R implying a potential role for the GH/IGF-I axis in the development of adaptive hypertrophy of the heart and vessels. Cardiovascular effects of GH in clinical studies include beneficial effects on contractility and exercise performance, and experimental studies suggest an increased Ca^{2+} responsiveness as one possible underlying cause, although effects of GH and IGF-I on apoptosis may possibly also play a role. However, it is clear that further basic and clinical studies are required to gain insight into the GH and IGF-I mechanisms of action and to monitor long-term effects of when GH is administered as substitution therapy or as an agent in the treatment of congestive heart failure.

REFERENCES

1 Courville C, Mason VR. The heart in acromegaly. Arch Intern Med 1938;61:704-713.

2 Bengtsson BÅ, Edén S, Ernest I, Odén A, Sjögren B. Epidemiology and long-term survival in acromegaly. A study of 166 cases diagnosed between 1955 and 1984. Acta Med Scand 1988;223:327-335.

3 Merola B, Cittadini A, Colao A, Longobardi S, Fazio S, Sabatini D, Saccà L, Lombardi G. Cardiac Structure and functional abnormalities in adult patients with growth hormone deficiency. J Clin Endocrinol Metab 1993;77:1658-1661.

4 Amato G, CarellaC, Fazio S, La Montagna G, Cittadini A, Sabatini D, Marciano-Mone C, Saccà L, Bellastella A. J Clin Endocrinol Metab 1993;77:1671-1676.

5 Caidahl K, Edén S, Bengtsson B-Å. Cardiovascular and renal effects of growth hormone. Clin Endocrinol (Oxf) 1994;40:393-400.

6 Fazio S, Sabatini D, Capaldo B, Vigorito C, Giordano A, Guida R, Pardo F, Biondi B Saccà L. A preliminary study of growth hormone in the treatment of dilated cardiomyopathy. N Engl J Med 1996;334:809-814.

7 Friberg P, Folkow B, Nordlander M. Structural adaptation of the rat left ventricle in response to changes in pressure and volume loads. Acta Physiol Scand 1985;125:67-79.

8 Schunkert H, Jahn L, Izumo S, Apstein CS, Lorell BH. Localization and regulation of c-fos and c-jun proto-oncogene induction by systolic wall stress in normal and hypertrophied rat hearts. Proc Natl Acad Sci USA 1991;88:11480-11484.

9 Simpson PC. Role of proto-oncogenes in myocardial hypertrophy. Am J Cardiol. 1988;62:13G-9.

10 Komuro I, Yazaki Y. Control of cardiac gene expression by mechanical stress. Annu Rev Physiol 1993;55:55-75.

11 Chien KR, Zhu H, Knowlton KU, Miller-Hance W, van Bilsen M, O'Brien TX, Evans SM. Transcriptional regulation during cardiac growth and development. Annu Rev Physiol 1993;55:77-95.

12 Friberg P, Isgaard J, Wåhlander H, Wickman A, Guron G, Adams MA. Cardiac hypertrophy and related growth processes: The role of Insulin-like growth factor I. Blood Pressure 1995;4 (suppl 2):22-29.

13 Murphy LJ, Bell GI, Duckworth L, Friesen G. Identification, characterization and regulation of a rat complimentary deoxyribonucleic acid which encodes insulin-like growth factor I. Endocrinology 1987;121:684-691.

14 Isgaard J, Nilsson A, Vikman K, Isaksson OGP. Growth hormone regulates the level of insulin-like growth factor I in rat skeletal muscle. J. Endocrinol. 1989;120:107-112.

15 D'Ercole AJ, Stiles AD, Underwood LE. Tissue concentration of somatomedin-C: Further evidence for multiple sites of synthesis and paracrine or autocrine mechanisms of action. Proc Natl Acad Sci USA 1984;81:935-939.

16 Han VKM, D'Ercole AJ, Lund KP. Cellular localization of somatomedin (insulin-like growth factor) messenger RNA in the human fetus. Science 1987;236:193-197.

17 Donohue TJ, Dworkin LD, Lango MN, Fliegner K, Lango RP, Benstein JA, Slater WR, Catanese VM. Induction of myocardial insulin-like growth factor I gene expression in left ventricular hypertrophy. Circulation 1994;89:799-809.

18 Wåhlander H, Isgaard J, Jennische E, Friberg P. Left ventricular insulin-like growth factor I increases in early renal hypertension. Hypertension. 1992;19:25-32.

19 Guron G, Friberg P, Wickman A, Brantsing C, Gabrielsson B, Isgaard J. Cardiac insulin-like growth factor I and growth hormone receptor expression in renal hypertension. Hypertension 1996;27 (part 2):636-642.

20 Mirsky J. Elastic properties of the myocardium: a quantitative approach with physiological and clinical applications. In: Handbook of Physiology. Chapter 14, Section 2, Volume I. American Physiological Society 1979:497-531.

21 Isgaard J, Wåhlander H, Adams MA, Friberg P. Increased expression of growth hormone receptor mRNA and insulin-like growth factor I mRNA in volume overloaded hearts. Hypertension 1994;23:884-888.

22 Russell-Jones DL, Leach RM, Ward JPT, Thomas CR. Insulin-like growth factor-I gene expression is increased in the right ventricle hypertrophy induced by chronic hypoxia in the rat. J. Endocrinol. 1993;10:99-102.

23 Reiss K, Meggs LG, Li P, Olivetti G, Capasso JM, Anversa P. Upregulation of IGF-I IGF-I receptor and late growth related genes in ventricular myocytes acutely after infarction in rats. J Cell Physiol 1994;158:160-168.

24 Delafontaine P. Insulin-like growth factor I and its binding proteins in the cardiovascular system. Card Vasc Res 1995;30:825-834.

25 Chen Y, Arner A, Bornfeldt KE, Uvelius B, Arnqvist HJ. Development of smooth muscle cell hypertrophy is closely associated with increased gene expression of insulin-like growth factor binding protein-2 and -4. Growth Regulation 1995;5:45-52.

26 Mathews LS, Engberg B, Norstedt G. Regulation of rat GH receptor gene expression. Biol Chem. 1989;264:9905-9910.

27 Ymer SI & Herington AC. Developmental expression of the growth hormone receptor gene in rabbit tissues. Mol Cell Endocrinol 1992;83:39-49.

28 Cittadini A, Strömer H, Katz SE, Clark R, Moses AC, Morgan JP, Douglas PS. Differential cardiac effects of growth hormone and insulin-like growth factor-I in the rat. A combined in vivo and in vitro evaluation. Circulation 1996;93:800-809

29 Yang R, Bunting S, Gillett N, Clark R, Jin H. Growth improves cardiac performance in experimental performance. Circulation 1995; 92:262-267.

30 Duerr RL, Huang S, Miraliakbar HR, Clark R, Chien KR, Ross Jr J. Insulin-like growth factor-1 enhances ventricular hypertrophy and function during the onset of experimental cardiac failure. J Clin Invest 1995;95:619-627.

31 Isgaard J, Kujacic V, Jennische E, Holmäng A, Sun XY, Hedner T, Hjalmarson Å, Bengtsson B-Å. Growth hormone improves cardiac function in rats with experimental myocardial infarction. Eur J Clin Invest 1997;27:517-525

32 Jin H, Yang R, Gillett N, Clark RG, Ko A, Paoni NF. Beneficial effects of growth hormone and insulin-like factor-1 in experimental heart failure in rats treated with chronic ACE inhibition. J Cardiovasc Pharmacol. 1995;26:420-425.

33 Shen YT, Wiedmann RT, Lynch JJ, Grossman W, Johnson RG. GH replacement fails to improve ventricular function in hypophysectomized rats with myocardial infarction. Am J Physiol 1996;271:H1721-H1727

34 Cittadini A, Grossman J, Napoli R, Katz S, Strömer H, Smith R, Clark R, Morgan JP, Douglas P. Growth hormone attenuates early ventricular remodeling and improves cardiac function in rats with large myocardial infarction. J Am College Cardiol 1997;29:1109-1116

35 Cuneo RC, Wilmshurst P, Lowy C, McGauley G, Sönksen PH. Cardiac failure responding to growth hormone. Lancet 1989;1:838-839

36 Frustaci A, Perrone GA, Gentiloni N, Russo MA. Reversible dilated cardiomyopathy due to growth hormone deficiency. Am J Clin Pathol 1992;97:503-511

37 O'Driscoll JG, Green DJ, Ireland M, Kerr D, Larbalestier RI. Treatment of end-stage cardiac failure with growth hormone. Lancet 1997;349:1068-1068

38 Volterrani M, Desenzani P, Lorusso R, d'Aloia A, Manelli F, Giustina A. Hemodynamic effects of intravenous growth hormone in congestive heart failure. Lancet 1997;349:1067-1068

39 Beer N, Tortoledo F, Beer R, Pinedo M. Beneficial effects of growth hormone in patients with congestive heart failure. Abstract "Heart Failure '97, Cologne Germany 1997.

40 Turner H, Wass JAH. Growth hormone in the treatment of dilated cardiomyopathy. Letter to the Editor, N Engl J Med 1996;335:672-672

41 Frustaci A, Gentiloni N, Russo MA. Growth hormone in the treatment of dilated cardiomyopathy. Letter to the editor, N Engl J Med 1996;335:672-673

42 Isgaard J, Bergh C-H, Caidahl K, Lomsky M, Hjalmarson Å, Bengtsson B-Å. A placebo controlled study of growth hormone in patients with congestive heart failure. Submitted Lancet

43 Mayoux E, Ventura-Clapier R, Timsit J, Béhar-Cohen F, Hoffman C, Mercadier J-J. Mechanical properties of rat cardiac skinned fibers are altered by chronic growth hormone hypersecretion. Circ Res 1993;72:57-64.

44 Strömer H, Cittadini A, Douglas PS, Morgan JP. Exogenously administered growth hormone and insulin-like growth factor-I alter intracellular Ca^{2+} handling and enhance cardiac performance. In vitro evaluation in the isolated isovolumic buffer-perfused rat heart. Circ Res 1996;79:227-236.

45 Ito H, Hiroe M, Hirata Y, Tsujino M, Adachi S, Shichini M, Koike A, Nogami A, Marumo F. Insulin-like growth factor-I induces hypertrophy with enhanced expression of muscle-specific genes in cultured rat cardiomyocytes. Circulation 1993;87:1715-1721.

46 Donath MY, Zapf J, Eppenberger-Eberhardt M, Froesch ER, Eppenberger HM. Insulin-like growth factor I stimulates myofibril development and decreases smooth muscle I-actin of adult cardiomyocytes. Proc Natl Acad Sci USA 1994;91:1686-1690.

47 Saraste A, Pulkki K, Kallajoki, M, Henriksen K, Parvinen M, Voipio-Pulkki LM. Apoptosis in human acute myocardial infarction. Circulation 1997;95:320-323

48 Narula J, Haider N, Virmani R, DiSalvo, TG, Kolodgie FD, Hajjar RJ, Schmidt U, Semigran MJ, Dec GW, Khaw BA. Apoptosis in myocytes in end-stage heart failure. N Engl J Med 1996;335-1182-1189

49 Itoh N, Yonehara S, Ishii A, Yonehara M, Mizushima SI, Sameshima M, Hase A, Seto Y, Nagata S. The polypeptide encoded by the cDNA for human cell surface antigen Fas can mediate apoptosis. Cell 1991;66:233-243

50 Nagata S, Golstein P. The Fas death factor. Science 1995;267:1449-1456

51 Kajstura J, Cheng W, Reiss K, Clark WA, Sonnenblick EH, Krajewski S, Reed JC, Olivetti G, Anversa P. Apoptotic and necrotic myocyte cell deaths are independent contributing variables of infarct size in rats. Lab Invest 1996;74:86-107

52 Buerke M, Mohara T, Skurk C, Nuss C, Tomaselli K, Lefer AM. Cardioprotective effect of insulin-like growth factor-I in myocardial ischemia followed by reperfusion. Proc Natl Acad Sci (USA) 1995;92:8031-8035

53 Tivesten Å, Matejka G, Caidahl K, Brantsing C, Sun XY, Bengtsson B-Å, Hedner T, Isgaard J. Growth hormone and IGF-I effects on cardiac function and regulation of the Fas receptor in rats with experimental myocardial infarction. Abstract # P1-126, 79th Annual Meeting of the Endocrine Society, June 11-14 1997, Minneapolis, USA

17 PSYCHOLOGICAL WELL-BEING IN GROWTH HORMONE DEFICIENCY AND CENTRAL NERVOUS EFFECTS OF GROWTH HORMONE

Jan-Ove Johansson and Bengt-Åke Bengtsson

Research Centre for Endocrinology and Metabolism, Sahlgrenska University Hospital, Göteborg, Sweden

Well-being in patients with untreated growth hormone deficiency

Although no widely accepted definition of the term "quality of life" exists, the close relationship between quality of life and health is accepted. In recent years, there has been a growing interest in the assessment of quality of life in both medical care and clinical trials. Measures which have often been used in evaluating quality of life are for example the two self-rating questionnaires the Nottingham Health Profile (NHP) and the Psychological General Well-Being Index (PGWB).

Children with untreated growth hormone deficiency (GHD) have been shown to be psychologically immature and to display personality disturbances (1). These findings may, however, be interpreted as a secondary effect of the decreased body height and not of the GHD per se. Long-term follow-up studies of adults treated with GH in childhood showed that these patients were more likely to be single and unemployed. Their educational level, however, did not differ from that of their close relatives (2, 3). In a study of 23 adults who had received GH therapy as children, Björk et al. observed that the patients with GHD had a poorer quality of life compared with a control group. According to NHP and PGWB patients with GHD were significantly more socially isolated, less physically mobile, slept less well and had a poorer emotional state. Fewer individuals in the patient group had a driving licence and were married than in the control group (4).

Up to 1989 little information about the quality of life in patients who acquired GHD in adult life was available. In a study comprising 24 adult patients with GHD, McGauley in using the NHP, found a significant increase in total score compared with matched controls, indicating a greater level of perceived health problems among the patients with GHD. The patients scored significantly higher than the controls in the areas of energy, emotional reaction and social isolation, implying that these patients perceived themselves as being more labile, more socially isolated and less energetic than the controls. However, the results in the study may have been influenced by the fact that nine of the 24 patients had GHD secondary to treatment of Cushing's disease (5). In a study from our unit, Rosén et al. compared 86 patients with GHD of adult onset with 86 controls from the Göteborg sample (1990) of the WHO MONICA Project. The controls were individually matched with the patients for sex, age, marital status and socioeconomic class. When using the NHP, it was found that the patients had a higher total score than the control group, thus indicating a higher level of perceived health problems. There were higher scores for energy ($p<0.001$), social isolation ($p<0.01$), emotional reaction ($p=0.056$) and sex life ($p<0.001$) among patients compared with the controls. The observed number of disability pensions among the patients tended to be higher than expected (19 vs. 12.4, $p=0.09$) (6) **(Table 1 and 2)**.

Table 1. Scores achieved in NHP part I in 86 GH-deficient subjects and in 86 controls. Higher scores indicate poor quality of life. Maximum score is 100. With the kind permission of *Clinical Endocrinology (Oxf)* (6).

	PATIENTS	CONTROLS	
Category	Mean SD	Mean SD	p
Emotional reaction	12.5 (±22.6)	6.9 (±14.6)	0.056
Sleep	15.7 (±23.0)	16.6 (±22.5)	ns
Energy	23.1 (±35.5)	7.8 (±19.4)	<0.001
Pain	12.4 (±23.8)	9.2 (±21.5)	ns
Physical mobility	9.0 (±15.5)	6.2 (±14.2)	ns
Social isolation	11.2 (±21.7)	3.0 (±11.2)	<0.01
Total score	14.0 (±17.0)	8.3 (±11.8)	<0.05

Table 2. Scores (% yes-responders) in NHP part II in 86 GH-deficient subjects and in 86 controls. Higher scores indicate poor quality of life. Maximum score is 100. With the kind permission of *Clinical Endocrinology (Oxf)* (6).

Category	PATIENTS	CONTROLS	p
Paid employment	8.1	10.8	ns
House work	19.8	9.3	0.08
Social life	15.1	7.0	ns
Family life	7.0	2.3	ns
Sex life	28.6	4.7	<0.001
Hobbies	22.1	17.4	ns
Holidays	14.0	12.8	ns

In another study from Japan (7), questionnaires were sent out to 1,116 adults with childhood onset GHD who ceased GH therapy before the age of 20, and to their physicians. The questionnaire used included present physical status, school and occupational career, socioeconomic and marital status, and the patients evaluation of the GH therapy. The Cornell Medical Index Health questionnaire was applied to assess physical and mental status. 709 patients (64%) answered the questionnaires anonymously. The distribution of occupation, job and income were similar to those in the general population. Patients with short stature, however, felt at a disadventage when finding jobs and marriage partners. Some psychological problems were reported among these patients but the conclusion was that most GH treated adults in Japan had adjusted themselves well to their psychosocial environment. The cultural difference between the European and Japanese society may in part explain the different outcome between the Japanese and the European studies. The fact that there were no tests done to confirm a GHD in adulthood must also be taken into account. However, additional recent studies have subsequently showed in a convincing way that adults with GHD have a poor quality of life (8-10). In one of these studies it was concluded that patients with childhood onset of GHD are significantly more socially disadvantaged than controls with respect to their living situation, partnership, profession and income (8). Furthermore, GH-deficient adults as a group have lower vigor scores and a higher state anxiety score than controls. Cognitive function, as indicated by a reduced memory performance, was also found to be subnormal in these subjects (10).

Effects of growth hormone treatment on well-being

Early anecdotal evidence has hinted at the favourable effect of GH on psychological well-being in adults with GHD. In 1962 Raben reported on a 35-year-old female school teacher with hypopituitatism who had been treated with conventional hormone substitution therapy for eight years. She was then treated with human GH, 3 mg three times a week, and within two months she noted increased vigor, ambition and sense of well-being (11). The beneficial effects of GH on psychological well-being are underlined by studies in recent years where treatment with recombinant human GH has been given to adult GH-deficient patients. In a study presented in 1986 five patients with GHD were given three weekly intramuscular injections of 8 IU GH during one month. Psychometric tests evaluating cognitive functions were performed before and after treatment. A beneficial effect of GH on certain cognitive functions, including memory, was suggested (12). In the previously mentioned study from McGauley the effect of daily subcutaneously administered GH (0.07 IU/kg body weight/day) on quality of life was assesed in a double-blind, placebo-controlled six-months trial. There were no differences between the GH- and placebo treated group after one month of the trial. After six months, however, the patients receiving active treatment experienced less perceived illness than the placebo group. They also noted significant psychological improvement in energy level and mood (5). We have reported similar findings in a double blind, placebo-controlled six-months cross-over trial. Ten patients with adult-onset GHD were given GH in a dose of 0.035-0.07 IU/kg body weight/day. The patients were examined by a psychiatrist and two rating scales were used - the Comprehensive Psychological Rating Scale (CPRS) and the Symptom Check List-90 (SCL-90). Complaints before treatment were mainly tiredness, low energy and lack of initiative, lack of concentration, memory difficulties and irritability. After six months of GH treatment, there was a significant improvement in the CPRS score while no significant change in the results of the SCL-90 was noted (13).

In one placebo-controlled cross-over study by Whitehead et al. the psychological well-being remained unchanged during 6 months of GH treatment (14). However, a multicentre study including 125 hypopituitary patients with adult onset GHD has confirmed that quality of life improves during GH treatment. The data from seven trials could be combined since they all had a similar design. The studies were double-blind, placebo-controlled trials for an initial six-month period, followed by an open period for an additional 6-12 months. Perceived health status was evaluated by the NHP and subjective well-being by the PGWB questionnaires. The energy section of NHP improved during the first 6 months of GH treatment. By 12 months the dimensions measuring emotional reaction and social isolation had also significantly improved compared with baseline. The changes observed indicate a normalization since similar levels are obtained in a healthy population. Moreover, the PGWB total score improved significantly after 6 months of GH treatment and reached normal values (15). In a recent study of 36 adults with GHD, Burman et al. also confirmed increased psychological well-being during treatment with GH, using the NHP, PGWB and Hopkins Symptom Check List (HSCL). With NHP, the dimensions of energy and emotions responded most to treatment. Further, the spouses observed their partners to be improved in several aspects of mood and behaviour (9). At the Centre for Endocrinology and Metabolism at Sahlgrenska University Hospital we have up to now treated almost 300 GH-deficient adults with GH. We have noted an improvement in psychological well-being in almost every patient. Some patients perceive this improvement, which is sometimes quite dramatic, relatively quickly after the initiation of treatment while others need almost a year before noting any change in quality of life. Interestingly, changes also occur in sleep-patterns during GH treatment, for example, there is no need to nap several hours during daytime. Thus the patients can work fulltime and still be able to have a social life after work.

The mechanisms underlying the beneficial effects of GH on psychological well-being are unclear. The low extracellular water volume noted among the patients with GHD has been suggested as a possible explanation for the tiredness. A restoration during GH therapy may in part explain the improvement in psychological well-being (6). The normalization of body composition with decreased body fat and increased lean body mass and the subsequent improvement in exercise capacity, may also contribute to this. However, it is also possible that GH may have direct neuroendocrine effects on the central nervous system (CNS). In fact, neural actions by GH were first documented by Zamenhof in the early 1940s

(16, 17). Other authors have later concluded that there are no studies indicating roles for GH within the CNS (18).

GH and GH receptors in the CNS

A possible direct effect of GH on the CNS is supported by the finding that GH-like material appears in the rat brain before its ontogenic appearance in the adenohypophysis (19). GH may thus play an important role in brain development, maturation and function. Several immunological studies have identified GH-like proteins throughout the CNS including the telencephalon (entire cerebral hemispheres), diencephalon (thalamus and hypothalamus), mid brain, metencephalon (pons and cerebellum) and spinal cord (19-26). In studies of rodents and primates, it has been shown that the greatest content of immunoreactive GH is in the amygdala, hippocampus and hypothalamus. However, the concentrations are less than 1% of those in the anterior pituitary gland (24). The long-term presence of immunoreactive GH in the brain following hypophysectomy also suggests a central site of synthesis, as does the continuous release of immunoreactive GH from dispersed CNS cells from the rat grown in tissue culture (24). This release of GH-like material into the growth medium can be suppressed with the addition of somatostatin (24). Moreover, the presence of GH mRNA in several areas of the brain supports the hypothesis of local synthesis of GH in the brain (27). A neurotrophic role for GH is also indicated by the severe deficits in brain development that occur in GH-deficient states (28-30) and that these defects are reversed by GH-treatment (30, 31). Zamenhof could clearly show that GH does increase the total number of cerebral neurons and glia in the developing tadpole (16) as well as in the offspring of female rats which receive GH during the time of neuronal proliferation (17, 32). Other studies have confirmed that the administration of GH to pregnant rats increases both the brain weight and the number of cortical neurons and enhances the subsequent learning ability in the offspring (33-35). It is now well established that GH stimulates neuronal and glial proliferation, myelination (28, 31, 36, 37) and RNA synthesis in the rat brain (38). It has also been shown that GH treatment of hypophysectomized rats influences the expression of the c-fos gene, which may serve as an indicator of neural activity, in some areas of the brain (39, 40). Furthermore, animal studies have shown that GH and GH releasing hormone (GHRH) modulate memory functions (41) as well as sleep (42). Although children with untreated GHD may show psychological immatutity (1) most of these patients are intellectually normal (3, 43). In patients with Laron-type dwarfism, i.e. a condition characterized by elevated serum levels of GH and decreased serum levels of IGF-1 caused by a deficiency of the cellular receptor for GH, intelligence has been reported as subnormal in the Israeli cohort but normal or even high in the European and Ecuadorian groups (44-47).

Recently, GH receptors have been found in many locations in the rodent and human brain, strongly suggesting CNS sites of GH action (48-53). It appears that the mRNA for the GH receptor and binding protein are expressed in neuronal and glia as well as endothelial cells (51, 54, 55). In fact, GH receptor mRNA in the rat brain decreases with postnatal age in contrast to its ontogeny in the liver. This finding supports the possibility of a role for GH in influencing neuronal maturation and glial cell formation (51, 56). The highest density of GH receptors in the adult human brain has been noted in the choroid plexus, the hippocampus, the hypothalamus and the pituitary gland (49) **(Fig. 1)**. While GH binding sites in the latter regions perform an obvious function in the regulation of pituitary GH secretion, the physiological relevance of GH binding sites in other brain regions is less clear. In most brain tissues examined, the number of GH receptors was higher in women and there was an age-related decrease in the number of GH receptors irrespective of sex (50). It seems likely that this loss of GH receptors in the brain of elderly people may have some physiological consequences. For examle, the decrease in GH binding at hypothalamic and pituitary levels may be of importance for the mechanisms behind the secretion of GH. A cellular localization of the GH binding protein/receptor has also been demonstrated in a variety of tissues including the central and periphereal nervous system (57, 58). As a result, GH may exert its central nervous effects using other mechanisms than stimulation of the extracellular part of the receptor (59).

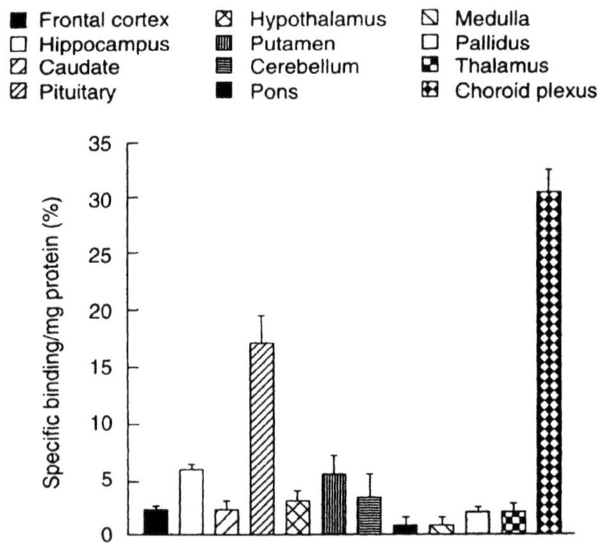

Fig. 1. Specific binding of ^{125}I-labelled GH to membranes from different parts of the human brain. The values are means ± SD (n=6). With the kind permission of *Brain Research* and the authors (49).

Does GH cross the blood-brain barrier?

The question of the permeability of the blood-cerebrospinal fluid (CSF) barrier to GH has been controversial (reviewed in 26). Acromegalics have higher CSF GH levels compared with controls (60). The infusion of large amounts of human GH to rhesus monkeys was not followed by a parallel increase in CSF GH levels (61). On the other hand, intra-peritoneal injections of ^{125}I-labelled GH to rats in another study resulted in an accumulation of radioactivity in several brain areas, thereby suggesting that GH crosses the blood-brain barrier in this species (62). It is also known that the systemic GH treatment of both normal and hypophysectomized rats influences the cerebral concentrations of monoamines. Monoamine changes were found as early as 15 min. after an injection of GH, thus suggesting that GH is rapidly taken up by brain tissue and quickly affects ongoing monoamine metabolism (63, 64).

In a one-month double-blind placebo-controlled study, we found that GH treatment in GH-deficient adults causes a mean ten-fold increase in GH in the CSF, thus indicating that GH does pass the human blood-CSF barrier (**Fig. 2**) (65). The possible passage of GH over the blood-CSF barrier is also supported by the dose-related increase in the GH concentration in CSF reported in a recent study (66). The concentration of GH in the CSF is still low compared with serum (CSF:serum ratio about 5%). The increase in GH concentrations in the CSF was similar in patients with and without a damaged blood-CSF barrier (65). One possible mechanism behind the increase in GH in CSF is ultrafiltration, since most of the proteins in CSF are derived by this process. Alternative explanations are a GH receptor-mediated transport in the choroid plexus (49), a carrier-mediated transport system or a diffusion via the incomplete blood-brain barrier in the median eminence of hypothalamus (26).

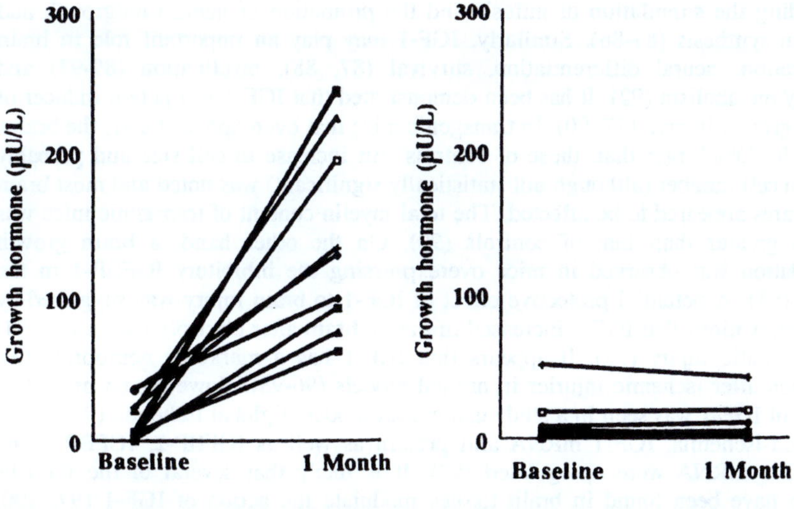

Fig. 2. Effects of GH (left) and placebo (right) on the GH concentration in the CSF in GH-deficient adults (ten per group). With the kind permission of *Neuroendocrinology* (65).

IGF-1 and IGF-1 receptors in the CNS

The variant form of IGF-1 (truncated IGF-1) lacks the tripeptide Gly-Pro-Glu at the N-terminus. Compared with the complete molecule, the truncated form is five to ten times more potent in stimulating DNA and protein synthesis, because of its lower binding affinity to the IGF binding proteins (IGFBPs) (67, 68). This variant form of IGF-1 and IGF-1 receptors have been found in a wide distribution in the human brain as well as in the rat brain (69-74). IGF-1 receptors are found in all regions of the adult human brain, including both neurones and glia, with high densities in the hippocampus, amygdala and parahippocampal gyrus and intermediate densities in the cerebellum, cerebral cortex and caudate nucleus (71).

IGF-1 mRNA is also expressed in both neurons and glia (75-77) and is eight to ten times more abundant in the cervical-thoracic spinal cord compared with the whole brain and three times more abundant in the midbrain and cerebellum (75). During normal brain development in the rat, the IGF-1 gene expression displays distinct developmental patterns in different regions of the brain (78). IGF-II and its mRNA have also been observed in normal rat brain in vivo at various developmental ages (75, 79, 80). However, only glia produce IGF-II mRNA during short-term cell culture (75). The choroid plexus and the meninges are a source of IGF-II (79) and it is likely that the IGF-II found in CSF is synthesized and secreted by the choroidal epithelial cells (79-81). The level of IGF-II in human CSF is usually higher than that of IGF-1 (82). However, the expression of IGF-II is not primarily regulated by GH (83).

There is now overwhelming evidence that IGFs have a definite biological role within the CNS. IGF-1 has a number of trophic effects on cultured neural tissue including the stimulation of mitosis and the promotion of neurite outgrowth and tubulin synthesis (84-86). Similarly, IGF-1 may play an important role in brain maturation, neural differentiation, survival (87, 88), myelination (89-91) and energy metabolism (92). It has been demonstrated that IGF-1 is a potent inducer of brain growth in vivo (37, 90). In transgenic mice that overexpress IGF-1, the brains were 50-70% larger than those of controls. An increase in cell size and probably also in cell number (although not statistically significant) was noted and most brain structures appeared to be affected. The total myelin content of transgenic mice was 130% greater than that of controls (90). On the other hand, a brain growth retardation was observed in mice overexpressing the inhibitory IGFBP-1 in the brain (93). A potential protective effect of IGF-1 to brain injury was suggested by the observations that IGF-1 increased in the rat brain after electrolytical lesion (94) or traumatic injury (95). It appears that IGF-1 has a markedly neuroprotective function after ischemic injuries in animal models (96-98). However, no protective effect of IGF-1 was seen in a study using a rat model of global ischemia (99). After hypoxia-ischemia, IGF-1 mRNA and protein as well as IGFBP-2, IGFBP-3 and IGFBP-5 mRNA were upregulated (97). It is likely that several of the IGFBPs which have been found in brain tissues modulate the action of IGF-1 (97, 100, 101). It has been proposed that different IGFBPs play different roles in repair and regeneration processes in the brain (97). Other growth factors such as the basic fibroblast growth factor and the nerve growth factor have been implicated as playing a role in the process of neural rescue after brain damage (102, 103).

In our study, we also observed that the mean increase in CSF IGF-1 concentrations was about 50% during one month of GH treatment (65) (**Fig. 3**). CSF IGFBP-3 concentrations also increased (65). IGF-1 may pass the blood-brain barrier in the rat (104) while an intravenous infusion of N-Met IGF-1 for 24 h to sheep resulted in an unchanged concentration of IGF-1 in the CSF (105). However, the question of the permeability of the human blood-CSF barrier to IGFs has not yet been fully answered (106). Previous clinical studies have indicated the existence of a blood-CSF barrier for IGFs and that IGF-1 present in the CSF is probably mainly derived from local CNS production (82). In line with these findings, we did not observe any correlation between the changes in plasma and CSF IGF-1 concentrations (65). In animal studies it has been shown that IGF-1 decreases both in serum and the CNS after hypophysectomy and the levels can be restored after administration of GH (107, 108). Both intracerebroventricular and systemic administration of GH increase mRNA and peptide levels of IGF-1 in the brain of adult rats (108, 109). It can be concluded that most of the IGF activity in the CNS is likely to be autocrine or paracrine rather than endocrine (83).

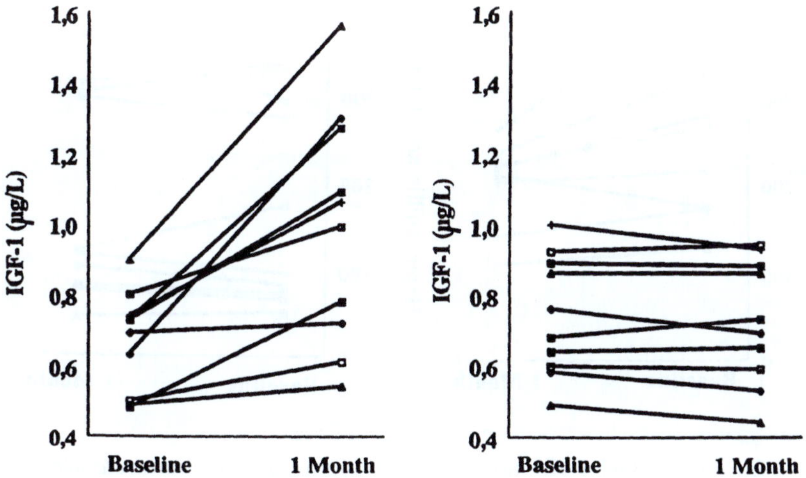

Fig. 3. Effects of GH (left) and placebo (right) on the IGF-1 concentration in the CSF in GH-deficient adults (ten per group). With the kind permission of *Neuroendocrinology* (65).

GH and neurotransmitters in the CSF

In our study, the CSF dopamine metabolite homovanillic acid (HVA) concentration decreased after one month of GH treatment in adults with GHD (**Fig. 4**) (65). This fall in the CSF HVA concentration indicates that GH affects the dopamine turnover in the CNS. A similar decline in CSF HVA concentration was also noted during treatment with GH for nine months (66). In that study there was a higher CSF concentration of the excitatory amino acid aspartate in the GH treated group suggesting that the lowering of HVA could be mediated by the N-methyl-D-aspartate receptor, which has been reported to affect dopamine release. Simultaneously, an increase in CSF free T4 concentration was observed (66). The results relating to CSF monoamine metabolites in depression and during treatment with antidepressants are equivocal (110-112). Although GH-deficient adults often have symptoms similar to those seen in depressed patients, the mechanism behind these symptoms may not necessarily be the same. In animal studies, region-dependent changes in dopamine, noradrenaline, serotonin and 5-HIAA levels during GH treatment have previously been reported (63, 64). It is possible that similar region-dependent changes also occur during GH treatment in GH-deficient patients although these changes are not necessarily mirrored in the CSF.

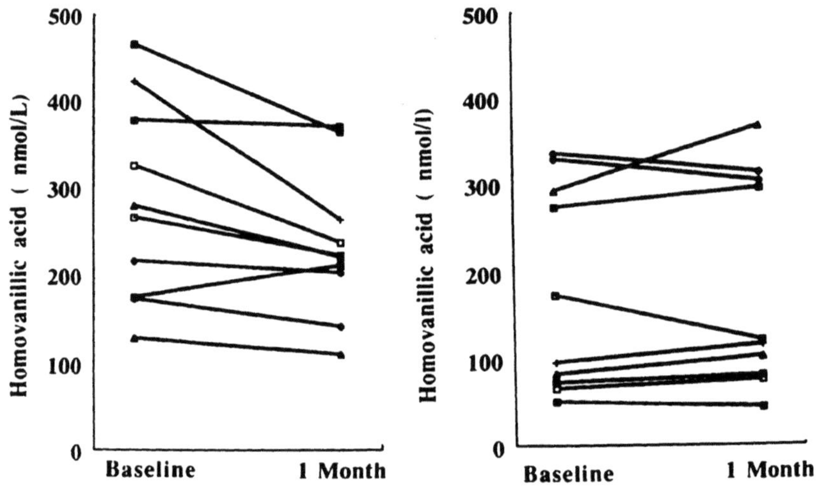

Fig. 4. Effects of GH (left) and placebo (right) on the homovanillic acid concentration in the CSF in GH-deficient adults (ten per group). With the kind permission of *Neuroendocrinology* (65).

A significant increase in CSF ß-endorphin immunoreactivity (i.r.) has been noted during GH treatment. Although the mean increase was limited (24.6% ± 7%), an increase was observed in all patients (**Fig. 5**) (65). ß-endorphin i.r. in the CSF decreases markedly after hypophysectomy (113) and hypopituitary patients may thus have a decreased concentration of CSF ß-endorphin i.r. The mechanism

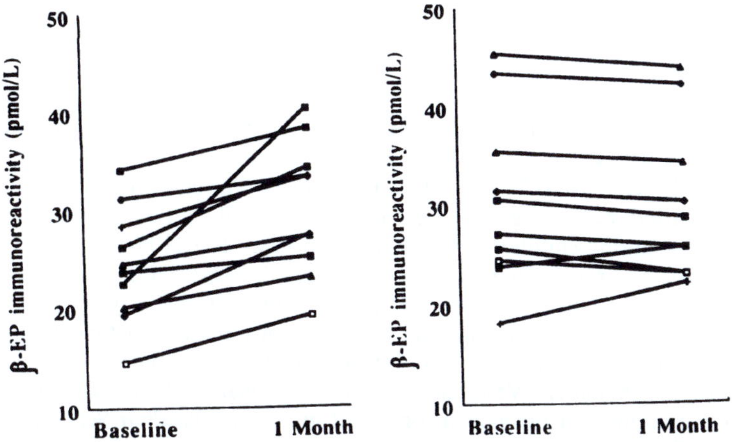

Fig. 5. Effects of GH (left) and placebo (right) on the β-endorphin (β-EP) immunoreactivity in the CSF in GH-deficient adults (ten per group). With the kind permission of *Neuroendocrinology* (65).

behind the increase in CSF β-endorphin i.r. is unclear, but it may be a direct effect by GH on cells producing peptides derived from pro-opiomelanocortin (POMC) or secondary to factors produced in the peripheral tissues that penetrate the blood-brain barrier. Recently, intracerebroventricular injections of GH to rats were found to increase the β-endorphin level in the pituitary and the levels of another opioid peptide in the spinal cord (114). Previously, it has been shown that opioid peptides are involved in the stimulation of GH release (115, 116) and affect the density of GH receptors in certain areas of the brain (117). Increased immunoreactivity of β-endorphin in plasma after exercise is well-documented and several studies support a role for β-endorphin in the emotional well-being that follows exercise (118, 119). The possibility that the increase in CSF β-endorphin immunoreactivity may be of importance for the improved psychological well-being noted during GH treatment in GH-deficient patients cannot be excluded.

Summary

Although GH is mainly synthesized in the pituitary there is also a widespread ectopic production in different areas of the brain. In addition, there is an abundant occurrence of GH receptors, IGF-1 and IGF-1 receptors. A clear understanding of the regulation and function of these factors and structures within the CNS has yet to be established. However, the possibility of an endogenous GH-IGF-1 axis involved in brain growth, maturation and function seems obvious. Furthermore, there are changes in CSF levels of GH, GH-dependent factors and neurotransmitters during GH treatment. Neuroendocrine mechanisms may very well be involved in the improvement in psychological well-being observed during GH treatment in GH-deficient adults.

References

1. Rotnem D, Genel M, Hintz RL, Cohen DJ. Personality development in children with growth hormone deficiency. J Am Acad Child Psychiatry 1977; 16:412-426.
2. Dean HJ, McTaggart TL, Fish DG, Friesen HG. The educational, vocational and marital status of growth hormone-deficient adults treated with growth hormone during childhood. Am J Dis Child 1985; 139:1105-1110.
3. Galatzer A, Aran O, Beit-Halachmi N, Nofar E, Rubitchek J, Pertzelan A et al. The impact of long-term therapy by a multidisciplinary team on the education, occupation and marital status of growt hormone deficient patients after termination of therapy. Clin Endocrinol 1987; 27:191-196.
4. Björk S, Jönsson B, Westphal O, Levin J-E. Quality of life of adults with growth hormone deficiency: a controlled study. Acta Paediatr Scand (Suppl) 1989; 356:55-59.
5. McGauley GA. Quality of life assessment before and after growth hormone treatment in adults with growth hormone deficiency. Acta Paediatr Scand (Suppl) 1989; 356:70-72.
6. Rosén T, Wirén L, Wilhelmsen L, Wiklund I, Bengtsson B-Å. Decreased psychological well-being in adult patients with growth hormone deficiency. Clin Endocrinol (Oxf) 1994; 40:111-116.
7. Takano K, Tanaka T, Saito T and the members of the committee for the study group of adult GH-deficiency. Psychosocial consequences of growth hormone deficiency (GHD) in adults. Clin Pediatr Endocrinol 1994; 3 (Suppl 5):79-85.
8. Rikken B, van Busschbach J, le Cessie S, Manten W, Spermon T, Grobbee R, Wit J-M. Impaired social status of growth hormone deficient adults as compared to controls with short or normal stature. Clin Endocrinol (Oxf) 1995;43:205-211.
9. Burman P, Broman JE, Hetta J, Wiklund I, Erfurth EM, Hägg E, Karlsson FA. Quality of life in adults with growth hormone (GH) deficiency: response to treatment with recombinant human GH in a placebo-controlled 21-month trial. J Clin Endocrinol Metab 1995;80:3585-3590.
10. Deijen JB, de Boer H, Blok GJ, van der Veen EA. Cognitive impairments and mood disturbances in growth hormone deficient men. Psychoneuroendocrinology 1996;21:312-322.
11. Raben MS. Clinical use of growth hormone. N Engl J Med 1962; 266:82-86.
12. Almqvist O, Thorén M, Sääf M, Eriksson O. Effects of growth hormone substitution on mental performance in adults with growth hormone deficiency: a pilot study. Psychoneuroendocrinology 1986; 11:347-352.
13. Bengtsson B-Å, Edén S, Lönn L, Kvist H, Stokland A, Lindstedt G, Bosaeus I, Tölli J, Sjöström L, Isaksson OGP. Treatment of adults with growth hormone (GH) deficiency with recombinant human GH. J Clin Endocrinol Metab 1993;76:309-317.
14. Whitehead HM, Boreham C, McIlrath EM, Sheridan B, Kennedy L, Atkinson AB, Hadden DR. Growth hormone treatment of adults with growth

hormone deficiency: results of a 13-month placebo controlled cross-over study. Clin Endocrinol 1992;36:45-52.
15. Mårdh G, Lundin K, Borg G, Jonsson B, Lindeberg A. Growth hormone replacement therapy in adult hypopituitary patients with growth hormone deficiency: combined data from 12 European placebo-controlled clinical trials. Endocrinol Metab 1994;1 (Suppl A):43-49.
16. Zamenhof S. Stimulation of the proliferation of neurons by the growth hormone: I. Experiments on tadpoles. Growth 1941;5:123-139.
17. Zamenhof S. Stimulation of cortical-cell proliferation by the growth hormone. III. Experiments on albino rats. Physiol Zool 1942;15:281-292.
18. Krieger DT. Pituitary hormones in the brain: what is their function ? Federation Proc 1980;39:2937-2941.
19. Hojvat S, Emanuele N, Baker G, Connick E, Kirsteins L, Lawrence AM. Growth hormone (GH), thyroid-stimulating hormone (TSH), and luteinizing hormone (LH)-like peptides in the rodent brain: non-parallel ontogenetic development with pituitary counterparts. Dev Brain Res 1982;4:427-434.
20. Pacold ST, Kirsteins L, Hojvat S, Lawrence AM. Biologically active pituitary hormones in the rat brain amygdaloid nucleus. Science 1978;199:804-806.
21. Van Houten M, Posner BI, Walsh RJ. Radioautographic identification of lactogen binding sites in rat median eminence using ^{125}I-human growth hormone: evidence for a prolactin "short-loop" feedback site. Exp Brain Res 1980;38:455-461.
22. Kyle CV, Evans MC, Odell WD. Growth hormone-like material in normal human tissues. J Clin Endocrinol Metab 1981;53:1138-1144.
23. Lechan RM, Nestler JL, Molitch ME. Immunohistochemical identification of a novel substance with human growth hormone-like immunoreactivity in rat brain. Endocrinology 1981;109:1950-1962.
24. Hojvat S, Baker G, Kirsteins L, Lawrence AM. Growth hormone (GH) immunoreactivity in the rodent and primate CNS: distribution, characterization and presence posthypophysectomy. Brain Res 1982;239:543-557.
25. Lechan RM, Molitch ME, Jackson IMD. Distribution of immunoreactive human growth hormone-like material and thyrotropin-releasing hormone in the rat central nervous system: evidence for their coexistence in the same neurons. Endocrinology 1983;112:877-884.
26. Harvey S, Hull KL, Fraser RA. Growth hormone: neurocrine and neuroendocrine perspectives. Growth Regul 1993;3:161-171.
27. Gossard F, Dihl F, Pelletier G, Dubois PM, Morel G. In situ hybridization to rat brain and pituitary gland of growth hormone cDNA. Neurosci Lett 1987;79:251-256.
28. Pelton EW, Grindeland RE, Young E, Bass NH. Effects of immunologically induced growth hormone deficiency on myelinogenesis in developing rat cerebrum. Neurology 1977;27:282-288.
29. Noguchi T, Sugisaki T, Takamatsu K, Tsukada Y. Factors contributing to the poor myelination in the brain of the snell dwarf mouse. J Neurochem 1982;39:1693-1699.

30. Noguchi T, Sugisaki T. Stimulatory effects of growth hormone and thyroxine on the concentration of gangliosides in the snell dwarf cerebrum. J Neurochem 1986;47:1785-1792.
31. Noguchi T, Sugisaki T, Watanabe M, Kohsaka S, Tsukada Y. Effects of bovine growth hormone on the retarded cerebral development induced by neonatal hydrocortisone intoxication. J Neurochem 1982;38:246-256.
32. Zamenhof S, Mosley J, Schuller E. Stimulation of the proliferation of cortical neurons by prenatal treatment with growth hormone. Science 1966;152:1396-1397.
33. Block JB, Essman WB. Growth hormone administration during pregnancy: a behavioural difference in offspring rats. Nature 1965;205:1136-1137.
34. Sara VR, Lazarus L, Stuart MC, King T. Fetal brain growth: selective action by growth hormone. Science 1974;186:446-447.
35. Sara VR, Lazarus L. Prenatal action of growth hormone on brain and behaviour. Nature 1974;250:257-258.
36. Rogers LJ, Schanberg SM, Fellows RE. Growth and lactogenic hormone stimulation of ornithine decarboxylase in neonatal rat brain. Endocrinology 1974;95:904-911.
37. Chen L, Lund PK, Burgess SB, Rudisch BE, McIlwain DL. Growth hormone, insulin-like growth factor I, and motoneuron size. J Neurobiol 1997;32:202-212.
38. Berti-Mattera LN, Gómez CJ, Krawiec L. Effects of bovine growth hormone on RNA synthesis in brain and liver of neonatal hypothyroid rats. Horm Metab Res 1983;15:286-289.
39. Minami S, Kamegai J, Sugihara H, Hasegawa O, Wakabayashi I. Systemic administration of recombinant human growth hormone induces expression of the *c-fos* gene in the hypothalamic arcuate and periventricular nuclei in hypophysectomized rats. Endocrinology 1992;131:247-253.
40. Kamegai J, Minami S, Sugihara H, Higuchi H, Wakabayashi I. Growth hormone induces expression of the *c-fos* gene on hypothalamic neuropeptide-Y and somatostatin neurons in hypophysectomized rats. Endocrinology 1994;135:2765-2771.
41. Schneider-Rivas S, Rivas-Arancibia S, Vázquez-Pereyra F, Vázquez-Sandoval R, Borgonio-Pérez G. Modulation of long-term memory and extinction responses induced by growth hormone (GH) and growth hormone releasing hormone (GHRH) in rats. Life Sci 1995;56:433-441.
42. Zhang J, Obál Jr. F, Fang J, Collins BJ, Krueger JM. Non-rapid eye movement sleep is supressed in transgenic mice with a deficiency in the somatotropic system. Neurosci Lett 1996;220:97-100.
43. Pollitt E, Money J. Studies in the psychology of dwarfism. I. Intelligence quotient and school achievement. J Pediatr 1964;64:415-421.
44. Laron Z, Galatzer A. Growth hormone, somatomedin and prolactin-relationship to brain function. Brain Dev 1985;7:559-567.
45. Rosenbloom AL, Guevara-Aguirre J, Rosenfeld RG, Fielder PJ. The little women of Loja - Growth hormone-receptor deficiency in an inbred population of southern Ecuador. N Engl J Med 1990;323:1367-1374.
46. Guevara-Aguirre J, Rosenbloom AL, Vaccarello MA, Fielder PJ, de la Vega A, Diamond Jr FB, Rosenfeld RG. Growth hormone receptor deficiency

(Laron Syndrome): clinical and genetic characteristics. Acta Paediatr Scand (Suppl) 1991;377:96-103.

47. Rosenfeld RG, Rosenbloom AL, Guevara-Aguirre J. Growth hormone (GH) insensitivity due to primary GH receptor deficiency. Endocr Rev 1994;15:369-390.

48. Di Carlo R, Muccioli G, Bellussi G, Pagnini G, Papotti M, Bussolati G. Lactogenic binding sites in the brain: regional distribution, species variation and characterization. Adv Biosci 1984;48:303-308.

49. Lai Z, Emtner M, Roos P, Nyberg F. Characterization of putative growth hormone receptors in human choroid plexus. Brain Res 1991;546:222-226.

50. Lai Z, Roos P, Zhai Q, Olsson Y, Fhölenhag K, Larsson C, Nyberg F. Age-related reduction of human growth hormone-binding sites in the human brain. Brain Res 1993;621:260-266.

51. Lobie PE, García-Aragón J, Lincoln DT, Barnard R, Wilcox JN, Waters MJ. Localization and ontogeny of growth hormone receptor gene expression in the central nervous system. Dev Brain Res 1993;74:225-233.

52. Mustafa A, Nyberg F, Bogdanovic N, Islam A, Roos P, Adem A. Somatogenic and lactogenic binding sites in rat brain and liver: quantitative autoradiographic localization. Neurosci Res 1994;20:257-263.

53. Nyberg F, Burman P. Growth hormone and its receptors in the central nervous system-location and functional significance. Horm Res 1996;45:18-22.

54. Mathews LS, Enberg B, Norstedt G. Regulation of rat growth hormone receptor gene expression. J Biol Chem 1989;264:9905-9910.

55. Burton KA, Kabigting EB, Clifton DK, Steiner RA. Growth hormone receptor messenger ribonucleic acid distribution in the adult male rat brain and its colocalization in hypothalamic somatostatin neurons. Endocrinology 1992;131:958-963.

56. Ilkbahar YN, Wu K, Thordarson G, Talamantes F. Expression and distribution of messenger ribonucleic acids for growth hormone (GH) receptor and GH-binding protein in mice during pregnancy. Endocrinology 1995;136:386-392.

57. Lobie PE, Garcia-Aragon J, Wang BS, Baumbach WR, Waters MJ. Cellular localization of the growth hormone binding protein in the rat. Endocrinology 1992;130:3057-3065.

58. Lobie PE, Ronsin B, Silvennoinen O, Haldorsén L-A, Norstedt G, Morel G. Constitutive nuclear localization of janus kinases 1 and 2. Endocrinology 1996;137:4037-4045.

59. Argetsinger LS, Carter-Su C. Mechanism of signaling by growth hormone receptor. Physiol Rev 1996;76:1089-1107.

60. Linfoot JA, Garcia JF, Wei W, Fink R, Sarin R, Born JL, Lawrence JH. Human growth hormone levels in cerebrospinal fluid. J Clin Endocrinol Metab 1970;31:230-232.

61. Belchetz PE, Ridley RM, Baker HF. Studies on the accessibility of prolactin and growth hormone to brain: effect of opiate agonists on hormone levels in serial, simultaneous plasma and cerebrospinal fluid samples in the rhesus monkey. Brain Res 1982;239:310-314.

62. Stern WC, Miller M, Resnick O, Morgane PJ. Distribution of ^{125}I-labelled rat growth hormone in regional brain areas and peripheral tissue of the rat. Am J Anat 1975;144:503-508.
63. Stern WC, Miller M, Jalowiec JE, Forbes WB, Morgane PJ. Effects of growth hormone on brain biogenic amine levels. Pharmacol Biochem Behav 1975;3:1115-1118.
64. Andersson K, Fuxe K, Eneroth P, Isaksson O, Nyberg F, Roos P. Rat growth hormone and hypothalamic catecholamine nerve terminal systems. Evidence for rapid and discrete reductions in dopamine and noradrenaline levels and turnover in the median eminence of the hypophysectomized male rat. Eur J Pharmacol 1983;95:271-275.
65. Johansson J-O, Larson G, Andersson M, Elmgren A, Hynsjö L, Lindahl A, Lundberg P-A, Isaksson OGP, Lindstedt S, Bengtsson B-Å. Treatment of growth hormone-deficient adults with recombinant human growth hormone increases the concentration of growth hormone in the cerebrospinal fluid and affects neurotransmitters. Neuroendocrinology 1995; 61:57-66.
66. Burman P, Hetta J, Wide L, Månsson J-E, Ekman R, Karlsson FA. Growth hormone treatment affects brain neurotransmitters and thyroxine. Clin Endocrinol (Oxf) 1996;44:319-324.
67. Tomas FM, Knowles SE, Owens PC, Read LC, Chandler CS, Gargosky SE, Ballard FJ. Effects of full-length and truncated insulin-like growth factor-I on nitrogen balance and muscle protein metabolism in nitrogen-restricted rats. J Endocrinol 1991;128:97-105.
68. Yamamoto H, Murphy LJ. Generation of des-(1-3) insulin-like growth factor-I in serum by an acid protease. Endocrinology 1994;135:2432-2439.
69. Sara VR, Hall K, von Holtz H, Humbel R, Sjögren B, Wetterberg L. Evidence for the presence of specific receptors for insulin-like growth factors 1 (IGF-1) and 2 (IGF-2) and insulin throughout the adult human brain. Neurosci Lett 1982;34:39-44.
70. Carlsson-Skwirut C, Jörnvall H, Holmgren A, Andersson C, Bergman T, Lundquist G, Sjögren B, Sara VR. Isolation and characterization of variant IGF-1 as well as IGF-2 from adult human brain. FEBS Lett 1986;201:46-50.
71. Adem A, Jossan SS, d'Argy R, Gillberg PG, Nordberg A, Winblad B, Sara V. Insulin-like growth factor 1 (IGF-1) receptors in the human brain: quantitative autoradiographic localization. Brain Res 1989;503:299-303.
72. Bondy CA, Lee W-H. Patterns of insulin-like growth factor and IGF receptor gene expression in the brain. Functional implications. Ann NY Acad Sci 1993;692:33-43.
73. de Keyser J, Wilczak N, de Backer J-P, Herroelen L, Vauquelin G. Insulin-like growth factor-I receptors in human brain and pituitary gland: an autoradiographic study. Synapse 1994;17:196-202.
74. Lesniak MA, Hill JM, Kiess W, Rojeski M, Pert CB, Roth J. Receptors for insulin-like growth factors I and II: Autoradiographic localization in rat brain and comparison to receptors for insulin. Endocrinology 1988;123:2089-2099.
75. Rotwein P, Burgess SK, Milbrandt JD, Krause JE. Differential expression of insulin-like growth factor genes in rat central nervous system. . Proc Natl Acad Sci USA, 1988;85:265-269.

76. Bondy CA. Transient IGF-I gene expression during the maturation of functionally related central projection neurons. J Neurosci 1991;11:3442-3455.
77. Aguado F, Sánchez-Franco F, Rodrigo J, Cacicedo L, Martínez-Murillo R. Insulin-like growth factor I-immunoreactive peptide in adult human cerebellar purkinje cells: co-localization with low-affinity nerve growth factor receptor. Neuroscience 1994;59:641-650.
78. Bach MA, Shen-Orr Z, Lowe Jr. WL, Roberts Jr. CT, LeRoith D. Insulin-like growth factor I mRNA levels are developmentally regulated in specific regions of the rat brain. Mol Brain Res 1991;10:43-48.
79. Stylianopoulou F, Herbert J, Soares MB, Efstratiadis A. Expression of the insulin-like growth factor II gene in the choroid plexus and the leptomeninges of the adult rat central nervous system. Proc Natl Acad Sci USA 1988;85:141-145.
80. Haselbacher G, Humbel R. Evidence for two species of insulin-like growth factor II (IGF II and "big"IGF II) in human spinal fluid. Endocrinology 1982;110:1822-1824.
81. Nilsson C, Mørck Hultberg B, Gammeltoft S. Autocrine role of insulin-like growth factor II secretion by the rat choroid plexus. Eur J Neurosci 1996;8:629-635.
82. Bäckström M, Hall K, Sara V. Somatomedin levels in cerebrospinal fluid from adults with pituitary disorders. Acta Endocrinol (Copenh) 1984;107:171-178.
83. Jones JI, Clemmons DR. Insulin-like growth factors and their binding proteins: biological actions. Endocr Rev 1995;16:3-34.
84. Aizenman Y, de Vellis J. Brain neurons develop in a serum and glial free environment: effects of transferrin, insulin, insulin-like growth factor-I and thyroid hormone on neuronal survival, growth and differentiation. Brain Res 1987;406:32-42.
85. DiCicco-Bloom E, Black IB. Insulin growth factors regulate the mitotic cycle in cultured rat sympathetic neuroblasts. Proc Natl Acad Sci USA 1988;85:4066-4070.
86. Caroni P, Grandes P. Nerve sprouting in innervated adult skeletal muscle induced by exposure to elevated levels of insulin-like growth factors. J Cell Biol 1990;110:1307-1317.
87. Bondy C, Werner H, Roberts Jr. CT, LeRoith D. Cellular pattern of type-I insulin-like growth factor receptor gene expression during maturation of the rat brain: comparison with insulin-like growth factors I and II. Neuroscience 1992;46:909-923.
88. Sara VR, Carlsson-Skwirut C. The role of the insulin-like growth factors in the regulation of brain development. Prog Brai Res 1988;73:87-99.
89. McMorris FA, Smith TM, DeSalvo S, Furlanetto RW. Insulin-like growth factor I/somatomedin C: a potent inducer of oligodendrocyte development. Proc Natl Acad Sci USA, 1986;83:822-826.
90. Carson MJ, Behringer RR, Brinster RL, McMorris FA. Insulin-like growth factor I increases brain growth and central nervous system myelination in transgenic mice. Neuron 1993;10:729-740.

91. Ye P, Carson J, D'Ercole AJ. In vivo actions of insulin-like growth factor-I (IGF-I) on brain myelination: studies of IGF-I and IGF binding protein-1 (IGFBP-1) transgenic mice. J Neurosci 1995;15:7344-7356.
92. Werner H, Raizada MK, Mudd LM, Foyt HL, Simpson IA, Roberts Jr. CT, LeRoith D. Regulation of rat brain/HepG2 glucose transporter gene expression by insulin and insulin-like growth factor-I in primary cultures of neuronal and glial cells. Endocrinology 1989;125:314-320.
93. D'Ercole AJ, Dai Z, Xing Y, Boney C, Wilkie MB, Lauder JM, Han VKM, Clemmons DR. Brain growth retardation due to the expression of human insulin-like growth factor binding protein-1 in transgenic mice: an in vivo model for the analysis of igf function in the brain. Dev Brain Res 1994;82:213-222.
94. Yamaguchi F, Itano T, Miyamoto O, Janjua NA, Ohmoto T, Hosokawa K, Hatase O. Increase of extracellular insulin-like growth factor I (IGF-I) concentration following electrolytical lesion in rat hippocampus. Neurosci Lett 1991;128:273-276.
95. Garcia-Estrada J, Garcia-Segura LM, Torres-Aleman I. Expression of insulin-like growth factor I by astrocytes in response to injury. Brain Res 1992;592:343-347.
96. Gluckman P, Klempt N, Guan J, Mallard C, Sirimanne E, Dragunow M, Klempt M, Singh K, Williams C, Nicolics K. A role for IGF-1 in the rescue of CNS neurons following hypoxic-ischemic injury. Biochem Biophys Res Comm 1992;182:593-599.
97. Gluckman PD, Guan J, Beilharz EJ, Klempt ND, Klempt M, Miller O, Sirimanne E, Dragunow M, Williams CE. The role of the insulin-like growth factor system in neuronal rescue. Ann NY Acad Sci 1993;692:138-148.
98. Johnston BM, Mallard EC, Williams CE, Gluckman PD. Insulin-like growth factor-1 is a potent neuronal rescue agent after hypoxic-ischemic injury in fetal lambs. J clin Invest 1996;97:300-308.
99. Bergstedt K, Wieloch T. Changes in insulin-like growth factor 1 receptor density after transient cerebral ischemia in the rat. Lack of protection against ischemic brain damage following injection of insulin-like growth factor 1. J Cereb Blood Flow Metab 1993;13:895-898.
100. Ocrant I. Insulin-like growth factor binding proteins in nervous-tissue-derived cells. Ann NY Acad Sci 1993;692:44-50.
101. Guan J, Williams CE, Skinner SJM, Mallard EC, Gluckman PD. The effects of insulin-like growth factor (IGF)-1, IGF-2, and des-IGF-1 on neuronal loss after hypoxic-ischemic brain injury in adult rats: evidence for a role for IGF binding proteins. Endocrinology 1996;137:893-898.
102. Kirschner PB, Henshaw R, Weise J, Trubetskoy V, Finklestein S, Schulz JB, Beal MF. Basic fibroblast growth factor protects against excitotoxicity and chemical hypoxia in both neonatal and adult rats. J Cereb Blood Flow Metab 1995;15:619-623.
103. Cheng B, Mattson MP. NGF and bFGF protect rat hippocampal and human cortical neurons against hypoglycemic damage by stabilizing calcium homeostasis. Neuron 1991;7:1031-1041.

104. Reinhardt RR, Bondy CA. Insulin-like growth factors cross the blood-brain barrier. Endocrinology 1994;135:1753-1761.
105. Hodgkinson SC, Spencer GSG, Bass JJ, Davis SR, Gluckman PD. Distribution of circulating insulin-like growth factor-I (IGF-I) into tissues. Endocrinology 1991;129:2085-2093.
106. Pardridge WM. Transport of insulin-related peptides and glucose across the blood-brain barrier. Ann NY Acad Sci 1993;692:126-137.
107. Van Wyk JJ. The somatomedins: biological actions and physiologic control mechanisms. In: Li CH (Ed). Hormonal proteins and peptides: growth factors. Academic press, New York. 1984;12:81-125.
108. Hynes MA, Van Wyk JJ, Brooks PJ, D'Ercole AJ, Jansen M, Lund PK. Growth hormone dependence of somatomedin-C/insulin-like growth factor-I and insulin-like growth factor-II messenger ribonucleic acids. Mol Endocrinology 1987;1:233-242.
109. López-Fernández J, Sánchez-Franco F, Velasco B, Tolón RM, Pazos F, Cacicedo L Growth hormone induces somatostatin and insulin-like growth factor I gene expression in the cerebral hemispheres of aging rats. Endocrinology 1996;137:4384-4391.
110. Ågren H, Mefford IN, Rudorfer MV, Linnoila M, Potter WZ. Interacting neurotransmitter systems. A non-experimental approach to the 5HIAA-HVA correlation in human CSF. J Psychiatr Res 1986;20:175-193.
111. Risby ED, Hsiao JK, Sunderland T, Ågren H, Rudorfer MV, Potter WZ. The effects of antidepressants on the cerebrospinal fluid homovanillic acid/5-hydroxyindoleacetic acid ratio. Clin Pharmacol Ther 1987:42:547-554.
112. Reddy PL, Khanna S, Subhash MN, Channabasavanna SM, Rao BSSR. CSF amine metabolites in depression. Biol Psychiatry 1992;31:112-118.
113. Schlachter LB, Wardlaw SL, Tindall GT, Frantz AG. Persistence of ß-endorphin in human cerebrospinal fluid after hypophysectomy. J Clin Endocrinol Metab 1983;57:221-224.
114. Johansson P, Ray A, Lai Z, Zhai Q-Z, Roos P, Nyberg F. Effects of intracerebroventricular injections of growth hormone on opioid peptides in the male rat. Analgesia 1995;1;4-6:481-485.
115. Bruhn TO, Tresco PA, Mueller GP, Jackson IMD. Beta-endorphin mediates clonidine stimulated growth hormone release. Neuroendocrinology 1989;50:460-463.
116. Delitala G, Tomasi PA, Palermo M, Ross RJM, Grossman A, Besser GM. Opioids stimulate growth hormone (GH) release in man independently of GH-releasing hormone. J Clin Endocrinol Metab 1989;69:356-358.
117. Zhai QZ, Lai Z, Yukhananov R, Roos P, Nyberg F. Decreased binding of growth hormone in the rat hypothalamus and choroid plexus following morphine treatment. Neurosci Lett 1995;184:82-85.
118. Janal MN, Colt EWD, Clark WC, Glusman M. Pain sensitivity, mood and plasma endocrine levels in man following long-distance running: effects of Naloxone. Pain 1984;19:13-25.
119. Morgan WP. Affective beneficence of vigorous physical activity. Med Sci Sports Exerc 1985;17:94-100.

18 DOSE TITRATION AND MONITORING GH TREATMENT IN THE ADULT.

SJ Howell, MRCP, A Rahim MRCP
and SM Shalet, MD FRCP

Department of Endocrinology, Christie Hospital NHS Trust, Withington, Manchester, U.K.

INTRODUCTION

Increasingly, growth hormone (GH) replacement is being offered to adults with severe GH deficiency (GHD). Initial reports of the benefits of GH therapy (23, 35) in the adult GHD population were based on trials using doses of GH comparable to those which had been used for years in paediatric practice. Whilst most studies demonstrated benefits of GH therapy, two observations suggested that in adults these doses were supraphysiological. Firstly, compared with children, the incidence of side effects in adults was greater and often responded to a dose reduction. Salomon et al (35) reported side effects of fluid retention, arthralgia or myalgia in more than half of the patients they treated with 0.07 IU/kg/day of GH. Furthermore a review by De Boer (10) of several studies published between 1989 and 1993 demonstrated that 61% of patients treated with GH at a dose of above 2.5 IU/ m^2/day reported side effects. Secondly a proportion of patients had IGF-1 levels above the normal range suggesting over-treatment. In 1993 Moller et al (27) demonstrated a dose-dependent rise in IGF-1 levels during treatment of GHD adults, with markedly elevated levels following 4 IU/ m^2/day of GH. More recent studies have confirmed raised IGF-1 levels in patients receiving high dose GH replacement (8, 11, 27).

Compared with the pioneering early adult GH replacement studies (23, 35), more recent trials of GH replacement in adults have often used lower doses. This has been associated with a reduction in the incidence of side effects. Despite the reduction in GH dose, significant improvements in GH-dependent parameters have still been observed (2, 29) with a reduction in fat mass following treatment with 1 IU/ m^2/day (27), 1.7 IU/day (30), and a variable dose adjusted to maintain the IGF-1 level within the normal range (13). Doses as low as 1.25 IU/day have been shown to increase markers of bone turnover (40) with no further demonstrable increase at a higher dose of 5 IU/day. Significant increases in bone mineral density (BMD) after 18 months (4) and 2 years of treatment (22) have also been observed with doses of 4 μg/kg/day and 7.8 μg/kg/day (0.01 IU/kg/day and 0.02 IU/kg/day) respectively.

A further problem with the doses of GH which were used in the initial adult studies was that they were calculated simply according to either body weight or surface area. This did not allow for the great inter-individual variability that is seen in the response to GH therapy in adults or for the possibility that the required dose may alter with time. Holmes et al (20) reported that side effects of treatment occurred more commonly in certain patients. Increasing age and weight, a peak GH response to provocative testing of greater than 1 mU/l, and a greater increment in IGF-1 during therapy were all associated with a higher incidence of side effects. Side effects were also more common in patients with adult onset GH deficiency compared with those who became GH deficient in childhood. This was confirmed by Attanasio et al (3), who also demonstrated a number of other differences between patients with adult and childhood onset GH deficiency, emphasising the heterogeneity of GH deficient adults.

The response to GH therapy of certain GH-dependent parameters also varies between individuals. Johannsson et al showed that changes in BMD in response to GH therapy were dependent on pretreatment BMD (13), as patients with significant pretreatment osteopaenia (i.e. a z-score < -1) achieved a greater increase in BMD compared with those with a z-score of > -1. Furthermore, improvements in body composition were greater in young, male patients with low pre-treatment GH binding protein levels (16). There is also evidence that there is a reduced requirement for GH with increasing age. A recent dose finding study in elderly GH deficient patients (39) reported that the ideal replacement dose in older patients is considerably less than that used in younger GH deficient adults. This is not surprising as GH secretion in healthy elderly adults is reduced compared with that in younger subjects (14), declining by approximately 14% per decade from the third decade of life (25). Furthermore, Johannsson et al (22) reported no change in IGF-1 despite a gradual reduction in mean GH dose of 20% over 18 months, suggesting that physiological GH requirements may decrease with time.

The dose of GH replacement therefore needs to be tailored to each patient individually, with monitoring continuing even after an optimal replacement dose has been achieved, as dose adjustments may still need to be made. To deliver therapy in this manner several questions need to be answered: What dose of GH should you start with? How do you assess whether a patient is receiving physiological GH replacement and how should you monitor patients in the long term?

DOSE SELECTION AND TITRATION

In view of the difficulty in predicting the ideal physiological replacement dose for a GH deficient adult, a reasonable strategy would be to start with a very low dose in all patients, and gradually increase this until an optimum replacement dose is reached. This should avoid over-replacement in patients whose GH requirements are relatively low and therefore reduce the incidence of side effects. At a recent consensus meeting in Australia, an arbitrary starting dose of 0.15 - 0.3 mg/day (0.45 - 0.9 IU/day) was agreed upon. This should then be increased at intervals of no less than one month on the basis of clinical and biochemical response.

Determining when an optimal replacement dose has been reached remains difficult. The ideal GH replacement dose is the one that corrects all the abnormalities associated with adult GH deficiency without causing any adverse effects. During dose titration, measurement of changes in the endpoints of therapy and the occurrence of side effects of treatment should ideally be used to guide replacement. In practice the monitoring of clinical response is supplemented by biochemical monitoring utilising the measurement of markers of GH status, such as IGF-1, to determine that physiological levels of GH have not been exceeded.

Measureable changes occurring after the institution of GH therapy include improvements in quality of life, bone mineral density and body composition. The usefulness of assessing alterations of these biological endpoints during the initial period of dose titration will depend to some extent on their rate of change. Significant improvements in BMD have only been demonstrated after at least 12 months of therapy in adult onset GHD, hence changes in this parameter do not help in the process of dose titration. Improvements in energy, mood and concentration, however, often occur in the first few weeks of treatment and can be quite marked (7, 26). Disease specific questionnaires for patients with GHD have been developed and validated. These may be used to assess these changes, although simple history-taking, ideally from both the patient and their partner, will often provide a good indication of the response to treatment. As improvement in psychological well-being occurs relatively quickly in response to GH therapy, it may provide useful guidance to the ideal replacement dose for an individual patient.

Changes in body composition may also occur rapidly. De Boer et al (9) demonstrated a reduction in extracellular water (ECW) in GH deficient adults based on a discrepancy between the expected bioimpedence, calculated by measurements of arm muscle area and skin fold thickness, and that assessed by a portable bioimpedance analyser (BIA). This finding was confirmed by Rosen et al (33) using an isotope dilution method and a four compartment model. De Boer et al (11) went on to show that GH deficient adults were relatively dehydrated, and in addition GH replacement rapidly increased extracellular water in a dose-dependent manner such that GH doses of above 2 IU/m^2/day resulted in overhydration. Normalisation of ECW may therefore be used as an indicator of optimal GH dose, and may thus be helpful in dose titration.

Several markers of GH status have been evaluated for their usefulness in assessing GH replacement. De Boer et al (12) studied insulin-like growth factor 1 (IGF-1), insulin-like growth factor binding protein 3 (IGFBP-3) and acid labile subunit (ALS) during GH therapy administered at several different doses in 46 young GH deficient adults (mean age 28.4 years). They reported a reduction in all three markers in most patients prior to treatment with GH, followed by a significant increase with GH replacement. The GH dose required to elevate mean IGF-1 levels into the normal range was lower than that needed to normalise IGFBP-3 or ALS concentrations. In addition, IGF-1 was the only marker to exceed the normal range during supraphysiological doses of GH. Several studies have demonstrated IGF-1 levels greater than 2 standard deviations (SD) above the mean in patients receiving high dose GH replacement, with higher IGF-1 values associated with an increased incidence of side effects (8, 11, 27). IGF-1, however, is not a reliable marker of under-replacement. Hoffman et al (18) found that 70% of patients with adult-onset GH deficiency had IGF-1 values within 2 SD of the mean of an age-matched normal population, whilst Holmes (19) found this figure to be 30% in a cohort of adults with GH deficiency of both adult- and childhood onset. Recent data (38) have also shown that with increasing age IGF-1 levels of normal and GH deficient patients demonstrate even more overlap, with 20 of a group of 24 elderly GH deficient patients with organic pituitary disease having normal IGF-1 levels. The relationship between changes in serum markers and alterations in the end-points of GH therapy have also been examined. Thoren et al (37) found a positive correlation between IGF-1 and beneficial changes in body composition following GH therapy, whilst no similar correlation existed with IGFBP-3 or ALS. Using BIA, De Boer et al (11) found that normalisation of ECW subsequent to GH therapy, was associated with normal IGF-1 levels, while patients who became overhydrated had abnormally high IGF-1 levels.

Thus, IGF-1 increases in a dose dependent manner following the introduction of GH therapy, reaches abnormally high values when supraphysiological doses of GH are administered and positively correlates with one of the end-points of therapy (body composition change). It would therefore appear to be an excellent marker of physiological GH replacement and in particular is a good marker of over-replacement. The use of IGF-1 as a marker of GH replacement is however limited by its inability to reliably identify under-treatment, as IGF-1 levels may be within the normal range prior to treatment in GH deficient adults. Thus when monitoring GH replacement a normal IGF-1 level does not exclude under-replacement, particularly in elderly patients. The inability of ALS and IGFBP-3 to distinguish between adequate and supraphysiological GH doses will limit their use in assessing optimal replacement doses.

Common acute side effects of GH treatment include fluid retention, myalgia and arthralgia (8). All have been shown to resolve on either discontinuation of treatment or dose reduction. Side effects are more common in older patients, those with adult-onset GH deficiency and those in whom a greater increment in IGF-1 occurs during GH replacement (20). Furthermore higher doses of GH are associated with a higher incidence of side effects (10). The occurrence of side effects is therefore a good indicator of over-treatment with GH and is thus useful in dose titration.

MONITORING

Once a stable replacement dose of GH has been achieved, regular monitoring needs to be continued for several reasons. Firstly, it is likely that the dose of GH required will decrease with increasing age of the patient so further dose adjustments may need to be made. This is best assessed by six- to 12-monthly measurements of IGF-1 with alterations of the GH dose calculated to keep the levels within the normal range. Secondly, it is important to monitor the effects of GH therapy to ensure that it is beneficial. Improvements in different aspects of the GH deficiency syndrome occur at different rates following the start of GH replacement. Changes in quality of life will often occur in the first few weeks of therapy and should certainly be evident by the end of six months treatment. Significant changes in body composition occur after as little as six weeks of GH therapy (5) and are sustained after three to four years of treatment (1, 24). Improvements in BMD do not take place as rapidly. Studies of 6 to 12 months of GH replacement have shown either no change or a reduction in BMD (17, 36). However, if GH replacement is continued for 18 to 24 months there is a resultant increase in bone density (4, 22). Beneficial effects on serum lipids have been reported after only 2 months of therapy (34) although there is some evidence that these may not be sustained during long term treatment (16). A reduction in cardiovascular mortality subsequent to GH therapy has yet to be established and will require studies with much longer periods of replacement than are currently available.

Thus, any strategy for assessing the benefits of GH treatment will depend to some extent on the original indication for treatment, and assessment of different biological endpoints will need to be undertaken at different times. If GH is prescribed primarily to improve quality of life then re-evaluation at six months by interview and disease-specific questionnaire is adequate to provide evidence of efficacy. Body composition can be monitored simply by serial measurement of weight and waist-hip ratio, and may be supplemented by BIA or dual energy x-ray absorptimetry (DEXA). Changes should be apparent by six months, but reassessment at 12 months would provide a better indication of the impact of therapy. For evidence of the effect of treatment on BMD a period of at least 2 years of GH replacement is required to discern whether patients have derived any benefit. A baseline DEXA scan should ideally be performed before the start of treatment followed by a repeat measurement after 2 years. If treatment is commenced on the basis of potentially reducing cardiovascular mortality then therapy should be considered to be life-long and assessment of benefit can only be made as part of a large trial. Fasting lipids should be measured at the start of treatment and then at least once yearly.

Monitoring is also necessary to identify side effects of treatment. Symptoms related to fluid retention, such as oedema or carpal tunnel syndrome have been frequently reported, as have mild arthralgia and myalgia. These are often transient and dose dependent and appear to be markers of over-treatment. They are uncommon if therapy is commenced at a low dose and gradually increased, but if they do occur will usually respond to a dose reduction. Because of the effect of GH on insulin sensitivity there is a possibility that GH therapy may be complicated by hyperglycaemia. However, although there is some evidence of an increase in fasting blood glucose and insulin levels (6, 35), the development of fasting hyperglycaemia, or a significant increase in glycosylated haemoglobin subsequent to GH therapy, is rare. Furthermore, the changes in body composition during GH therapy, and in particular the reduction in truncal obesity, are likely to improve insulin sensitivity (15). Nonetheless, markers of glycaemia should be measured at intervals during GH replacement.

GH influences the metabolism and action of many substances. Coexisting therapy therefore needs to be monitored during GH replacement as dose adjustments may need to be made. GH replacement increases the peripheral conversion of T4 to

T3, so a decrease in the dose of concurrent treatment with thyroxine may be necessary. In addition, the use of T4 estimation alone in the monitoring of thyroxine replacement may be misleading. A decrease in total cortisol levels may occur following GH replacement, but there is evidence that this is due to a decrease in cortisol binding protein, with free cortisol levels remaining unaltered (31). Thus, changes in cortisol levels in patients on glucocorticoid replacement after the introduction of GH need to be interpreted with care.

The risk of colonic cancer is increased in acromegaly (32). Although it is inappropriate to directly extrapolate from this to physiological GH replacement in GH deficient adults, there is a theoretical risk of increased tumour growth during GH therapy. However, the evidence to date has demonstrated no increase in the incidence of new tumours or the likelihood of tumour recurrence (28) following the institution of GH replacement. Nevertheless, good clinical practice requires regular imaging of residual pituitary tumours and GH therapy should not affect this or any other standard cancer screening practice. Before the start of GH treatment, a baseline scan is recommended for all patients previously treated for a piuitary adenoma or other intracranial tumours.

SUMMARY

There are many potential benefits of treatment of GH deficient adults with GH replacement. Administering the correct, physiological dose of GH is important to maximise the benefit of therapy while reducing side effects to a minimum and ensuring that treatment is as cost-effective as possible. We recommend that patients be started on a dose of 0.15 - 0.3 mg/day. During the initial phase of dose titration, changes in GH dose should be made at intervals of 1 to 2 months on the basis of the clinical response, as demonstrated by improvements in well-being or the occurrence of side effects, and biochemical criteria. The dose should be adjusted to elevate or maintain the IGF-1 level to within the age-matched normal range, and IGF-1 levels of greater than 2 SD above the mean would be an indication for dose reduction. The use of BIA measurements may expedite the process of dose titration and will be particularly useful in patients whose IGF-1 levels are normal prior to treatment. The occurrence of side effects is usually an indication of over-replacement and should result in a dose reduction of 25 to 50%. However, the use of a low starting dose followed by a gradual increase will minimise the incidence of side effects.

Monitoring should continue throughout continued GH replacement at six to 12 month intervals. It should include the measurement of IGF-1 levels, fasting lipid levels and a marker of glycaemia. In addition, assessment of the efficacy of treatment can be undertaken. Quality of life is best assessed by interview and disease-specific questionnaires, while changes in body composition can be estimated by serial measurements of weight and waist:hip ratio. This can be supplemented by BIA and DEXA., the latter of which can also be used to assess changes in BMD. The strategy for monitoring the efficacy of treatment and for deciding the duration of therapy will depend on the indication for treatment. The dose of GH required to maintain physiological replacement will probably gradually reduce with increasing age and the dose should be adjusted to maintain the IGF-1 level within the normal range.

REFERENCES

1. Al-Shoumer-KA; Page-B; Thomas-E; Murphy-M; Beshyah-SA; Johnston-DG. Effects of four years' treatment with biosynthetic human growth hormone (GH) on body composition in GH-deficient hypopituitary adults. Eur-J-Endocrinol. 1996; 135(5): 559-67.

2. Amato-G; Carella-C; Fazio-S; La-Montagna-G; Cittadini-A; Sabatini-D; Marciano-Mone-C; Sacca-L; Bellastella-A. Body composition, bone metabolism, and heart structure and function in growth hormone (GH)-deficient adults before and after GH replacement therapy at low doses. J-Clin-Endocrinol-Metab. 1993; 77(6): 1671-6.

3. Attanasio-AF; Lamberts-SW; Matranga-AM; Birkett-MA; Bates-PC; Valk-NK; Hilsted-J; Bengtsson-BA; Strasburger-CJ. Adult growth hormone (GH)-deficient patients demonstrate heterogeneity between childhood onset and adult onset before and during human GH treatment. Adult Growth Hormone Deficiency Study Group. J-Clin-Endocrinol-Metab. 1997; 82(1): 82-8.

4. Baum-HB; Biller-BM; Finkelstein-JS; Cannistraro-KB; Oppenhein-DS; Schoenfeld-DA; Michel-TH; Wittink-H; Klibanski-A. Effects of physiologic growth hormone therapy on bone density and body composition in patients with adult-onset growth hormone deficiency. A randomized, placebo-controlled trial . Ann-Intern-Med. 1996; 125(11): 883-90.

5. Bengtsson-BA; Eden-S; Lonn-L; Kvist-H; Stokland-A; Lindstedt-G; Bosaeus-I; Tolli-J; Sjostrom-L; Isaksson-OG. Treatment of adults with growth hormone (GH) deficiency with recombinant human GH. J-Clin-Endocrinol-Metab. 1993; 76(2): 309-17.

6. Binnerts-A; Swart-GR; Wilson-JH; Hoogerbrugge-N; Pols-HA; Birkenhager-JC; Lamberts-SW. The effect of growth hormone administration in growth hormone deficient adults on bone, protein, carbohydrate and lipid homeostasis, as well as on body composition. Clin-Endocrinol-Oxf. 1992; 37(1): 79-87.

7. Burman-P; Broman-JE; Hetta-J; Wiklund-I; Erfurth-EM; Hagg-E; Karlsson-FA. Quality of life in adults with growth hormone (GH) deficiency: response to treatment with recombinant human GH in a placebo-controlled 21-month trial. J-Clin-Endocrinol-Metab. 1995; 80(12): 3585-90.

8. Chipman-JJ, Attanasio-AF, Birkett-MA, Bates-PC, Webb-S, Lamberts-SWJ. The safety profile of GH replacement therapy in adults. Clin-Endocrinol-Oxf. 1997; 46 (4): 473-481.

9. De-Boer-H; Blok-GJ; Voerman-HJ; De-Vries-PM; Van-der-Veen-EA. Body composition in adult growth hormone-deficient men, assessed by anthropometry and bioimpedance analysis. J-Clin-Endocrinol-Metab. 1992, 75(3): 833-7.

10. De Boer-H, Blok-GJ, Van Der Veen-EA. Clinical aspects of growth hormone deficiency in adults. Endocrine Reviews. 1995; 16 (1), 63-86.

11. De-Boer-H; Blok-GJ; Voerman-B; de-Vries-P; Popp-Snijders-C; van-der-Veen-E. The optimal growth hormone replacement dose in adults, derived from bioimpedance analysis. J-Clin-Endocrinol-Metab. 1995; 80(7): 2069-76.
12. De-Boer-H; Blok-GJ; Popp-Snijders-C; Stuurman-L; Baxter-RC; van-der-Veen-E. Monitoring of growth hormone replacement therapy in adults, based on measurement of serum markers. J-Clin-Endocrinol-Metab. 1996 ; 81(4): 1371-7.
13. De-Boer-H; Blok-GJ; Voerman-B; Derriks-P; van-der-Veen-E. Changes in subcutaneous and visceral fat mass during growth hormone replacement therapy in adult men. Int-J-Obes-Relat-Metab-Disord. 1996; 20(6): 580-7.
14. Finkelstein-JW; Roffwarg-HP; Boyar-RM; Kream-J; Hellman-L. Age-related change in the twenty-four-hour spontaneous secretion of growth hormone. J-Clin-Endocrinol-Metab. 1972 Nov; 35(5): 665-70
15. Fowelin-J; Attvall-S; Lager-I; Bengtsson-BA. Effects of treatment with recombinant human growth hormone on insulin sensitivity and glucose metabolism in adults with growth hormone deficiency. Metabolism. 1993; 42(11): 1443-7.
16. Garry-P; Collins-P; Devlin-JG. An open 36-month study of lipid changes with growth hormone in adults: lipid changes following replacement of growth hormone in adult acquired growth hormone deficiency. Eur-J-Endocrinol. 1996; 134(1): 61-6.
17. Hansen-TB; Brixen-K; Vahl-N; Jorgensen-JO; Christiansen-JS; Mosekilde-L; Hagen-C. Effects of 12 months of growth hormone (GH) treatment on calciotropic hormones, calcium homeostasis, and bone metabolism in adults with acquired GH deficiency: a double blind, randomized, placebo-controlled study. J-Clin-Endocrinol-Metab. 1996; 81(9): 3352-9.
18. Hoffman-DM; O'Sullivan-AJ; Baxter-RC; Ho-KK. Diagnosis of growth-hormone deficiency in adults. Garvan Institute of Medical Research, St Vincent's Hospital, Sydney, NSW, Australia. Lancet. 1994; 343(8905): 1064-8.
19. Holmes-SJ. Growth hormone replacement in adults with growth hormone deficiency. MD thesis - University of London. 1995.
20. Holmes-SJ; Shalet-SM. Which adults develop side-effects of growth hormone replacement? Clin-Endocrinol-Oxf. 1995 Aug; 43(2): 143-9
21. Johannsson-G; Bjarnason-R; Bramnert-M; Carlsson-LM; Degerblad-M; Manhem-P; Rosen-T; Thoren-M; Bengtsson-BA. The individual responsiveness to growth hormone (GH) treatment in GH-deficient adults is dependent on the level of GH-binding protein, body mass index, age, and gender. J-Clin-Endocrinol-Metab. 1996; 81(4): 1575-81.
22. Johannsson-G; Rosen-T; Bosaeus-I; Sjostrom-L; Bengtsson-BA. Two years of growth hormone (GH) treatment increases bone mineral content and density in hypopituitary patients with adult-onset GH deficiency. J-Clin-Endocrinol-Metab. 1996; 81(8): 2865-73.
23. Jorgensen-JO; Pedersen-SA; Thuesen-L; Jorgensen-J; Ingemann-Hansen-T; Skakkebaek-NE; Christiansen-JS. Beneficial effects of growth hormone treatment in GH-deficient adults. Lancet. 1989; 1(8649): 1221-5.
24. Jorgensen-JO; Thuesen-L; Muller-J; Ovesen-P; Skakkebaek-NE; Christiansen-JS. Three years of growth hormone treatment in growth hormone-deficient adults: near normalization of body composition and physical performance. Eur-J-Endocrinol. 1994; 130(3): 224-8.

25. Lizarralde-G; Veldhuis-JD. Age and relative adiposity are specific negative determinants of the frequency and amplitude of growth hormone (GH) secretory bursts and the half-life of endogenous GH in healthy men. Iranmanesh-A; J-Clin-Endocrinol-Metab. 1991 Nov; 73(5): 1081-8.
26. McGauley-GA. Quality of life assessment before and after growth hormone treatment in adults with growth hormone deficiency. Acta-Paediatr-Scand-Suppl. 1989; 356: 70-2.
27. Moller-J; Jorgensen-JO; Lauersen-T; Frystyk-J; Naeraa-RW; Orskov-H; Christiansen-JS. Growth hormone dose regimens in adult GH deficiency: effects on biochemical growth markers and metabolic parameters . Clin-Endocrinol-Oxf. 1993; 39(4): 403-8.
28. Ogilvy-Stuart-AL; Ryder-WD; Gattamaneni-HR; Clayton-PE; Shalet-SM. Growth hormone and tumour recurrence. BMJ. 1992; 304(6842): 1601-5.
29. O'Halloran-DJ; Tsatsoulis-A; Whitehouse-RW; Holmes-SJ; Adams-JE; Shalet-SM. Increased bone density after recombinant human growth hormone (GH) therapy in adults with isolated GH deficiency. J-Clin-Endocrinol-Metab. 1993; 76(5): 1344-8.
30. Orme-SM; Sebastian-JP; Oldroyd-B; Stewart-SP; Grant-PJ; Stickland-MH; Smith-MA; Belchetz-PE. Comparison of measures of body composition in a trial of low dose growth hormone replacement therapy. Clin-Endocrinol-Oxf. 1992; 37(5): 453-9.
31. Rodriguez-Arnao-J; Perry-L; Besser-GM; Ross-RJ. Growth hormone treatment in hypopituitary GH deficient adults reduces circulating cortisol levels during hydrocortisone replacement therapy. Clin-Endocrinol-Oxf. 1996 Jul; 45(1): 33-7
32. Ron-E; Gridley-G; Hrubec-Z; Page-W; Arora-S; Fraumeni-JF Jr. Acromegaly and gastrointestinal cancer. Cancer. 1991; 68(8): 1673-7
33. Rosen-T; Bosaeus-I; Tolli-J; Lindstedt-G; Bengtsson-BA. Increased body fat mass and decreased extracellular fluid volume in adults with growth hormone deficiency. Clin-Endocrinol-Oxf. 1993 38(1): 63-71.
34. Russell-Jones-DL; Watts-GF; Weissberger-A; Naoumova-R; Myers-J; Thompson-GR; Sonksen-PH. The effect of growth hormone replacement on serum lipids, lipoproteins, apolipoproteins and cholesterol precursors in adult growth hormone deficient patients. Clin-Endocrinol-Oxf. 1994; 41(3): 345-50.
35. Salomon-F; Cuneo-RC; Hesp-R; Sonksen-PH. The effects of treatment with recombinant human growth hormone on body composition and metabolism in adults with growth hormone deficiency. N-Engl-J-Med. 1989; 321(26): 1797-803
36. Sartorio-A; Ortolani-S; Conti-A; Cherubini-R; Galbiati-E; Faglia-G. Effects of recombinant growth hormone (GH) treatment on bone mineral density and body composition in adults with childhood onset growth hormone deficiency. J-Endocrinol-Invest. 1996 ; 19(8): 524-9.
37. Thoren-M; Hilding-A; Baxter-RC; Degerblad-M; Wivall-Helleryd-IL; Hall-K. Serum insulin-like growth factor I (IGF-I), IGF-binding protein-1 and -3, and the acid-labile subunit as serum markers of body composition during growth hormone (GH) therapy in adults with GH deficiency. J-Clin-Endocrinol-Metab. 1997; 82(1): 223-8.

38. Toogood-AA; O'Neill-PA; Shalet-SM. Beyond the somatopause: growth hormone deficiency in adults over the age of 60 years. J-Clin-Endocrinol-Metab. 1996; 81(2): 460-5.
39. Toogood-AA, Wieringa-GE, Adams-JE and Shalet-SM. Growth hormone replacement therapy in the elderly with organic growth hormone deficiency: Do they benefit? Abstract in 79th annual meeting of the Endocrine society,june11-14, 1997.
40. Wollmann-HA; Schonau-E; Blum-WF; Meyer-F; Kruse-K; Ranke-MB. Dose-dependent responses in insulin-like growth factors, insulin-like growth factor-binding protein-3 and parameters of bone metabolism to growth hormone therapy in young adults with growth hormone deficiency. Horm-Res. 1995; 43(6): 249-56.

19 SIDE EFFECTS OF GROWTH HORMONE TREATMENT

John P Monson

Department of Endocrinology, St Bartholomew's Hospital London EC1A 7BE, UK

INTRODUCTION

Human recombinant growth hormone (GH) replacement therapy is now an established component of the management of adult hypopituitarism [1,2] in addition to its traditional role in children with isolated GH or multiple pituitary hormone deficiencies. Apart from the tragic consequences for some recipients of cadaveric GH, replacement therapy in the child has not been associated with any significant adverse effects. In the adult, achieving physiological GH replacement might, at first sight, appear a relatively straightforward exercise to be undertaken in parallel with the established routine replacement of thyroid, glucocorticoid and sex hormones. However, the precision with which these hormones are replaced is far from perfect and this is particularly the case with glucocorticoid replacement for which there is no reliable marker of adequate tissue exposure [3]. The situation with GH replacement in the hypopituitary adult shares these difficulties and is further complicated by the fact that in most instances treatment is commenced against a background of prolonged deficiency which will have resulted in reduction in total body water in addition to various metabolic perturbations. Some of these secondary phenomena reverse rapidly with GH replacement and are responsible for the classical early treatment related symptoms consistent with fluid retention whereas others, including adverse changes in insulin sensitivity, may not reverse immediately and may paradoxically be exacerbated if the dose of GH employed is excessive. Recent experience employing lower doses of GH than were used in the original trials of adult GH replacement are remarkable for the paucity of adverse symptoms and events. For these reasons it is clear that consideration of the side effects of GH therapy must be undertaken in terms of the dose schedules employed.

In this chapter I shall review the important adverse events associated with GH replacement therapy in the hypopituitary adult focussing on those events occurring immediately after commencement of treatment and subsequent metabolic consequences. The potential effects on other hormone replacements and the question of tumour recurrence and neoplasia will also be considered.

Early Adverse Events During GH Replacement

The earliest trials of GH replacement therapy in adults utilised weight or surface area derived doses which were essentially an adaptation of paediatric practice [4-10]. Furthermore, the greater sensitivity of adults to GH and the wide interindividual variation in response was not fully appreciated and could not be incorporated into the design of those pioneering studies. The consequence was the potential for substantially supra-physiological doses, particularly in obese patients, and a very high incidence of initial side effects of therapy at least some of which were clearly related to the antinatriuretic effect of GH.

The most frequently described initial adverse symptom was oedema which affected between 50 and 100% of patients receiving GH in doses exceeding 0.4 IU/kg/week (2.4 IU/m^2/day) in six individual early studies [4,6,7,8,9,10]. The mechanism of oedema formation presumably includes renal sodium and water retention but a contributory metabolic effect similar to the phenomenon of refeeding oedema remains possible. Somewhat against the latter is the fact that the oedema responded rapidly to withdrawal of GH therapy, a feature consistent with a renal mechanism [11]. The mechanism of GH-induced antinatriuresis remains poorly defined. It is partly explained by activation of the renin-angiotensin system [11] although intact aldosterone secretion is not a prerequisite; it has been postulated that enhanced renal tubular sodium reabsorption may also contribute. The reported incidence of severe oedema on lower doses of GH is variable. Thus, Jørgensen et al described a single subject, out of a total study population of 22 patients treated with GH (0.33 IU/kg/week), who developed oedema and headache necessitating withdrawal of therapy [5]. In contrast, Mårdh et al [12], reporting combined data from 233 patients in 12 placebo controlled clinical trials using identical protocols and a lower GH dose of 0.25 IU/kg/week (maximum 4 IU/day), documented a prevalence of oedema of 37.4% in the active group as opposed to 3.4% in the placebo group (Table 1). It should be pointed out that the study population reported by Jørgensen et al were adults with childhood onset GH deficiency whereas that of Mårdh et al were predominantly adult onset. Furthermore, it is noteworthy that the earliest studies placed understandable emphasis on symptoms which were sufficient to necessitate a reduction in dose or cessation of therapy and might therefore have reflected different severity of symptoms compared with subsequent studies using lower GH doses.

Of the other early adverse symptoms described by Mårdh et al [12], arthralgia, myalgia, paraesthesia and carpal tunnel syndrome were substantially more common in GH treated than in placebo treated patients (Table 1). These symptoms were a uniform finding in early reports and were not associated with evidence of soft tissue or joint inflammation. They were usually self-limiting and it seems probable that that they also reflected the sodium and water retaining effects of GH replacement rather than a direct effect of GH on articular structures or myocytes. Interestingly,

GH replacement was not associated with an increased incidence of hypertension as might be predicted on the basis of its antinatriuretic effects; this aspect is considered further below in relation to the effects of GH on cardiovascular function. Importantly, both oedema and arthralgia tended to affect the same individuals and were more likely to occur *de novo* in the initial phase of therapy (Fig 1) with subsequent improvement. Verhelst et al [13] and Holmes and Shalet [14] have demonstrated a relationship between symptoms of fluid retention and the increment in serum IGF-I observed over 6 months. In keeping with these findings, the dose ranging studies of de Boer et al [15] clearly demonstrated that symptoms of fluid retention were dose dependent for the study population as whole while exhibiting wide interindividual variation.

Table 1
Adverse events occurring during 6 months of GH replacement therapy (n=115) or placebo (n=118). The dose of GH was 0.25 IU/kg/week (1.5 IU/m^2 /day) to a maximum of 4 IU/day. Data from Mårdh et al [12] with permission.

Adverse events	**GH group (%)**	**Placebo group (%)**
Oedema	37.4	3.4
Arthralgia	19.1	1.7
Myalgia	15.7	3.4
Paraesthesia	7.8	0.8
Carpal tunnel syndrome	1.7	0
Headache	2.6	1.7
Upper resp tract infection	11.3	8.5
Nausea	2.6	2.5
Diarrhoea	1.7	3.4
Dizziness	0.9	3.4

From these preliminary data it became clear that sensitivity to GH replacement therapy was highly variable between individuals and could not be predicted from baseline characteristics. Subsequent work [16] has highlighted the contribution of gender to individual susceptibility with males demonstrating a greater dose response consistent with the known higher physiological GH production rate in females. In addition, serum concentrations of GH binding protein (GHBP) may also provide a marker for individual susceptibility [16]. For these reasons, dosing regimens based on weight or body surface area are unlikely to prove satisfactory for the generality of patients and it is now becoming increasingly common to adopt a titration policy [17] in which an initial low dose of GH is slowly increased on the basis of serum IGF-I and body composition measurements; this approach would predict a reduction in those adverse symptoms attributable to sodium and water retention.

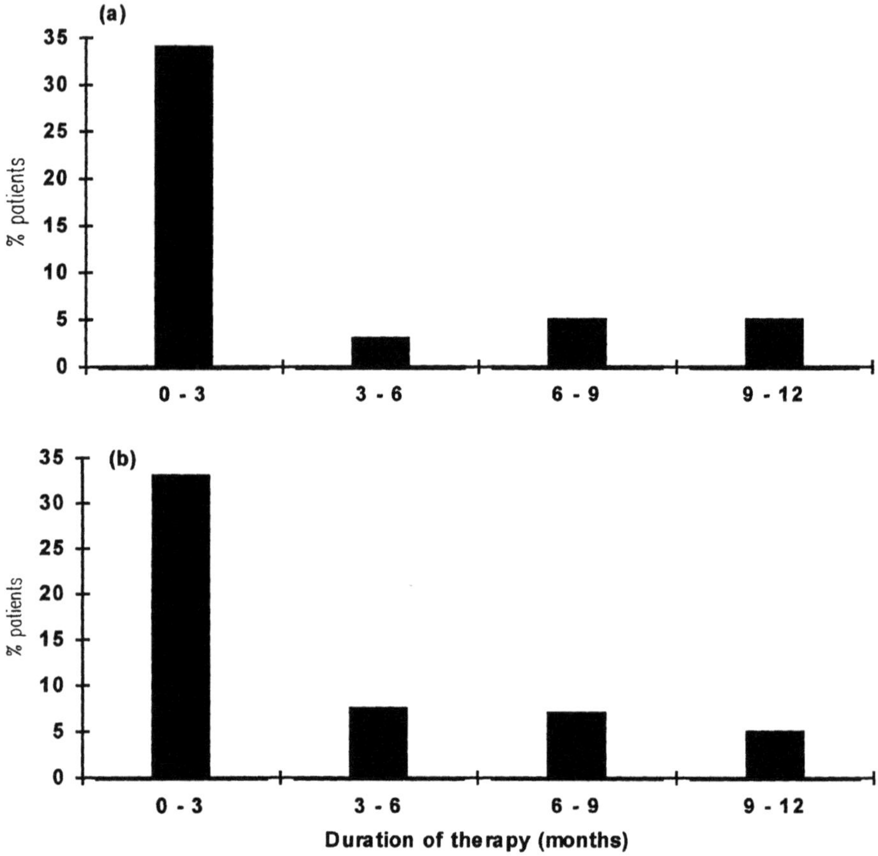

Figure 1
Incidence of (a) oedema and (b) arthralgia and myalgia during GH replacement therapy in 115 hypopituitary, GH-deficient adults. Data from Mårdh et al [12] with permission.

In support of this, data from KIMS, an international safety surveillance database of GH treated hypopituitary adults, demonstrates a dramatic reduction in the rate of adverse events indicative of fluid retention in patients commencing treatment with dose titration compared with rates observed in the original multicentre clinical trials (0.29/year v. 2.26/year) [18] - (Fig. 2).

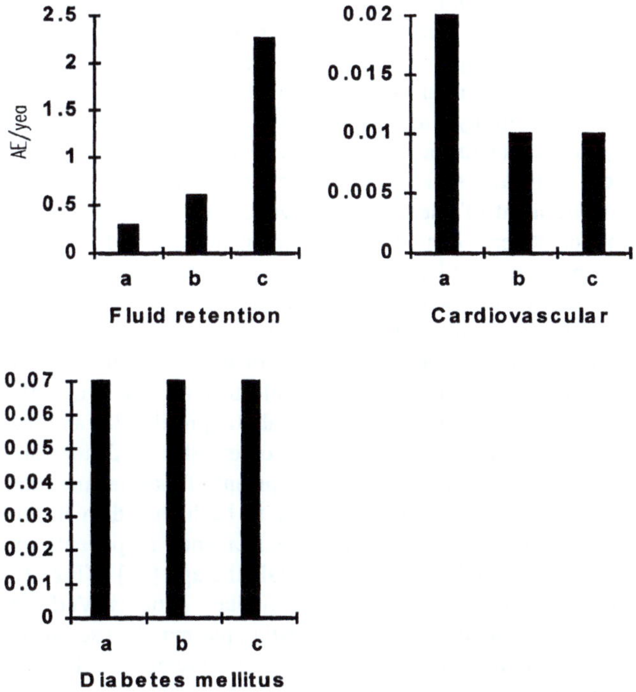

Figure 2
Incidence of adverse events per treatment year in patients enrolled into (a) KIMS (n=275) versus patients from previous placebo-controlled trials receiving (b) placebo or (c) GH.

Metabolic Consequences

Although GH deficiency states are associated with a decreased counterregulatory response to hypoglycaemia, sensitivity to insulin is reduced [19,20]. This phenomenon may partly explain the increased prevalence of impaired glucose tolerance in hypopituitary adults [21] and is likely to be related to the abnormal body composition, with increased central adiposity, which characterises hypopituitarism. The inverse correlations between body fat and its distribution on the one hand and various indices of insulin sensitivity remain evident in GH-deficient hypopituitary patients [20] - (Table 2). Furthermore, this change in body fat distribution is accompanied by additional adverse changes in other cardiovascular risk factors, including increments in LDL [22,23,24] and a variable decrement in HDL cholesterol [25] and alterations in serum fibrinogen and plasminogen activator inhibitor type 1 [26]; the summation of these effects may explain, at least in part, the increased cardiovascular morbidity and mortality associated with hypopituitarism [27,28].

With the exception of the relative insulin insensitivity, the constellation of adverse cardiovascular risk factors outlined are substantially and rapidly improved by GH replacement therapy. However, despite impressive and highly reproducible reductions in central adiposity, the initial consequence of GH replacement is a further decline in insulin sensitivity, as determined by homeostatic model assessment [20] or euglycaemic clamp techniques [27]. In keeping with this, the correlation between indices of insulin sensitivity and fat distribution increase after commencement of GH replacement (Table 2). This adverse change in net insulin responsiveness is evident for at least 6 months after commencement of GH and subsequently an improvement in insulin sensitivity to pretreatment levels is observed [27]. To date, no study has documented a net increase in overall insulin responsiveness above baseline measurements despite apparently complete normalisation of body composition. The potential impact of this on cardiovascular risk remains to be determined. However, it is improbable that insulin is directly atherogenic rather than indicating an overall adverse metabolic profile [28] and the situation in question is unique in that the majority of other adverse factors are reversed by therapy. Most importantly, the rate of development of diabetes mellitus in patients on GH replacement therapy does not appear to be higher than in the general population and is certainly no higher than the incidence in patients on placebo therapy during the original multicentre trials of GH therapy [18] - (Fig. 2). In a 2 year study of GH replacement in 148 adult hypopituitary patients, Verhelst and colleagues [13] demonstrated subtle increments in fasting plasma glucose (mean 0.365 mM) over the duration of the study but a marginal increment in HbA1c evident at 6 and 12 months subsequently reduced at 18 and 24 months; two patients developed biochemical evidence of diabetes mellitus but in retrospect this had been present in both prior to commencement of the study. It is likely that there will be major differences between individuals in terms of the effect of GH deficiency on central fat distribution, insulin sensitivity and genetic predisposition to diabetes mellitus so that the summation of these effects and the outcome of GH replacement on insulin action and glucose homeostasis will be variable.

Table 2
Associations between indices of insulin sensitivity (IS by homeostatic model assessment, fasting specific insulin and total insulin) and body fat and its distribution (W/H - waist/hip ratio; BMI - body mass index, kg/m^2; central fat, dual energy X ray absorptiometry) in 22 adult hypopituitary subjects before and after 6 months of GH replacement (mean dose 0.2 IU/kg/week). Data from Weaver et al [20], with permission.

	IS		**Specific insulin**		**Total insulin**	
	Before	**After**	**Before**	**After**	**Before**	**After**
Total fat						
P	0.02	0.001	0.001	0.002	0.068	<0.001
r	-0.45	-0.63	0.63	0.63	0.4	0.68
% Fat						
P	0.12	0.02	0.07	0.09	0.07	0.015
r	-0.39	-0.48	0.39	0.37	0.4	0.51
W/H						
P	0.07	0.08	0.01	0.01	0.05	0.06
r	-0.39	-0.37	0.51	0.55	0.42	0.41
BMI						
P	0.064	0.007	<0.001	0.003	0.02	0.006
r	-0.61	-0.55	0.72	0.62	0.49	0.56
Central fat						
P	0.062	<0.001	0.003	<0.001	0.03	<0.003
r	-0.44	-0.66	0.6	0.69	0.45	0.7

The adverse consequences of GH deficiency in increasing total and LDL cholesterol levels are reversed by GH replacement therapy. However, a number of studies in both normal [29] and hypopituitary adults [20,30,31] have documented an increment in serum lipoprotein(a) (Lp(a)), a quantitatively minor lipoprotein species containing apo B100 and apo(a), the concentration of which is largely genetically determined and which appears to be an independent risk factor for macrovascular disease. However, not all studies have shown this effect and in one, the increment in Lp(a) was only observed when levels were already elevated at baseline [31]. Furthermore, one study documenting an increment in Lp(a) also demonstrated favourable alterations in HDL cholesterol in the same individuals [30]. The long term significance of these alterations is uncertain at present but should be viewed in the context of an overall favourable change in cardiovascular risk profile.

Cardiovascular Consequences
The effects of GH on the heart are mediated by local IGF-I production and possibly also by a direct action of GH. Rat cardiac myocytes contain GH receptors [32] and express IGF-I mRNA which increases in response to GH administration [33]. GH

deficiency is therefore likely to influence cardiac morphology and function, a phenomenon which is most clearly evident in childhood onset GH deficiency. The latter is associated with relative cardiac hypokinesia, reduced left ventricular wall mass, cardiac output and blood pressure [34]. In contrast, some studies in adult onset GH deficiency have documented normal ventricular dimensions [35,36]. These apparent anomalies may be due to heterogeneity in adult onset hypopituitarism both in terms of duration of disease and the variable presence of hypertension which may be increased in adult onset disease.

GH replacement therapy increases LV wall mass, cardiac output and specifically stroke volume [34,36]. Fractional shortening is also increased and diastolic function improved [36]. Importantly, GH therapy results in a reduction in peripheral vascular resistance [37] which may partly explain the improvement in blood pressure control in some hypopituitary subjects with essential hypertension and the small but significant reductions in systolic and diastolic pressure observed in normotensive patients in some studies [13]. Clearly, normalisation of cardiac dimensions in patients with childhood onset hypopituitarism is desirable. However the longer term consequences in adult onset disease are less clear cut and the possibility of inducing left ventricular hypertrophy should be borne in mind. Some parallels may be drawn with the effects of endogenous GH excess on cardiac morphology [38]. Early acromegaly is associated with a hyperkinetic cardiac state with increased cardiac output and decreased peripheral resistance. This subsequently gives way to cardiac hypertrophy with impaired diastolic relaxation, the latter being due to fibrotic change within the hypertrophied ventricles. Progression of fibrosis eventually results in cardiac decompensation and a clinical picture of dilated cardiomyopathy. This is, of course, the natural history of severe continuing GH excess extending over many years. However, it serves to emphasise the importance of optimum dose schedules in GH replacement and in particular the avoidance of over-replacement. The overall effects of GH replacement on cardiac structure and function and on the peripheral vascular system are likely to be beneficial for the individual patient but clearly, continuing detailed assessments of cardiac morphology in representative subsets of patients are desirable. Currently available multicentre surveillance data indicates that there is no increase in adverse cardiac events in patients on maintenance growth hormone replacement (Fig. 2).

Benign Intracranial Hypertension

The development of papilloedema and increased cerebrospinal fluid pressure in the absence of radiological evidence of space occupying lesions has been reported in one adult and 22 children receiving high dose GH therapy [39]. The mechanisms underlying this phenomenon are unclear but it is probable that they are at least partly dose related. Headache is a frequently reported symptom in the early phase of GH replacement, occurring significantly more commonly than during placebo therapy [12]. The relationship between self-limiting headache and intracranial hypertension is speculative but the frequency of the former symptom emphasises the need for awareness of the possibility of increased intracranial pressure and fundoscopy should be performed in all instances.

Changes in Serum Thyroid Hormone Concentrations

Changes in thyroid hormone concentrations during GH therapy in children were first described over 20 years ago. The initial observation of reduction in serum thyroxine (T4) concentration [40] was followed by reports of increased triiodothyronine (T3) concentrations [41] with an additional artefactual effect of changes in thyroid hormone binding globulin concentrations [43]. The possible mechanisms contributing to these changes include reduction in TSH secretion because of increased somatostatin tone secondary to GH therapy [40,43], increased peripheral conversion of T4 to T3 [41,44,45] and enhanced negative feedback on TSH secretion by T3 [45]. Clearly the relative contribution of these factors depends on the presence or absence of residual endogenous TSH secretion.

Studies in adults [4,46,47] have indicated essentially similar findings as those in children and were evident in both euthyroid patients and in those with TSH deficiency suggesting that increased conversion of T4 to T3 is a major factor in these observations. The changes have been predominantly subclinical although in one instance the onset of atrial fibrillation may have been triggered by an increase in serum T3 concentration after commencement of GH. Our own observations suggest that increments in T3 are most uniform in patients with intact TSH reserve in whom serum thyroid hormone concentrations remain predictably in the normal range [47]. In addition, changes in T4 reverse between 6 and 12 months. Furthermore, more recent investigations have documented less perturbation in serum thyroid hormone levels, perhaps because of lower GH doses [13]. Nonetheless, all patients commencing GH therapy should have regular monitoring of thyroid hormone concentrations, particularly if they are on high replacement doses of thyroxine. The observations to date do not indicate a significant increase in thyroid function as a result of a direct trophic effect of GH on the thyroid gland.

Changes in Cortisol Metabolism and Cortisol Binding Globulin

Two studies have documented decreased serum concentrations of cortisol binding globulin (CBG) in hypopituitary adults treated with GH [48,49]. The consequence of this is an artefactual reduction in total serum cortisol concentrations which would not predict any alteration in glucocorticoid activity but should be borne in mind when assessing serum cortisol measurements in these patients. However, a potentially more significant change occurs in cortisol metabolism and specifically in cortisol-cortisone interconversion under the influence of 11ß-hydroxysteroid dehydrogenase (11ßHSD). The renal (type 2) isoform of this enzyme functions as an NAD-dependent dehydrogenase converting cortisol to cortisone and therefore protects renal mineralocorticoid receptors from exposure to cortisol. The type 1 isoform has its major effect in liver and functions predominantly as a reductase, converting cortisone to cortisol. GH therapy is associated with partial inhibition of the type 1 enzyme as determined by measurements of the ratio of 11-hydroxy to 11-oxo cortisol metabolites which reduce by approximately 20% [48] but without alteration in activity of the renal isoenzyme as determined by the urine free cortisol to cortisone ratio [50]. This phenomenon is accompanied by modest reduction in the bioavailability of oral hydrocortisone [48] and the combined effects of these factors might theoretically compromise glucocorticoid action in patients on lower hydrocortisone replacement doses. Clearly, serum cortisol measurements are a less reliable indicator in this situtation because of changes in CBG and therefore clinical awareness is particularly important. Patients with intact ACTH reserve demonstrate similar changes in cortisol metabolites but are not at risk of glucocorticoid insufficiency because of intact negative feedback.

Neoplasia and Tumour Recurrence

The increased risk of neoplasia, particularly colonic, in patients with endogenous GH excess is now clearly established [53,54]. This and the probable overall reduction in incidence of malignancy in adult hypopituitary patients [27] raise obvious concerns about the incidence of neoplasia during GH replacement. The theoretical risk of tumour formation or an increased rate of relapse of treated tumours has not been evident in follow up of adult hypopituitary patients on GH replacement for periods in excess of 5 years [18]. A recent analysis of 1034 patients on the KIMS database, treated for a total of 1232 patient years documented 3 deaths, none of which was related to GH therapy. Recurrence of pituitary or CNS tumours was reported in 6 patients; clinically non-functioning pituitary tumour in 4, a gonadotrophinoma in one and a hypothalamic-pituitary dysgermonoma in one patient. The expected recurrence rate of pitutary tumours is difficult to quantify because of other factors including the variable use of post-operative irradiation in different centres but is likely to be in the region of 2% annually [55]. This would suggest, on the basis of the present evidence which is preliminary in terms of duration of therapy, that an increased rate of tumour relapse does not occur; clearly longterm surveillance is essential. In the same study, new tumours were reported in 5 patients; three were basal cell carcinomas in 3 Australian patients treated with GH for between 1.0 and 2.8 years, one patient developed a testicular teratoma and one patient had an abnomal cervical smear.

An increased growth rate of cutaneous naevi has been observed in children with either GH deficiency or Turner's syndrome while on GH therapy [56]. Although this has not resulted in an overtly increased incidence of malignant melanoma it is an additional important area for increased awareness in the context of adult GH replacement.

Miscellaneous Reported Side Effects

The development of Dupuytren's contracture 3 months after commencent of GH therapy has been described in a 14 year old female patient; the specific relationship to GH remains unclear [57]. Reversible gynaecomastia has been reported in elderly male patients on GH [58].

The identification of rare adverse events and the need to establish their relationship (or otherwise) with GH therapy further emphasises the importance of large scale, physician-managed databases for the long term surveillance of this treatment.

REFERENCES

1. Cuneo RC, Salomon F, McGauley GA, Sönksen PH. The growth hormone deficiency syndrome in adults. *Clin Endocrinol* 1992; 37: 387-97.
2. De Boer H, Blok G-J, Van der Veen EA. Clinical aspects of growth hormone deficieincy in adults. *Endocrine Reviews* 1995; 16: 63-86.
3. Monson JP. The assessment of glucocorticoid replacement therapy. *Clin Endocrinol* 1997; 46: 269-70.
4. Salomon F, Cuneo RC, Hesp R, Sönksen PH. The effects of treatment with recombinant human growth hormone on body composition and metabolism in adults with growth hormone deficiency. *N Engl J Med* 1989; 321: 1797-803.
5. Jørgensen JOL, Pedersen SA, Thuesen L, Jørgensen J, Ingemann-Hanssen T, Skakkebæk NE, Christiansen JS. Beneficial effects of growth hormone treatment in GH-deficient adults. *Lancet* 1989; i: 1221-5.
6. Van der Veen EA, Netenblos JC. Growth hormone replacement therapy in adults: bone and calcium metabolism. *Horm Res* 1990; 33(suppl 4): 65-8.
7. Whitehead HM, Boreham C, McIlrath EM, Sheridan B, Kennedy L, Atkinson AB, Hadden DR. Growth hormone treatment of adults with growth hormone deficiency: results of a 13 month placebo controlled cross over study. *Clin Endocrinol* 1992; 36: 45-52.
8. Degerblad M, Elgindy N, Hall K, Sjöberg, Thoren M. Potent effect of recombinant growth hormone on bone mineral density and body composition in adults with panhypopituitarism. *Acta Endocrinol (Copenh)* 1992; 126: 387-93.
9. Binnerts A, Swart GR, Wilson JH, Hoogerbrugge N, Pols HAP, Birkenhäger JC, Lamberts SWJ. The effect of growth hormone administration in growth hormone deficient adults on bone, protein, carbohydrate and lipid homeostasis, as well as on body composition. *Clin Endocrinol* 1992; 37: 79-87.
10. Bengtsson B-Å, Edén S, Lönn L, Kvist H, Stokland A, Lindstedt G, Bosaeus I, Tölli J, Sjöström A, Isaksson OGP. Treatment of adults with growth hormone (GH) deficiency with recombinant human GH. *J Clin Endocrinol Metab* 1993; 76: 309-17.
11. Ho K, Weissberger AJ. The antinatriuretic action of biosynthetic growth hormone in man involves activation of the renin-angiotensin system. *Metabolism* 1990; 39: 133-7.
12. Mårdh G, Lundin K, Borg G, Jonsson B, Lindeberg A on behalf of the investigators. Growth hormone replacement therapy in adult hypopituitary patients with growth hormone deficiency: combined data from 12 European placebo-controlled clinical trials. *Endocrinol Metab* 1994; 1(suppl A): 43-9.
13. Verhelst J, Abs R, Vandeweghe M, Mockel J, Legros J-J, Copinschi G, Mahler C, Velkeniers B, Vanhaelst L, Van Aelst A, De Rijdt D, Stevenaert A, Beckers A. Two years of replacement therapy in adults with growth hormone deficiency. *Clin Endocrinol* 1997 (in press).
14. Holmes SJ, Shalet SM. Which adults develop side effects of growth hormone replacement? *Clin Endocrinol* 1995; 43: 143-9.

15. De Boer H, Blok G-J, Voerman B, De Vries P, Popp-Snijders C, Van der Veen EA. The optimal growth hormone replacement dose in adults, derived from bioimpedance analysis. *J Clin Endocrinol Metab* 1995; 80: 2069-76.
16. Johansson G, Bjarnason R, Bramnert M, Carlsson LM, Degerblad M, Manhem P, Rosén T, Thoren M, Bengtsson B-Å. The individual responsiveness to growth hormone (GH) treatment in GH-deficient adults is dependent on the level of GH-binding protein, body mass index, age and gender. *J Clin Endocrinol Metab* 1996; 81: 1575-81.
17. Johansson G, Bengtsson B-Å. Individualised GH dose titration in GH-deficient, hypopituitary adults. *Clin Endocrinol* 1997 (in press).
18. Abs R, Bengtsson B-Å, Hernberg-Stahl E, Monson JP, Tauber J-P, Wilton P, Wüster C. GH replacement in 1034 growth hormone deficient adults: demographic and clinical characteristics, dosing and safety. *Proceedings of the European Neuroendocrine Association* 1997.
19. Salomon F, Cuneo RC, Sönksen PH. Glucose metabolism in adults with growth hormone deficiency. *Acta Paediatr Scand* 1991; suppl 377: 64-8.
20. Weaver JU, Monson JP, Noonan K, John WG, Edwards A, Evans KA, Cunningham J. The effect of low dose recombinant human growth hormone replacement on regional fat distribution, insulin sensitivity, and cardiovascular risk factors in hypopituitary adults. *J Clin Endocrinol Metab* 1995; 80: 153-9.
21. Markussis V, Beshyah SA, Fisher C, Sharp P, Nicolaides AN, Johnston DG. Detection of premature atherosclerosis by high resolution ultrasonography in symptom-free hypopituitary adults. *Lancet* 1992; 340: 1188-92.
22. Angelin B, Rudling M. Growth hormone and hepatic lipoprotein metabolism. *Curr Opin Lipidol* 1994; 5: 160-165.
23. Cuneo RC, Salomon F, Watts GF, Hesp R, Sönksen PH. Growth hormone treatment improves serum lipids and lipoproteins in adults with growth hormone deficiency. *Metabolism* 1993; 42: 1519-23.
24. De Boer H, Blok G-J, Voerman HJ, Phillips M, Schouten JA. Serm lipid levels in growth hormone deficient men. *Metabolism* 1994; 43: 199-203.
25. Rosén T, Edén S, Larsson G, Wilhelmsen L, Bengtsson B-Å. Cardiovascular risk factors in adult patients with growth hormone deficiency. *Acta Endocrinol (Copenh)* 1993; 129: 195-200.
26. Johansson J-O, Landin K, Tengborn L, Rosén T, Bengtsson B-Å. High fibrinogen and plasminogen activator inhibitor activity in growth hormone-deficient adults. *Arterioscler Thromb* 1994; 14: 434-7.
27. Rosén T, Bengtsson B-Å. Premature mortality due to cardiovascular disease in hypopituitarism. *Lancet* 1990; 336: 285-8.
28. Bülow B, Hagmar L, Mikoczy Z, Nordström CH, Erfurth EM. *Clin Endocrinol* 1997; 46: 75-81.
29. Fowelin J, Attvall S, Lager I, Bengtsson B-Å. Effects of treatment with recombinant human growth hormone on insulin sensitivity and glucose metabolism in adults with growth hormone deficiency. *Metabolism* 1993; 42: 1443-7.
30. Elliot TG, Viberti GC. Relationship between insulin resistance and coronary artery disease in diabetes mellitus and the general population: a critical

appraisal (review). *Baillière's Clinics in Endocrinology and Metabolism* 1993; 7: 1079-1103.

31. Edén S, Wiklund O, Oscarsson J, Rosén T, Bengtsson B-Å. Growth hormone treatment of growth hormone-deficient adults results in a marked increase in Lp(a) and HDL cholesterol concentrations. *Arterioscer Thromb* 1993; 13: 296-301.
32. Olivecrona H, Ericsson S, Berglund L, Angelin B. Incresaed concentrations of serum lipoprotein (a) in response to growth hormone treatment. *B Med J* 1993; 306: 1726-7.
33. Modigliani E, Kerchouni R, Uzzan B, Valensi P, Chanson P, Caron J. Modification of blood lipids and lipoproteins after human growth hormone treatment in adults with growth hormone deficiency: a preliminary report. *Endocrinol Metab* 1994; 1(suppl A): 31-35.
34. Mathews LS, Enberg B, Norstedt G. Regulation of rat growth hormone receptor gene expression. *J Biol Chem* 1989; 17: 9905-10.
35. Isgaard J, Nisson A, Vikman K, Isaksson OGP. Growth hormone regulates the level of insulin-like growth factor-I mRNA in rat muscle. *J Endocrinol* 1989; 120: 107-12.
36. Amato G, Carella C, Fazio S, La Montagna G, Cittadini A, Sabatini D, et al. Body composition, bone metabolism, heart structure and function in growth hormone deficient adults before and after growth hormone replacement therapy at low doses. *J Clin Endocrinol Metab* 1993; 77: 1671-6.
37. Shahi M, Beshyah SA, Hackett D, Sharp PS, Johnston DG, Foale RA. Myocardial dysfunction in treated adult hypopituitarism: a possible explanation for increased cardiovascular mortality. *Br Heart J* 1992; 67: 92-6.
38. Fort S, Weaver JU, Monson JP, Mills P. The effects of low-dose recombinant human growth hormone on cardiovascular structure and function in hypopituitary growth hormone deficient adults. *Endocrinol Metab* 1995; 2: 119-26.
39. Caidahl K, Edén S, Bengtsson B-Å. Cardiovascular and renal effects of growth hormone. *Clin Endocrinol* 1994; 40: 393-400.
40. Sacca L, Cittadini A, Fazio S. Growth hormone and the heart. *Endocrine Reviews* 1994; 15: 555-73.
41. Malozowski S, Tanner LA, Wysowski D, Fleming GA. Growth hormone, insulin-like growth factor I, and benign intracranial hypertension. *N Engl J Med* 1993; 329: 665-6.
42. Lippe BM, Van Herle AJ, LaFranchi SH, Uller RP, Lavin N, Kaplan SA. Reversible hypothyroidism in growth hormone-deficient children treated with human growth hormone. *J Clin Endocrinol Metab* 1975; 40: 612-8.
43. Sato T, Suzuki Y, Taketani T et al. Enhanced peripheral conversion of thyroxine to triiodothyronine during hGH therapy in GH deficient children. *J Clin Endocrinol Metab* 1977; 45; 324-9.
44. Gacs G, Banos C. The effect of growth hormone on the plasma levels of T4, free-T4, T3, reverse T3 and TBG in hypopituitary patients. *Acta Endocrinol (Copenh)* 1981; 96: 475-9.

45. Ho KY, Weissberger AJ, Stuart MC, Day RO, Lazarus L. The pharmacokinetics, safety and endocrine aspects of authentic biosynthetic human growth hormone in normal subjects. *Clin Endocrinol* 1989; 30: 335-45.
46. Rezvani I, DiGeorge AM, Dowshen SA, Bourdony CJ. Action of human growth hormone (hGH) on extrathyroidal conversion of thyroxine (T4) to triiodothyronine (T3) in children with hypopituitarism. *Pediatr Res* 1981; 15: 6-9.
47. Grunfeld C, Sherman BM, Cavalieri RR. The acute effects of human growth hormone administration on thyroid function in normal men. *J Clin Endocrinol Metab* 1988; 67: 1111-4.
48. Jørgensen JOL, Pedersen SA, Laurberg P, Weeke J, Skakkebæk NE, Christiansen JS. Effects of growth hormone therapy on thyroid function of growth hormone-deficient adults with and without concomitant thyroxine-substituted central hypothyroidism. *J Clin Endocrinol Metab* 1989; 69: 1127-32.
49. Monson JP, Weaver JU, Noonan K, Burrin JM. Effect of growth hormone on serum thyroid hormone concentrations and desmopressin requirements in adults with hypopituitarism ands central diabetes insipidus. *Endocrinol Metab* 1994; 1 (suppl A): 51-5.
50. Weaver JU, Thaventhiran L, Noonan K, Burrin JM, Taylor NF, Norman MR, Monson JP. The effect of growth hormone replacement on cortisol metabolism and glucocorticoid sensitivity in hypopituitary adults. *Clin Endocrinol* 1994; 41: 639-48.
51. Rodriguez-Arnao J, Perry L, Besser GM, Ross RJM. Growth hormone treatment in hypopituitary GH deficient adults reduces circulating cortisol levels during hydrocortisone replacement. *Clin Endocrinol* 1996; 45, 33-7.
52. Gelding SV, Taylor NF, Weaver JU, Wood P, Monson JP. The effect of growth hormone replacement therapy on cortisol-cortisone interconversion in hypopituitary adults: evidence for growth hormone modulation of extrarenal 11β-hydroxysteroid dehydrogenase activity. *Clin Endocrinol* 1997; (in press).
53. Bengtsson B-Å. Acromegaly and neoplasia. *J Pediatr Endocinol* 1993; 6: 73-8.
54. Jenkins PJ, Fairclough P, Lowe D, Grossman AB, Monson JP, Besser GM, Wass JHW. Acromegaly and colonic neoplasia. *Clin Endocrinol* 1997; 47, 17-22.
55. Bradly KM, Adams CB, Potter CP, Wheeler DW, Anslow PJ, Burke CW. An audit of selected patients with non-functioning pituitary adenoma treated by transsphenoidal surgery without irradiation. *Clin Endocrinol* 1994; 41: 655-9.
56. Bourgignon JP, Piérard GE, Ernould C, Heinrichs C, Craen M, Rochiccioli P, Arrese JE, Franchimont C. Effects of human growth hormone therapy on melanocyte naevi. *Lancet* 1993; 341: 1505-6.
57. Kiess W, Butenandt O. Development of Dupuytren's contracture during growth hormone therapy. *Lancet* 1993; 342: 181-2.
58. Cohn L, Feller AG, Draper MW, Rudman IW, Rudman D. Carpal tunnel syndrome and gynaecomastia during growth hormone treatment of elderly men with low circulating IGF-I concentrations. *Clin Endocrinol* 1993; 39: 417-25.

New frontiers in adult GH therapy

20 GROWTH HORMONE IN HEART FAILURE

Antonio Cittadini
Serafino Fazio
Luigi Sacca

Department of Medicine
Federico II Medical School, Naples, ITALY

INTRODUCTION

Heart failure represents the final common pathway of most cardiovascular disorders. Its annual incidence amounts to 400.000 new cases in the United States and recent estimates indicate that two million patients are currently being treated for this condition (1). The only drugs that have demonstrated to reduce mortality in controlled clinical trials are angiotensin converting enzyme (ACE) inhibitors (2) and, recently, the β-blocker carvedilol (3), whereas digitalis only reduces the hospitalization rate (4). Despite the proven efficacy of these drugs, the prognosis of heart failure still remains poor.

The clinical usefulness of ACE inhibitors and their ability to reduce hypertrophy in several experimental settings have led to the view that activation of myocardial growth is an unfavorable component of left ventricular (LV) pathologic remodeling in heart failure. Therefore, therapeutic agents capable of blunting LV hypertrophy and dilatation are potential candidates to new strategies aimed at preserving and/or improving myocardial function in heart failure.

An alternative strategy would be to activate myocardial growth, by inducing a type of physiological and adequate hypertrophy. In this regard, GH and IGF-1 may represent candidate molecules because of their proven ability to trigger a growth response in several organs including the heart (5). In the following sections, we will describe: 1) evidence linking GH and the heart, 2) the mechanistic impact of this interaction, 3) the rationale for the use of GH and IGF-1 in heart failure, and 4) finally we will discuss some recent studies performed in experimental and human models of heart failure.

GH/IGF-1 TARGET THE HEART

The first description of cardiac involvement in acromegaly was made in 1895 (6). Several reports have subsequently documented cardiac abnormalities in acromegaly (5). A series of elegant studies performed by Beznak in the fifties demonstrated the essential role of GH in the maintenance of normal LV mass and an efficient hypertrophic response to pressure overload (7,8). Nevertheless, the importance of the link between GH and the heart has been largely overlooked. Until a few years

ago, GH was considered to be essential only for linear growth, maintenance of lean body mass, and energy metabolism. Its target tissues included classically the liver (IGF-1 synthesis, insulin antagonism), the growth plate (growth), the adipose tissue (lipolysis), and the skeletal muscle (insulin antagonism, anabolism) (9).
The advent of echocardiographic imaging and, more recently, of modern molecular techniques, has shed new light on the modulation of cardiovascular morphology and function by the GH/IGF-1 axis. In addition, relevant information has been derived from cross-sectional studies performed in conditions of GH excess or deficiency.

GH excess
The "trophic" role of GH has been suggested by the demonstration of LV hypertrophy of the concentric type in acromegalic patients (10,11,12). This finding was confirmed even after exclusion of known causes of cardiac disease, such as arterial hypertension, coronary artery disease, and diabetes mellitus (13). From a functional standpoint, the increased LV mass is associated with impaired diastolic filling, as shown by both Doppler echocardiographic and scintigraphic techniques (13,14). Treatment with the somatostatin analogue octreotide reduces LV mass and improves diastolic filling in parallel with the inhibition of GH activity (15). The ejection phase indexes are normal at rest in acromegalic patients, while cardiac dysfunction becomes evident after physical exercise (14).

Studies performed in animal models of GH excess confirmed and expanded on these notions. Rats treated with high doses of exogenous GH and/or IGF-1 develop cardiac hypertrophy associated with increased myocyte size (16). Interestingly, no alterations of capillary density or increase in interstitial tissue was observed in GH treated rats, at variance with what is observed in other models of LV hypertrophy. Diastolic function is normal in GH treated rats, or even enhanced as shown by invasive indexes of relaxation. Systolic function is increased as demonstrated in the intact animal and in the isolated whole heart (16,17). The concentric growth and the increased myofilament calcium responsiveness both contribute to the enhanced contractility (*vide infra*) (17,18).

Taken together, studies conducted in conditions of GH excess demonstrate that induction of cardiac growth is a hallmark of GH overactivity. The different functional adaptations observed in the human and rat model of GH excess may be reconciled considering that: 1) short-term administration of GH to normal humans increases cardiac function (19); 2) most of the clinical studies of GH excess are performed in patients with longstanding acromegaly with consequent relevant changes in myocardial structure; 3) acromegaly of brief duration (< 5 years) is associated with a hyperkinetic syndrome and no abnormalities of diastolic filling (20); and 4) both chronic (4 weeks) and acute GH/IGF-1 treatment enhance myocardial contractility in rats (16,17). Therefore, it may be speculated that the initial impact of GH excess on the heart is to increase its performance. After years of exposure to elevated GH and IGF-1 levels, interstitial fibrosis usually occurs and abnormalities of diastolic and subsequently of systolic function eventually develop. The observation that short-term acromegaly does not impair diastolic and systolic function is somewhat reassuring in view of the potential therapeutic use of GH.

GH deficiency
GH deficiency is a powerful model with which to explore the question whether GH plays any role in adult life after completion of linear growth. The data generated by this approach have greatly enhanced our understanding of the importance of GH in cardiovascular physiology. Cardiac morphology in childhood onset GH deficiency is markedly altered and it is characterized by cardiac atrophy, with decreased LV mass and relative wall thickness (21,22). Furthermore, GH deficient patients display a hypokinetic syndrome whose main features are low LV ejection fraction, low cardiac output, and increased peripheral vascular resistance (23). Most of these abnormalities reverse or improve after GH substitutive therapy (21-23).

These findings have been confirmed in animal models of GH deficiency. In the dwarf rat, there is a reduced myocyte size and cardiac mass, and impaired myocardial contractility *in vitro* (24). These alterations are associated with reduced myofilament responsiveness to both calcium and isoproterenol (25). The impaired β-adrenergic responsiveness does not appear to depend upon reduced β-adrenergic receptor density, affinity, or coupling to the adenylyl cyclase system (24).

MECHANISMS of GH ACTION

Heart performance can be regulated by shifting moment among three different types of controls: organ physiology, cell biochemistry and biophysics, and gene expression (26). We will examine the mechanisms of GH action from this perspective, describing the GH interaction with the heart at molecular, cell, and organ level.

Molecular level
The molecular mechanisms governing GH and IGF-1 interrelationship is still a matter of debate. GH receptor is expressed in several tissues and its relative abundance in the heart is higher than in other target organs (27). On the other hand, IGF-1 is a ubiquitous protein, present with its receptors in a high number of tissues, including the heart (28,29). Several lines of experimental evidence support the hypothesis that GH acts by stimulating the local production of IGF-1, which exerts its growth-promoting effects through paracrine/autocrine mechanisms. Moreover, it has been demonstrated that IGF-1 may also act as a classical endocrine factor (30). In fact, it is produced by the liver under GH control, and circulates in the blood stream reaching the target tissues with levels 1000 times higher than insulin. To further complicate this scenario, the presence of GH receptor in several tissues and the differential effects of GH and IGF-1 in several organs raise the possibility that GH may have direct effects independent of IGF-1 mediation (27,16).

Extensive evidence also supports the concept that IGF-1 is involved in the intricate cascade of events leading to cardiac hypertrophy. In response to pressure or volume hemodynamic overload, IGF-1 expression increases in parallel to hypertrophy (31-33). More importantly, exogenous infusion of IGF-1 and GH in different *in vivo* and *in vitro* model systems stimulates LV hypertrophy (34,16,17). Additional evidence that GH and IGF-1 are direct cardiac growth factors is that GH exerts an anabolic effect on skeletal and cardiac muscle consisting of activation of protein synthesis (35,36).

The intracellular events downstream of the GH receptor-ligand complex are largely unknown. There is evidence of activation of the tyrosine kinase JAK2, MAP kinase, and STAT transcription factors (37). Two recent investigations offer some insight into the physiological relevance of GH signaling in the heart (34,38). Stimulation of isolated cardiac preparations with exogenous GH does not evoke any contractile response, nor does it activate protein synthesis and myocyte growth. In contrast, exogenous IGF-1 augments contractility and the incorporation of amino acids, leading to cell hypertrophy. These observations are consistent with the view that IGF-1 mediates most, if not all, of the effects of GH on the heart, at least in acute settings. GH receptors might mediate other long-term adaptations or modulate IGF-1 local production.

The intracellular signaling pathways activated by the IGF-1-receptor complex are the object of active investigation. Experimental evidence has identified several major substrate proteins activated by IGF-1 binding to its cognate receptor: 1) the insulin related substrate-1 (IRS-1); 2) the Shc proteins; and 3) Crk, a cellular homologue of v-crk. IRS-1 in turn binds the p85 ß-subunit domain of PI3-kinase (39). In addition to the activation of IRS-1, Shc, and Crk, other second messengers may be generated by IGF-1 receptor activation, including the phospholipase C pathway. In fact, rapid elevation of intracellular inositol phosphates levels has been reported in cardiac cells (40).

A major breakthrough has recently come from the observation that IGF-1 rapidly activates multiple intracellular pathways in isolated rat cardiomyocytes, including the phosphatidylinositol 3-kinase and the phospholipase C (41). Linkage of this latter signaling cascade to the modulation of cardiac contractility is provided by the observation that its bifurcated end products, inositol phosphate and diacylglycerol and protein kinase C activation, are known to influence myocardial inotropy in several ways. Considering that the number of genes which are growth regulated is very high, exceeding 500, elucidation of the intracellular pathways activated by GH and IGF-I is in its infancy.

Cellular level

Recent experimental evidence supports the notion that GH possesses the ability to modify the role of intracellular calcium in the excitation contraction coupling. In the tumour model of GH excess, Mayoux *et al.* demonstrated that the augmented contractility of the papillary muscle is paralleled by increased calcium responsiveness of the myofilament apparatus (18). In other words, for each calcium concentration, the heart was capable of developing more force. A subsequent study by Strömer *et al.* found that intracellular concentrations of calcium did not change in rat whole hearts after 4 weeks of treatment with high doses of recombinant GH (17). The main mechanism of positive inotropism was an increased responsiveness of myocardial myofilament to calcium.

Diametrical abnormalities were found when calcium handling was examined in the heart of dwarf rats, an animal model of GH deficiency. Papillary muscles of dwarf rats developed less force than control rats and these abnormalities were coupled to a major decrease of myofilament calcium responsiveness, with unchanged intracellular calcium levels (25). Impaired calcium responsiveness was reversible after 3 weeks of GH therapy. A more recent investigation by Cittadini and coworkers has shown that acute infusions of IGF-1 in the isovolumic, isolated rat whole heart slightly decrease intracellular calcium levels while sensitizing the myofilament apparatus to calcium (38). Interestingly, in this latter study, GH had no acute effects on myocardial contractility. Taken together, these studies demonstrate that GH excess augments myocardial contractility not by raising the amount of calcium available to the contractile machinery, but by increasing the responsiveness of the myofilament apparatus to calcium. This calcium sensitising effect takes place also independently of cardiac growth. At least in an acute setting, IGF-1 seems to mediate the effect of GH on contractility.

Another interesting feature of the GH and heart interrelationship is GH 's ability to affect the thermodynamic status of the contractile apparatus. Papillary muscle of rats chronically exposed to high levels of circulating GH develop more force than controls, despite myosin pheconversion towards the low ATPase activity V3 isoform (42,18). Furthermore, the content of high energy phosphates following acute IGF-1 application in the whole heart does not change in parallel with the increase of the developed tension (38). Therefore, at variance with other types of cardiac hypertrophy, the enhanced contractility evoked by the activation of the GH/IGF-1 axis does not seem to require more energy. The myocardium appears to work at a lower energetic cost.

Organ level

In addition to myocardial contractility, another major regulator of cardiac performance is the ventricular load. One mechanism responsible for the GH favorable impact on ventricular mechanics is the reduced afterload (*i.e.* the force limiting LV ejection). A reduced afterload, in turn, is expected to improve cardiac performance, as anticipated by the force-velocity relation.

End-systolic stress, which can be derived from Laplace's law (wall stress = [pressure x radius]/wall thickness), is regarded as an accurate index of afterload, and is consistently reduced in conditions of GH excess as a consequence of the concentric pattern of LV growth (increased wall thickness and unchanges radius). The decreased peripheral vascular resistance may provide a further contribution to maintaining the afterload low.

There is convincing evidence that conditions of GH excess induce cardiac hypertrophy. However, it is important to stress that such hypertrophy is concentric, and therefore characterized by increased values of relative wall thickness (*i.e.* ventricular thickness normalized by ventricular radius). Relative wall thickness in turn is a major determinant of wall stress, as predicted by Laplace's law. In other words, the higher the relative wall thickness, the lower wall stress and consequently the higher the shortening velocity and stroke volume.

The other mechanism by which GH may contribute to keeping afterload low and therefore to increase cardiac function is the reduction of peripheral vascular resistance. This effect has been consistently documented both in human and animal studies and seems related to the release of nitric oxide from the endothelium (43-47). Indeed, direct infusions of GH and IGF-1 into the brachial artery induce vasodilatation that is prevented by L-NAME, an inhibitor of the nitric oxide pathway. Furthermore, nitric oxide formation is reduced in GH deficient patients

(48). Substitutive therapy with GH reverses this abnormality and in parallel it decreases peripheral vascular resistance (48).

In addition to reducing afterload, GH may indirectly affect cardiac function also by increasing preload. This effect is due to its well recognized sodium retaining effect, which causes volume expansion. It has been reported that some subjects treated with exogenous GH develop joint swelling or ankle edema. The cardiac consequences of volume expansion are reflected by the increased LV end-diastolic dimension.

This brief overview of GH's mechanisms of action has focused on its cardiac effects. However, before considering the rationale of GH use in heart failure, it is worth noting that some of its extracardiac effects can be relevant to its therapeutic use. For example, GH improves exercise capacity, by increasing skeletal muscle mass and strength (5). This may in turn reverberate on cardiac performance through a conditioning mechanism. Furthermore, GH may also induce a sense of well-being, which might contribute to improving the clinical status of patients with heart failure.

RATIONALE for GH / IGF-1 USE in HEART FAILURE

The rationale for GH and IGF-1 use in heart failure appears evident when one considers the main morphological features of heart failure, on the one hand, and GH's effects on cardiovascular function, on the other hand. From a structural and functional standpoint, the principal abnormalities in heart failure are reduced contractility, LV cavity dilation, and increased peripheral vascular resistance. The consequent deterioration of cardiac performance and the increased LV volumes, according to Laplace's law, lead to a major increase of wall stress. Cardiac dilation, which helps maintaining an adequate stroke volume, initiates a vicious cycle whereby "dilation begets dilation".

The experimental evidence provided in the previous sections indicates that GH may be beneficial at all steps of heart failure. The activation of cardiac growth with a concentric pattern of remodeling reduces the augmented afterload by reducing wall stress, according to Laplace's law, as previously discussed. The vasodilating properties of GH may also attenuate the pathologic remodeling by decreasing peripheral vascular resistance. In addition to the improvement of cardiac function mediated by the unloading mechanism, GH' s positive inotropic effect provides an additional mechanism that may directly counteract the impaired contractility, which represents the *primum movens* of the vicious cycle responsible for pathologic remodeling. (Figure 1)

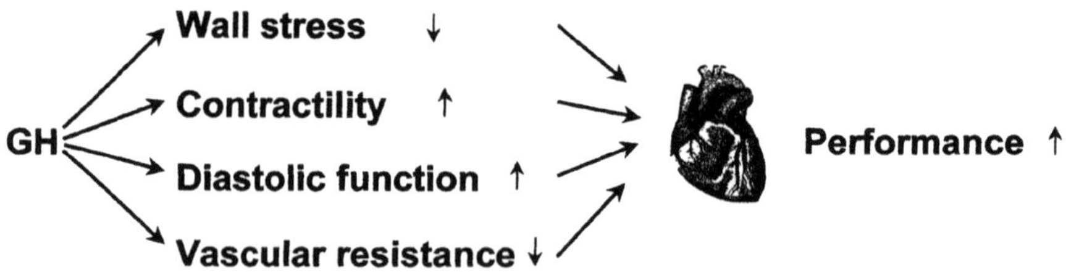

Table 1. GH/IGF-1 in experimantal heart failure (adapted from Cittadini *et al.* Endocrinol Metab 1997;4:27-31)

REFERENCE	*DRUG and DOSE*	*TREATMENT DURATION*	*EXPERIMENTAL SETTING*	*RESULTS*
Castagnino *et al.*, 1992 (49)	GH, 0.07 IU/kg/day	3 days	Post-infarction heart failure (early phase of remodelling)	Preservation of myocardial collagen framework. Reduced incidence of aneurysms
Ambler *et al.*, 1993 (50)	IGF-I, 0.8 mg/kg/day	3 weeks	Doxorubicin-induced cardiomyopathy (late phase of remodelling)	Increased cardiac output
Duerr *et al.*, 1995 (51)	IGF-I, 3 mg/kg/day	2 weeks	Post-infarction heart failure (early phase of remodelling)	Increased LV/tibial length Increased cardiac output Increased ejection fraction
Jin *et al.*, 1995 (52)	GH, 2 mg/kg/day,+ IGF-I, 2 mg/kg/day (in rats receiving chronic ACE inhibitors)	2 weeks	Post-infarction heart failure (late phase of remodelling)	Increased cardiac output Reduced systemic vascular resistance Reduced LV end-diastolic pressure
Yang *et al.*, 1995 (54)	GH, 2 mg/kg/day	2 weeks	Post-infarction heart failure (late phase of remodelling)	Increased cardiac output Reduced systemic vascular resistance Reduced LV end-diastolic pressure
Duerr *et al.*, 1996 (53)	GH, 2 mg/day (≈0.73 mg/kg/day)+ IGF-I, 3 mg/kg/day	4 weeks	Post-infarction heart failure (late phase of remodelling)	Increased cardiac output Reduced systemic vascular resistance Reduced LV dilatation
Shen *et al.*, 1996 (56)	GH, 3.2 mg/kg/day	4 weeks	Post-infarction heart failure (early phase of remodelling)	No significants changes
Cittadini *et al.*, 1997 (55)	GH, 3 mg/kg/day,	3 weeks	Post-infarction heart failure (early phase of remodelling)	Increased LV/tibial length Increased cardiac output Reduced systemic vascular resistance Reduced LV dilatation Increased contractility *in vitro*

GH / IGF-1 in EXPERIMENTAL HEART FAILURE

Several studies have investigated the effects of the activated GH/IGF-1 axis in experimental models of heart failure (Table 1). The first report dates back to 1992, when Castagnino *et al.* demonstrated that a 3-day treatment with GH (0.07 IU/Kg/day) decreased the incidence of post-infarction aneurysms from 25% to 7% (49). The rationale for this investigation was the previous finding that GH and IGF-1 activated fibroblast growth and proliferation. A subsequent study was performed in the doxorubicin-induced cardiomyopathy model. Ambler *et al.* demonstrated that 0.8 mg/kg/day of IGF-1 increased cardiac output as well as reduced histologically detected myocyte damage (50). In another investigation performed in the rat model of myocardial infarction, Duerr *et al.* provided evidence that exogenous administration of IGF-1 increases cardiac output and LV ejection fraction, in particular in animals with large infarctions, along with an induction of cardiac growth indicated by the higher LV/tibial length ratio (51).

Two subsequent investigations by Jin *et al.* and Duerr *et al* (52,53) addressed the question whether a combination therapy with GH and IGF-1 was able to improve cardiac function in a rat model of myocardial infarction. In both studies, therapy was initiated in the late phase of remodeling, 4 weeks after myocardial infarction, when most of pathologic remodeling had already occurred. The rationale for using GH and IGF-1 in combination was to potentiate the anabolic properties of the hormones while limiting their potential side effects on carbohydrate metabolism. Both studies yielded positive results with regard to cardiac function, reduction of peripheral vascular resistance, and reduced LV dilatation. Interestingly, the study by Jin et *al.* was performed with a background therapy using ACE inhibition (57). While the hypertrophic response of GH/IGF-1 therapy was blunted, their effects on remodeling and function were still evident, pointing to a GH induced functional benefit independent of myocardial growth.

Yang *et al.* and Cittadini *et al.*, using GH alone in late and early phases of post-infarction experimental heart failure, respectively, provided more straightforward evidence of GH' s ability to attenuate pathologic remodeling and to improve cardiac function (54.55). In both studies LV filling pressures recorded by high fidelility catheters were significantly reduced with increased cadiac output and decreased systemic vascular resistance. Moreover, hypertrophy of non-infarcted myocardium was documented both histologically and echocardiographically after 3-week therapy with 3 mg/kg/day of GH by Cittadini *et al* (55).

A recent animal study by Shen *et al.* reported results that were at variance with earlier studies (56). GH was ineffective in improving the hemodynamic profile in rats with myocardial infarction. This inconsistency may be related to the differences in the methodology employed to induce myocardial infarction. In fact, the authors made several deep ligatures into myocardial walls rather than occluding the left coronary artery, probably creating an unphysiological constraint to the remodeling process and to its modulation by GH.

An important piece of evidence supporting the utility of IGF-1 as cardioprotective agent has been provided by Lefer *et al.* in the ischemia-reperfusion rat model (57). Systemic administration of low dose IGF-1 significantly prevented CPK loss by reducing the apoptosis process and neutrophile accumulation which follows the ischemic process. This observation appears promising in view of the recent evidence showing the key role of the apoptotic process in the development of heart failure.

Taken together, studies performed in experimental models of heart failure are encouraging and provide a solid background for the implementation of studies in human heart failure of various etiologies.

GH / IGF-1 in HUMAN HEART FAILURE

Despite the large body of information from animal models of heart failure, few studies have addressed the question whether the activation of GH/IGF-1 axis may be beneficial in human heart failure. Three case reports have reported an impressive improvement of cardiac function following GH therapy in patients with GH deficiency who developed severe heart failure during the course of the disease and did not respond to conventional therapy (58-60). Given the very high doses of GH employed, the documented beneficial effect may depend not only on GH replacement, but also on a pharmacological effect. The only preliminary study conducted so far has been reported by Fazio *et al.* in seven patients with idiopathic dilated cardiomyopathy (61). Three month treatment with low doses of GH improved exercise tolerance, hemodynamic profile, myocardial energetics, and mechanical efficiency. There was a hypertrophic response indicated by transthoracic echocardiography, with a reshaped LV geometry, and a dramatic decrease of wall stress. Cardiac ouput rose, in particular during exercise, as well as ventricular work, despite reductions in myocardial oxygen consumption and energy production. Moreover, clinical symptoms, exercise capacity, and patients' quality of life improved. Altough promising, these results should be interpreted with caution in view of the lack af an adequate control group, and of the small number of patients studied.

IGF-1 may have positive inotropic actions independent of cardiac growth. Acute administration of IGF-1 have been recently demonstrated to increase cardiac output in patients with heart failure in a double-blind placebo-controlled study by Sütsch *et al* (62). This observation is congruent with the increase in contractility demonstrated by IGF-1 in a variety of isolated cardiac muscle preparations, ranging from isolated myocyte to papillary muscle and isolated whole heart. Another support to the hypothesis that GH/IGF-1 may advantageously affect patients suffering from heart failure comes from two recent reports (63,64). In one study, GH was administered at high doses (10-14 U/day) to two patients with end-stage heart failure and cardiac cachexia (63). Both subjects achieved a significant improvement in their clinical status and functional capacity: one was successfully bridged to cardiac transplantation and the second improved to such a degree that he was delisted. The second study describes beneficial hemodynamic effects of intravenous GH in 12 patients with congestive heart failure (64). After 24 hours of GH infusion, mean pulmonary artery pressure decreased by approximately 25% while cardiac index increased by more than 50%.

FUTURE DIRECTIONS

The demonstration that GH and IGF-1 target the heart and have beneficial effects in experimental heart failure and in preliminary human studies raises a number of questions.

Basic research studies should ideally parallel clinical studies trying to unravel more mechanistic issues. Although convincing, the studies performed with the rat model of myocardial infarction lack a long-term proof of efficacy. Therefore, the availability of recombinant rat GH should foster the initiation of survival studies in this model system. Secondly, the mechanism of action of both GH and IGF-1 on myocardial function is unknown. Changes in myofilament calcium responsiveness appear to play a prominent role in the direct effects of GH and IGF-1 on contractility. A thorough examination of the events downwards of the hormone-receptor ligand is necessary to identify the intracellular signaling pathways activated by both GH and IGF-1 in the cardiac cell, and responsible for the myofilament calcium sensitization.

The potential of GH and IGF-1 in other experimental settings such as cardiac apoptosis and myocardial stunning deserves further attention considering the enormous clinical impact of both conditions.

Further clinical trials are needed to confirm the preliminary findings in a larger patient population with a placebo double-blind design. Once this information will become available, larger prospective trials should address whether GH induced short-term beneficial effects will translate into improvement of prognosis in patients suffering from heart failure. The data from previous trials have shown that a clinical and functional improvement in the short term does not necessarily imply that the long term prognosis is improved as well. Studies performed using phosphodiesterase inhibitor drugs have indicated that the opposite may be true. Long-term clinical studies will also answer the fundamental question whether a prolonged GH therapy in patients with heart failure will induce the same adverse effects frequently encountered in acromegalic patients, such as interstitial remodeling and diastolic dysfunction.

A final consideration relates to the catabolic syndrome that in some instances is associated with heart failure, in particular during its end-stage, a condition of wasting also termed cardiac cachexia. Recently, wasting has been demonstrated to be an independent predictor of mortality in heart failure (65). Cardiac cachexia is also associated with impaired secretion of anabolic hormones and with some degree of GH resistance that could be overcome by exogenous GH administration (66). Therefore, clinical studies in patients with end-stage heart failure and cardiac cachexia are needed to test the potential efficacy of GH in this clinical setting.

REFERENCES

1. Braunwald E, Grossman W. Clinical aspects of heart failure. In *Heart Disease*. Ed. Braunwald E. Philadelphia: WB Saunders Co; 1992:444-463.
2. The SOLVD Investigators. Effect of enalapril on survival in patients with reduced left ventricular ejection fraction and congestive heart failure. *N Engl J Med 1991; 325:293-302*
3. Packer M, Bristow MR, Cohn JN, Colucci, WS, Fowler MB, Gilbert EM, Shusterman NH. The effect of carvedilol on morbidity and mortality in patients with chronic heart failure. *N Engl J Med 1996 ; 334 :1349-55*
4. The Digitalis Investigation Group. The effect of digoxin on mortality and morbidity in patients with heart failure. *N Engl J Med 1997; 336:525-533*
5. Sacca' L, Cittadini A, Fazio S. Growth hormone and the heart. *Endocrine Reviews 1994 ;15:555-573*
6. Huchard H. Anatomie pathologique, lesions et trouble cardiovasculaires de l'acromegalie. *J Praticiens 1895;9:249-251*
7. Beznàk M. The restoration of cardiac hypertrophy and blood pressure in hypophysectomized rats by large doses of lyophilized anterior pituitary and growth hormone. *J Physiol 1954;124:64-74*
8. Beznàk M. The effect of pituitary growth hormone on the blood pressure and on the ability of the heart to hypertrophy. *J Physiol 1952a;116:74-83*
9. Thorner MO, Vance ML, Horvath E, Kovacs K. The anterior pituitary. In: Wilson JD, Foster DW, eds. Williams Textbook of Endocrinology (8th edition). Philadelphia: WB Saunders Co; 1992:221-310
10. Mc Guffin Jr. WL, Sherman BM, Roth J, Gorden P, Kahn CR, Roberts WC, Frommer PL. Acromegaly and cardiovascular disordes. A prospective study. *Ann Intern Med 1974;81:11-18*
11. Smallridge RC, Rajfer S, Davia J, Schaaf M. Acromegaly and the heart. An echocardiographic study. *Am J Med 1979;66:22-27*
12. Morvan D, Komajda M, Grimaldi A, Turpin G, Grosgogeat Y. Cardiac hypertrophy and function in asymptomatic acromegaly. *Eur Heart J 1991;12:666-672*
13. Fazio S, Cittadini A, Sabatini D, Merola B, Colao AM, Biondi B, Lombardi G, Sacca' L. Evidence for biventricular involvement in acromegaly: a Doppler echocardiographic study. *Eur Heart J 1993;14:26-33*
14. Fazio S, Cittadini A, Cuocolo A, Merola B, Sabatini D, Nicolai E, Colao AM, Biondi B, Lombardi G, Sacca' L. Impaired cardiac performance is a distinct feature of uncomplicated acromegaly, *J Clin Endocrinol Metab 1994;79:441-446*
15. Merola B, Cittadini A, Colao A, Ferone D, Fazio S, Sabatini D, Biondi B, Sacca' L, Lombardi G. Chronic treatment with octreotide improves cardiac abnormalities in patients with acromegaly. *J Clin Endocrinol Metab 1993;77:790-793*
16. Cittadini A, Strömer H, Katz SE, Clark R, Moses AC, Morgan JP, Douglas PS.Differential cardiac effects of growth hormone and IGF-I in the rat: a combined in *vivo* and in *vitro* evaluation. *Circulation 1996;93:800-809*

17. Strömer H, Cittadini A, Douglas PS, Morgan JP. Exogenously administered growth hormone and IGF-I alter intracellular calcium handling and enhance cardiac performance: in *vitro* evaluation in the isolated isovolumic buffer perfused rat heart. *Circ Res 1996;79:227-236*
18. Mayoux E, Ventura-Clapier R, Timsit J, Hoffmann Bèhar-Choen C, Mercadier J. Mechanical properties of rat cardiac skinned fibers are altered by chronic growth hormone hypersecretion. *Circ Res 1993;72:57-64*
19. Thuesen L, Christiansen JS, Sorensen KE, Jorgensen JOL, Orskov H, Henningsen P. Increased myocardial conctractility following growth hormone administration in normal man. *Dan Med Bull 1988;35:193-196*
20. Fazio S, Colao A, Sabatini D, Merola B, Riccio G, Ferone D, Biondi B, Lombardi G, Sacca' L. Effects of short-term acromegaly on cardiac morphology and function. In: Abstracs of the International Congress of Endocrinology, San Francisco, June 12-15, 1996. Bethesda, MD: Endocrine Society 1996: P2-P272
21. Merola B, Cittadini A, Colao A, Longobardi S, Fazio S, Sabatini D, Sacca' L, Lombardi G. Cardiac structural and functional abnormalities in adult patients with growth hormone deficiency. *J Clin Endocrinol Metab 1993;77:1658-1661*
22. Amato G, Carella C, Fazio S, La Montagna G, Cittadini A, Sabatini D, Marciano-Mone C, Sacca' L, Bellastella A. Body composition, bone metabolism, heart structure and function in growth hormone deficient adult before and after growth hormone replacement therapy at low doses. *J Clin Endocrinol Metab 1993;77:1671-1676*
23. Cittadini A, Cuocolo A, Merola B, Fazio S, Sabatini D, Nicolai E, Colao A, Longobardi S, Lombardi G, Sacca' L. Impaired cardiac performance in growth hormone deficient adults and its improvement after growth hormone replacement. *Am J Physiol 1994;267:E219-E225*
24. Cittadini A, Strömer H, Vatner DE, Grossman JD, Katz ES, Clark R, Morgan JP, Douglas PS. Consequences of GH deficiency on cardiac structure, function, and β-adrenergic pathway: Studies in mutant dwarf rats.*Endocrinology* 1997: *in press*
25. Strömer H, Cittadini A Grossman JD, Morgan JP. Normal cardiac function requires growth hormone.*Circulation 1996;94:I-421*
26. Regulation of myocardial conctractility. In: *Physiology of the Heart.* Ed. Katz AM. *New York, Raven press: 1992:319*
27. Mathews LS, Enberg B, Norstedt G. Regulation of rat growth hormone receptor gene expression. *J Biol Chem 1989;17:9905-9910*
28. D'Ercole AJ, Stiles AD, Underwood LE. Tissue concentrations of somatomedin C: further evidence for multiple sites of synthesis and paracrine or autocrine mechanisms of action. *Proc Natl Acad Sci USA 1984;81:935-939*
29. Engelmann GL, Bohem KD, Haskell JF, Khairallah PA, Ilan J. Insulin-like growth factors and neonatal cardiomyocyte development: ventricular gene expression and membrane receptor variations in normotensive and hypertensive rats. *Mol Cell Endocrinol 1989;63:1-14*
30. Guler H-P, Zapf J, Scheiwiller E, Froesch ER. Recombinant human insulin-like growth factor I stimulates growth and has distinct effects on organ size in hypophysectomized rats. *Proc Natl Acad Sci USA 1988;85:4889-4893*

31. Wahlander H, Isgaard J, Jennische E, Friberg P. Left ventricular insulin-like growth factor I increases in early renal hypertension. *Hypertension 1992;19:25-32*
32. Hanson MC, Kenneth AF, Alexander RW, De Lafontaine P. Induction of cardiac insulin-like growth factor I gene expression in pressure overload hypertrophy. *Am J Med Sci 1993;306:69-74*
33. Isgaard J, Wahlander H, Adams MA, Friberg P. Increased expression of growyh hormone receptor mRNA and insulin-like growth factor-I mRNA in volume-overloaded hearts. *Hypertension 1994;23:884-888*
34. Ito H, Hiroe M, Hirata Y, Motoyoshy T, Adachi S, Masayoshi S, Koike A, Nogami A, Marumo F. Insulin-like growth factor-I induces hypertrophy with enhanced expression of muscle specific genes in cultured rat cardyomicytes. *Circulation 1993;87:1715-1721*
35. Fong Y, Rosenbaum M, Tracey KJ, Raman G, Hesse DG, Matthews DE, Leibel RL, Gertner JM, Fischman DA, Lowry SF. Recombinant growth hormone enhances muscle myosin heavy-chain mRNA accumulation and amino acid accrual in humans. *Proc Natl Acad Sci USA* 1989;86:3371-3374
36. Hjalmarson A, Isaksson O, Ahren K. Effects of growth hormone and insulin on amino acid transport in perfused heart rat. *Am J Physiol 1969;217:1795-1802*
37. Sotiropoulos A, Perrot-Applanat M, Dinerstein H, et al. Distinct cytoplasmic regions of the growth hormone receptor are required for activation of JAK2, mitogen-activated protein kinase, and transcription. *Endocrinology 1994;135:1289-1291*
38. Cittadini A, Spindler M, Strömer H, Ishguro Y, Travers KE, Ingwall JS, Morgan JP. Mechanisms of the IGF-I positive inotropic action. *J Am Coll Cardiol 1997;29:404A (abstract)*
39. Jones JI, Clemmons DR. Insulin-like growth factors and their binding proteins: biological actions. *Endocrin Rev 1995;16:3.34*
40. Guse AH, Kiess W, Funk B, Kessler U, Berg I, Gercken G. Identification and characterization of insulin-like growth factor receptors on adult rat cardiac myocytes: linkage to inositol 1,4,5-triphosphate formation. *Endocrinology 1992;130:145-151*
41. Foncea R, Andersson M, Ketterman A, Blakesley V, Sapag-Hagar M, Sugden PH, LeRoith D, Lavandero S. Insulin-like growth factor-I rapidly activates multiple signal transduction pathways in cultured rat cardiac myocytes. *J Bioch Chem 1997;272:19115-19124*
42. Timsit J, Riou B, Bertherat J, Wisnewsky C, Kato NS, Weisberg AS, Lubetzki J, Lecarpentier Y, Winegrad S, Mercadier JJ. Effects of chronic growth hormone hypersecretion on intrinsic contractility, energetic, isomyosin pattern, and myosin adenosine triphosphatase activity of rat left ventricle. *J Clin Invest 1990;86:507-515*
43. Fryburg DA, Gelfand RA, Barrett JB. Growth hormone acutely stimulates forearm muscle protein synthesis in normal humanes. *Am J Physiol 1991;260:E499-E504*
44. Fryburg DA. Insulin-like growth factor-I exerts growth hormone and insulin-like actions on human muscle protein. *Am J Physiol 1994;267:E331-E336*

45. Wu HY, Yue CL, Chyu KY, Hsueh WA, Chan TM. Endothelial-dependent vascular effects of insulin-like factor I in the perfused rat mesenteric artery and aortic ring. *Diabetes 1994;43:1027-1032*
46. Haylor J, Singh I, El Nahas AM. Nitric oxide inhibitor prevents vasodilatation by insulin-like growth factor I. *Kidney Int 1991;39:333-335*
47. Tsukahara H, Gordienko DV, Tonshoff B, Gelato MC, Goligorsky MS. Direct demostration of insulin-like factor-I-induced nitric oxide production by endothelial cells. *Kidney Int 1994;45:598-604*
48. Böger RH, Skamira C, Bode-Boger SM, et al. Nitric oxide may mediate the hemodynamic effects of recombinant growth hormone in patients with acquired growth hormone deficiency. A double-blind, placebo-controlled study. *J Clin Invest 1996;98:2706-2713*
49. Castagnino HE, Toranzos FA, Milei J, Weiss V, Beigelman R, Sarchi MI, Bordenave CA, Azcoaga R. Preservation of the myocardial collagen framework by human growth hormone in experimental infarction and reduction in the incidence of ventricular aneurysms. *Int J Cardiol 1992;35:101-114*
50. Ambler GR, Johnston BM, Maxwell L, Gravin JB, Gluckman PD. Improvement of doxorubicin induced cardiomyopathy in rats treated with insulin-like growth factor I. *Cardiovasc Res 1993;27:1368-1373*
51. Duerr RL, Huang S, Miraliakbar HR, Clark R Chien KR, Ross J Jr. Insulin-like growth factor I enhances ventricular hypertrophy and function during the onset of experimental cardiac failure. *J Clin Invest 1995;95:619-627*
52. Jin H, Yang R, Gillet N, Clark R, Ko A, Paoni NF. Beneficial effects of growth hormone and insulin-like growth factor-I in experimental heart failure in rats with chronic ACE inhibition. *J Cardiovasc Pharmacol 1995,26:420-425*
53. Duerr RL, McKirnan D, Gim RD, Clark R, Chien KR, Ross J Jr. Cardiovascular effects of Insulin-like growth factor I and growth hormone in chronic left ventricular failure in the rat. *Circulation 1996,93:2188-2196*
54. Yang R, Bunting S, Gillet N, Jin H. Growth hormone improves cardiac performance in experimental heart failure. *Circulation 1995;92:262-267*
55. Cittadini A, Grossman JD, Napoli R, Katz SE, Strömer H, Smith RJ, Clark R, Morgan JP, Douglas PS. Growth hormone attenuates early left ventricular remodeling and improves cardiac function in rats with large myocardial infarction. *J Am Coll Cardiol 1997;29:1109-1116*
56. Shen Y-T, Wiedmann RT, Lynch JJ, Grossman W, Johnson G. GH replacement fails to improve ventricular function in hypophysectomized rats with myocardial infarction. *Am J Phsiol 1996;271:H1721-H1727*
57. Burke M, Murohara T, Skurk C, Nuss C, Tomaselli K, Lefer AM. Cardioprotective effect of insulin-like growth factor I in myocardial ischemia followed by reperfusion. *Proc Natl Acad Sci USA 1995;92:8031-8035*
58. Cuneo RC, Wilmshurst P, Lowy C, McGauley G, Sonksen PH. Cardiac failure responding to growth hormone. *Lancet 1989;1;838-839*
59. Frustaci A, Perrone GA, Gentiloni N, Russo MA. Reversible dilated cardiomyopathy due to growth hormone deficiency. *Am J Clin Pathol 1992;97:503-511*
60. Fazio S, Biondi B, Sabatini D, et al. Long-term growth hormone deficiency as a cause of cardiomyophathy and its reversibility with specific replacement therapy. *J Clin Endocrinol Metab 1996;81:887-890*

61. Fazio S, Sabatini D, Capaldo B, Vigorito C, Giordano A, Guida R, Pardo F, Biondi B, Sacca' L. A preliminary study of growth hormone in the treatment of heart failure in patients with dilated cardiomyopathy *N Eng J Med 1996;334:809-814*
62. Sütsch G, Donath M, Yan XW, Brunner HP, Follath F, Froesch ER, Kiowski W. Beneficial hemodynamic effects of acute IGF-I in patients with chronic heart failure. *J Am Coll Cardiol 1997;29:285A (abstract)*
63. O'Driscoll JG, Green DJ, Ireland M, Kerr D, Larbalestier RI. Treatment of end-stage cardiac failure with growth hormone. *Lancet 1997;349:1068*
64. Volterrani M, Desenzani P, Lorusso R, D'Aloia A, Manelli F, Giustina A. Haemodynamic effects of intravenous growth hormone in congestive heart failure. *Lancet 1997;349:1067-1068*
65. Anker SD, Ponikowski P, Varney S, Chua TP, Clark AL, Webb-Peploe KM, Harrington D, Kox WJ, Poole-Wilson PA, Coats AJS. Wasting as independent risk factor for mortality in chronic heart failure. *Lancet 1997;349:1050-1053*
66. Anker SD, Chua TP, Ponikowski P, Harrington D, Swan JW, Kox WJ, Poole-Wilson PA, Coats AJS. Hormonal changes and catabolic/anabolic imbalance in chronic heart failure and their importance for cardiac cachexia. *Circulation 1997;96:526-534*

21 GH AND CATABOLIC STATES

Douglas Wilmore, M.D.

From the Laboratories for Surgical Metabolism and Nutrition, Department of Surgery, Brigham and Women's Hospital, Harvard Medical School

Accelerated weight loss, associated with a decrease in body fat and skeletal muscle mass characterizes the metabolic response to catabolic illness. While these events are commonly observed in patients who have sustained extensive injury or severe infection, similar responses have been described in individuals with other disorders such as multiple organ systems failure, patients with chronic obstructive lung disease, those with AIDS and individuals undergoing extensive chemotherapy and radiation treatment for cancer.[1]

Adipose tissue represents a form of stored energy and use of this body component as a fuel source is appropriate during times of stress. Body protein, which is a critical component of the lean body mass, is not stored but comprises structural or functional tissue. Loss of body protein from this compartment is associated with loss of function. While these changes may not be obvious in the short term, (5-7 days in a normally nourished individual), unchecked catabolism for more than 7-10 days results in increased morbidity and prolonged convalescence. Such responses impede the patient's ability to return to a healthy life.

The accelerated net breakdown of body mass can be attenuated in critically ill patients by the administration of adequate nutrients. However, serial measurements of body composition[2] and substrate-flux studies[3] indicate that it is extremely difficult to maintain or replenish body protein during the catabolic state. Many factors may contribute to this response including bed rest, sedation, mechanical ventilation and curarization, inotropic support and inadequate exercise. However, all of these factors commonly effect patients in most modern critical care units and these measures contribe to the survival of almost all seriously ill patients. In view of these practices, it can be concluded that aggressive nutritional support does not prevent substantial loss of body protein during these severe catabolic illnesses;[4] such depletion is related to increased morbidity and mortality.

The mediators of these complex catabolic events are associated with the inflammation that follows infection and injury. These mediators can be grouped into three broad categories: hormones, cytokines, and lipid mediators. A simple cascade of events does not usually occur after major inflammatory stimuli; the response patterns depend on the stimulus, and release of multiple mediators often occurs. Moreover, these three catabolic response systems are interconnected by amplification loops and negative and positive feedback signals. These responses have proven extremely difficult to block or significantly attenuate as a strategy to decrease catabolism.

Another approach, however, is to counterbalance the catabolic mediators with anabolic factors. In an effort to understand the endogenous metabolism of growth hormone (GH) in catabolic patients, a variety of studies have recently been performed. In one report, GH levels were 4.5 mU/L, greater than age match and gender match controls, but similar to the same controls when fasted.[5] Using single sample analysis, mean GH levels were found to be elevated in burn patients and one study reported that concentrations remained greater than controls for 24 days.[6] Both nutritional state and pharmacological therapy effect the hypothalamic-pituitary response and it has been shown in very detailed studies that inotropic support with dopamine reduces GH elaborations.[7] When examining the effect of dopamine on GH secretion in the same patient, the mean GH concentration was 6.2 mU/L in the control period and 4.2 mU/L during dopamine infusion, demonstrating that this commonly used ionotrope suppressed GH elaboration.

Almost all studies support the contention that there is increased GH secretion associated with critical illness (the exception being the response during dopamine infusion) and that this is due in large part to the raised interpulse GH levels, similar to those observed during fasting.[8] However, in this catabolic setting, insulin-like growth factor-1 (IGF-1) concentrations have been found to be in the low or low-normal range.[5,9]

This reduction in IGF-1 level is related to a number of factors: reduced IGFBP-3, the main carrier protein;[10] clevage of IGF-1 from its binding proteins by circulating proteases;[11] and increased clearance of IGF-1 from the plasma compartment.[12,13] In addition, there is decreased GH receptor number on skeletal muscle following operation.[14] These changes account for the apparent GH resistance observed in critically ill patients; that is, for any level of GH the IGF-1 level will be substantially lower in seriously-ill patients when compared to IGF-1 concentrations in normals. To overcome this effect, larger doses of GH need to be given to catabolic patients to achieve the same anabolic endpoints observed when GH is administered to normals or those without a hypercatabolic disease process.

GH AND PERIOPERATIVE METABOLISM

The usual metabolic response to an operation is increased excretion of nitrogen in the urine. While the extent of nitrogen loss varies depending on the nutritional status of the host, the underlying disease processes, the type of anesthesia and the duration and extent of the surgical procedure, the postoperative nitrogen loss has been so predictable and reproducible that investigators have frequently utilized these patients as a model of the catabolic state. There has been a long-term interest in altering the postoperative nitrogen loss following an operation by using anabolic agents such as testosterone, anabolic steroids, or extracts of the anterior pituitary gland. However, it was not until the development and availability of bioengineered GH, that studies of sufficient duration containing significant numbers of patients were possible.

Ward et al administered GH (0.3U/kg.day or 0.12 mg/kg.day) or placebo to postoperative patients receiving approximately 400 calories from 5% dextrose solutions.[15] Nitrogen excretion over 6 days was reduced from 42.7g to 24.0 and fat oxidation increased with GH treatment. There was also a rise in serum glucose, insulin and IGF-1. A similar study was performed by Carli et al, who confirmed the GH effect on enhancing postoperative protein synthesis.[16]

Manson and Wilmore demonstrated that positive nitrogen balance could be achieved in normal volunteers by providing adequate nitrogen but only 50% of the required energy.[17] This approach was applied by Jiang and associates who studied a group of postoperative surgical patients following gastrectomy or colectomy.[18] The patients received parenteral nutrition containing 20 kcal/kg day and 1 gram protein/kg day and were randomized in a blinded manner to receive placebo or GH (0.06 mg/kg day or 0.15 U/kg.day). The growth hormone subjects lost significantly less weight (1.3 vs. 3.3 kg); cumulative nitrogen loss over 8 days was only 7.1 g. vs. 32.6 g in controls. Amino acid flux studies across the forearm demonstrate that the GH treated patients had increased uptake of amino acids into forearm skeletal muscle while the control subjects released amino acids from their forearm. In the control patients, hand grip force decreased approximately 10% following operation while the patients who received GH maintained their grip strength throughout the postoperative period.

This maintenance of strength was the first demonstration that this anabolic therapy could effect a positive functional outcome in catabolic patients. These effects on handgrip and respiratory muscle strength have been confirmed by others.[19] One group has extended this concept and presented evidence that the administration of a low dose of GH (8 U/day) along with postoperative intravenous feedings will significantly reduce postoperative fatigue.[20] The applicability and cost-effectiveness of GH following operation in elective surgical patients will need to be determined to justify its use over the coming years.

An alternative approach to the patient requiring an elective operation is to treat the individual before their procedure. In the depleted patient, this strategy should replete a portion of their lean tissue mass, and may be useful to enhance postoperative anabolism and boost immunological responses.

To quantify this effect, Byrne et al fed depleted surgical patients more than maintenance quantities of calories and protein in order to restore body mass, with targeted levels being 50 kcal/kg body weight and 2 g protein/kg.[21] Of the 14 patients studied, four received standard nutritional support and ten received similar support but with the addition of GH (0.14 mg/kg). Balance studies were performed daily, along with the measurement of respiratory gas exchange. Body composition was analyzed before and at the conclusion of the 3-week treatment period. GH-treated patients gained significantly more lean mass and protein-containing tissue ($P < 0.05$) than those receiving conventional therapy. This increase was not associated with a significant gain in body fat or with an inappropriate expansion of body water. The patients on standard therapy had increased fat deposition and a disproportionate increase in extracellular water (Table I). Patients receiving conventional nutritional support used a greater proportion of the available energy for storage as adipose tissue. The GH-treated patients gained 1.66 times more body protein per available calories than did those receiving standard therapy.

How do we apply this approach to other patients? A reasonable goal in patients with weight loss would be to restore body protein to within 10% of normal predicted protein mass. This data suggests that the time needed to achieve this level of repletion would be about 66% longer if GH was not administered. In a 60 kg individual, repletion of 20% of body protein to a satisfactory level would require approximately 3-4 weeks of nutritional therapy. Utilizing GH, this same therapeutic effect could be accomplished in 13-16 days. A reduction of 9-11 days in the length of time spent in hospital would achieve significant cost savings, which would be more than sufficient to pay for the GH therapy.

Table I Changes in body composition following 3 weeks of therapy

	ABSOLUTE (KG OR L)		RELATIVE (%)	
	STANDARD	GH	STANDARD	GH
WEIGHT	3.06	4.65	5.1	8.2
FAT	0.97	0.26	26.1	0.4
LEAN BODY MASS	1.99	4.31*	70.5	98.6
BODY PROTEIN	0.09	1.42*	4.3	45.5*
TOTAL BODY WATER	1.90	2.89+	66.3	53.1

*P < 0.05, + approximately one-third of the gain in total body water occurred in the intracellular compartment in the GH group, but was all extracellular water in the standard group.

What dose of GH should be used in postoperative patients? In the literature, doses range from 0.03 - 0.14 mg/kg.day (0.075 - 0.35 U/kg.day). Jauch and associates performed a dose response study in postoperative patients and saw an increasing effect of dose on nitrogen retention over a wide range.[22] In general, administration of 0.1 mg/kg day approaches the upper limit of effects on protein metabolism without causing significant side effects and this is an appropriate dose if contraindications are not present.

GH FOLLOWING ACCIDENTAL INJURY AND BURNS

Early studies performed independently by Liljedahl, Soroff and Wilmore all established that improved nitrogen balance could be achieved in injured patients following GH administration.[23] With the availability of recombinant GH, more extensive trials were performed, confirming these findings [24,25]; these studies also suggested that GH might confirm benefits in patients by enhancing respiratory muscle strength and shortening convalescent recovery.[25] It should be realized that these effects were achieved by administering reasonably large doses of GH (0.15 and 0.13 mg/kg.day) to injured patients and were accompanied by the increased concentration of IGF-1 and insulin with no significant effects on the elaboration of catabolic hormones such as glucocorticoids.

In 1989, Shernan and associates[26] and Herndon, et al[27] reported that GH enhanced wound healing in adult and pediatric burn patients, respectively. The study in children is notable: young children (average age 9 years) with extensive burns (average 59% of total body surface) received GH (0.2 mg/kg.day) or placebo injections until their burns were healed or adequately skin grafted. The rate of wound healing was increased in the GH group by approximately 20%, which reduced the length of hospital stay by about two weeks (Table II). This and a subsequent study[28] appear to document clear cut efficacy of GH in injured patients with large open wounds.

Table II The effect of GH in children who sustained major burns

	PLACEBO	GH
Number of patients	18	10
Age (years)	10.2 ± 1.2	7.0 ± 1.3
Total surface area of burn (%)	60 ± 3	57 ± 5
Percentage third-degree burns	43 ± 5	41 ± 7
Healing time of first donor site (days)	9.1 ± 0.4	7.4 ± 0.5[a]
Length of hospitalization per % total body surface burn (days per % total burns)	0.80 ± 0.09	0.54 ± 0.04[a]

[a] $P < 0.05$ vs placebo (Student's t-test)

GH IN SEPSIS, CRITICAL ILLNESS AND MULTIORGAN FAILURE

Acute infection is usually superimposed on other illnesses and this complication enhances the body's catabolic response. Because infection increases morbidity and prolongs recovery, GH has been considered a potential agent to attenuate these responses. In one study, net protein catabolism was determined by stable isotopic techniques in septic surgical patients receiving intravenous feedings with and without GH (give at a dose of 0.3U/kg.day twice daily).[29] In the control group, net protein catabolism gradually fell from 1.12 to 0.61 g protein/kg-day. In the GH treated group, this change was significantly greater; 0.93 to -0.20. Although this group moved into positive protein balance, no significant difference in outcome was reported.

It has been suggested that GH may enhance respiratory muscle function in critically ill patients. Pape et al tested this hypothesis by studying patients with stable chronic obstructive pulmonary disease (COPD).[30] They demonstrated an enhancement in primary function, which was related to improved respiratory muscle strength in patients receiving GH; this effect was not observed in the placebo group. Others, however, have not reported an advantage in administering GH to patients with COPD.[31]

In a phase I trial, Knox and associates administered GH to 53 patients in a surgical intensive care unit.[32] These patients were specifically selected because of decreased respiratory muscle strength, not COPD, and had failed ventilator weaning protocols. The patients were on average 63 years old and had a wide variety of co-existing morbidities. The average number of operative procedures they underwent during the period of GH administration was 3.9 per patient. GH was given for an average of 38 days; 81% of this population was weaned from mechanical ventilation. The overall survival rate of 76%. This 24% mortality rate compares quite favorably with the overall predicted mortality of 42% using the APACHE II scoring system, or 44% using the organ failure scoring system. The apparent improvement in survival occurred even though the patients had an average of 3.4 failed organs per individual. This study suggests that carefully selected ICU patients may benefit from GH therapy; a randomized, blinded trial is required to confirm this concept.

Finally it should be noted that complications have been noted in critically ill patients, particularly with high dose GH administration. These included hyperglycemia,[24, 25, 29] fluid retention and hypercalcemia.[33] Physicians giving GH in the intensive care unit setting should realize the potential for these side effects, and patients should be monitored and treated accordingly.

OTHER AREAS

GH therapy has been evaluated in a variety of other groups who have diseases associated with more chronic catabolic changes. Beneficial effects have been reported with GH administration in patients with AIDS wasting,[34] those on chronic hemodialysis,[35] and patients with the short bowel syndrome.[36] Rehabilitation following major operations such as total hip replacement may also be positively effected by such treatment, and studies are now being performed in this area. In addition, IGF-1 has been given with GH in catabolic conditions to enhance the anabolic signal. In all cases, cost must be balanced with clinical efficiency and the clinical benefit as perceived by the patient.

REFERENCES

1.Wilmore DW. Catabolic illness, strategies for enhancing recovery. NEJ Med 1991; 325:695-702.

2. Streat SJ, Beddoe AH, Hill GL. Aggressive nutritional support does not prevent protein loss despite fat gain in septic intensive care patients. J Trauma 1987;27:262-6.

3.Loder PB, Smith RC, Kee AJ, et al. What rate of infusion of intravenous nutrition solution is required to stimulate uptake of amino acids by peripheral tissues in depleted patients? Ann Surg 1990;211:360-8.

4.Warnold I, Eden E, Lundholm K. The inefficiency of total parenteral nutrition to stimulate protein synthesis in moderately malnourished patients.Ann Surg 1982;208:M3-49.

5. Ross RJM, Miell J, Freeman E, et al. Critically ill patients have high basal growth hormone levels with attenuated oscillatory activity associated with low levels of insulin like growth factor-1. Clin Endocrinol 1991;35:47-54.

6. Nygren J, Sammann M, Malm M, et al. Disturbed anabolic hormonal patterns in burned patients: the relation to glucagon. Clin Endocrinol 1995;43:491-500.

7. Van Den Berghe G, De Zegher F, Lauwers P, Veldhuis JD. Growth hormone secretion in critical illness: effect of dopamine. J Clin Endocrinol Metab 1994;79:1141-1146.

8. Ho KY, Veldhuis JD, Johnson ML, et al. Fasting enhances growth hormone secretion and amplifies the complex rhythms of growth hormone secretion in man. J Clin Endocrinol Metab 1988;81:968-975.

9. Ross RJM and Chew SL. Acquired growth hormone resistance. European J Endo 1995;132:655-60.

10. Ross RJM, Miell JP, Holly JMP, et al. Levels of GH and IGF binding proteins, insulin, blood glucose and cortisol in intensive care patients. Clin Endocrinol 1991;35:361-67.

11. Holly JMP, Claffey DCP, Cwyfan-Hughes SC, et al. Proteases acting on IGFBPs; their occurrence and physiological significance. Growth Reg 1993;3:88-91.

12. Miell JP, Taylor AM, Jones J, et al. Administration of human recombinant insulin- like growth factor-1 to patients following major gastrointestinal surgery. Clin Endocrinol 1992;37:542-51.

13. Herndon DN, Barrow RE, Kunkel KR. Effect of recombinant human growth hormone on donor-site healing in severely burned children. Ann Surg 1990;212:424-29.

14. Hermansson M, Wickelgren RB, Hammargvist F, et al. Measurement of human growth hormone receptor messenger ribonucleic acid by a quantitative polymerase chain reaction-based assay: demonstration of reduced expression after elective surgery. J Clin Endo and Metab 1997;82:421-8.

15. Ward HC, Halliday D, Sim JW. Protein and energy metabolism and biosynthetic human growth hormone after gastrointestinal surgery. Ann Surg 1987;206:56-61.

20. Vara-Thorbeck R, Guerrero JA, Ruiz-Requena E, Garcia-Carriazo M. Can the use of growth hormone reduce the postoperative fatigue syndrome. World J. Surg 1996;20:81-6.
21. Byrne TA, Morrissey TB, Gatzen, et al. Anabolic therapy with growth hormone accelerates protein gain in surgical patients requiring nutritional rehabilitation. Ann Surg 1993;218:400-18.
22. Jauch KW, Hermann A, Hart W, et al. Dose-dependent effects of human growth hormone (rHGH) on postoperative substrate metabolism. Clin Nutr 1992;11:9 (Supp, Abstr).
23. Wilmore DW, Moylan JA, Briston BP, et al. Anabolic effects of human growth hormone and high caloric feedings following thermal injury. Surg Gynec Obst 1974;138:875-84.
24. Peterson SR, Haladay NJ, Jeevanandam M. Enhancement of protein synthesis efficiency (PSE) in parenterally fed trauma victims by adjuvant recombinant human growth hormone. J Trauma 1994;36:726-33.
25. Ziegler TR, Young LS, Ferrari-Balioiera E. Use of human growth hormone combined with nutritional support in a critical care unit. JPEN 1990;14:574-81.
26. Shernan SK, Demling RH, LaLonde C, et al. Growth hormone enhanced re-epithelialization of human split-thickness skin graft donor sites. Surg. Forum 1989;40:37-39.
27. Herndon DN, Barrow RE, Kunkle KR, et al. Effects of recombinant human growth hormone on donor site healing in severely burned children. Ann Surg 1990;212:424-29.
28. Herndon DN, Hawkins HK, Nguyen TT et al. Characterization of growth hormone enhanced donor site healing in patients with large cutaneous burns. Ann Surg 1995;221:649-56.
29. Koea JB, Breier BH, Douglas RG, et al. Anabolic and cardiovascular effects of recombinant human growth hormone in surgical patients with sepsis. Brit J Surg 1996;83:196-202.
30. Pope GS, Friedman M, Underwood LE, et al. The effect of growth hormone on weight gain and pulmonary function in patients with chronic obstructive lung disease. CHEST 1991;99:1495-1500.
31. Pichard C, Jolliet P, Chevrolet JC, et al. Recombinant human growth hormone in chronic and acute respiratory insufficiency. Hormonal Research 1996;46:222-9.
32. Knor JP, Wilmore DW, Demling RH, et al. Use of growth hormone for postoperative respiratory failure. Am J Surg 1996;171:576-80.
33. Knox JB, Demling RH, WIlmore DW, et al. Hypercalcemia associated with the use of human growth hormone in an adult surgical intensive care unit. Arch Surg 1995;130:442-45.
34. Schambelan M, Mulligan K, Grunfeld C, et al. Recombinant human growth hormone in patients with HIV-associated wasting. Ann Intern Med 1996;125:873-82.
35. Garibotto G, Barreca A, Russo R, et al. Effects of recombinant human growth hormone on muscle protein turnover in malnourished hemodialysis patients. J Clin Invest 1997;99:97-105.
36. Wilmore DW, Lacey JM, Soultanakis RP, et al. Factors predicting a successful outcome after pharmacologic bowel compensation. Ann Surg, in press.

33. Knox JB, Demling RH, WIlmore DW, et al. Hypercalcemia associated with the use of human growth hormone in an adult surgical intensive care unit. Arch Surg 1995;130:442-45.
34. Schambelan M, Mulligan K, Grunfeld C, et al. Recombinant human growth hormone in patients with HIV-associated wasting. Ann Intern Med 1996;125:873-82.
35. Garibotto G, Barreca A, Russo R, et al. Effects of recombinant human growth hormone on muscle protein turnover in malnourished hemodialysis patients. J Clin Invest 1997;99:97-105.
36. Wilmore DW, Lacey JM, Soultanakis RP, et al. Factors predicting a successful outcome after pharmacologic bowel compensation. Ann Surg, in press.

22 GHRP-GH SECRETAGOGUES

Cyril Y. Bowers

Tulane University Medical Center
New Orleans, Louisiana, USA

I. INTRODUCTION

GH secretion is not infrequently low in normal elderly men and women and may compromise normal health (1,2). In the past, when a hormonal deficiency(s) was considered to be responsible for this decreased GH secretion in these subjects, a hypothalamic disorder rather than a pituitary disorder was proposed and a decreased secretion of GHRH or an increased secretion of SRIF was envisioned. One of our more long term objectives has been to utilize GHRP (GH releasing peptide) in order to further understand the pathophysiology of decreased GH secretion as a function of age, especially in elderly men and women, and to distinguish when decreased secretion of GH in elderly subjects is due to normal aging versus a specific hormonal deficiency(s).

Further investigation of age related decreased GH secretion has been investigated by determining the acute GH response to GHRH, GHRP-2 and GHRH+GHRP-2 in normal younger and older men and women. GHRP-2 is the hexapeptide, DAlaDßNalAlaTrpDPheLysNH_2. It is now well established that GHRPs act on both the hypothalamus and pituitary to release GH.

II. OVERVIEW

Even though 3 different chemical classes of GHRPs have been developed including peptides, partial peptides and nonpeptides as listed in Table 1 almost all evidence indicates that they release GH by the same molecular mechanism via the same receptor and intracellular signal transduction pathway. This makes it possible to directly compare both the basic and clinical results obtained even with the different chemical types of GHRPs. A seminal accomplishment by the Merck group in 1992 was the development of the first nonpeptidyl peptidomimetic GHRP/GHS (GH secretagogue), the benzolactam, L-692,429 (3).

In 1996, another milestone in the GHRP saga was the cloning of the GHRP/GHS receptor by the Merck group (4). This is a new 7 transmembrane G-protein coupled receptor. Genomic analysis of the receptor supports the presence of a single highly conserved gene in humans, chimpanzees, pigs, cows, rats and mice DNA. Previously, except in mice, which is still undetermined, GHRP has been reported to release GH in each of the above species. Highly relevant to a putative physiological GHRP system that is possibly involved in the regulation of GH secretion is the anatomic localization of the GHRP/GHS receptor on somatotroph cells of the pituitary, the arcuate and ventral medial nucleus of the hypothalamus as well as an extra hypothalamic site, the hippocampus (4,5). Most of these CNS areas are known or thought to be involved in the regulation of GH secretion.

Table 1: GHRP/GHS CHEMISTRY MILESTONES

			Peptide # residues
		FIRST GENERATION	
1976	1.	First GHRP	5
1977-80	2.	Intermediate Classes of GHRPs	4-5
1980	3.	GHRP-6	6
		SECOND GENERATION	
1989	1.	GHRP-1	7
1990	2.	GHRP-2	6
		THIRD GENERATION	
1992	1.	Europeptide's GHRP	6
1993-95	2.	Merck's GHRP/GHS	nonpeptide
1994	3.	Genentech's GHRP	partial peptide
1994-95	4.	Novo Nordisk's GHRP	partial peptide

Because of the availability and data on GHRP GH release in humans, a didactic, and somewhat euphemistic hypothesis has been proposed that the decreased secretion of GH in some elderly subjects is due to a hormonal deficiency, i.e., a decreased secretion of the putative hypothalamic endogenous GHRP-like hormone rather than a consequence of normal aging (6-8). Any involvement of decreased secretion of GHRH and/or excess secretion of SRIF would be considered a secondary rather than a primary disorder.

To release GH from the pituitary, GHRP activates the PI-PKC (phosphoinositol-protein kinase-C) pathway and GHRH activates the cAMP-PKA (cyclic AMP-protein kinase A) pathway (9). Nevertheless, dependent as well as independent actions of GHRP and GHRH on GH release are induced by these 2 peptides. Depending on the experimental conditions, the GH release induced by GHRP and GHRH can be demonstrated to be not only independent and dependent, but also additive and synergistic as well as permissive (10). Specific examples of these relationships are the following. Each peptide independently releases GH *in vitro*. GHRP's release of GH depends on endogenous GHRH *in vivo*. The two peptides together release GH essentially additively *in vitro* but synergistically *in vivo*. The synergistic release of GH induced by a low dose of GHRP plus a high dose of GHRH *in vivo* reveals the permissive relationship of the 2 peptides.

Several conclusions emanate from the relationship between GHRP and GHRH. Both the additive and independent action of the 2 peptides support that GHRP and GHRH release GH by different mechanisms. Furthermore, the synergistic release of GH induced by the 2 peptides together supports an independent and complementary action. This synergistic release of GH by GHRP-2+GHRH *in vivo,* but not *in vitro,* reveals that GHRP has a dual action on the pituitary and hypothalamus. Even though the pituitary action of GHRP has been elucidated in major part, except for an effect on GH synthesis, the hypothalamic action is still incompletely understood. From *in vivo* results which clearly demonstrate that the release of GH by GHRP is markedly dependent on the presence of endogenous GHRH in animals as well as humans, it is logical to conclude that GHRP releases GH *in vivo* by stimulating the release of GHRH via a hypothalamic action of GHRP. However, what appears to occur is that GHRP releases endogenous GHRH at high but not low dosages. Noteworthy is that the release of GH by low dose GHRP depends on the presence of endogenous GHRH but is not mediated via GHRH (11,12). For this reason, the relationship between low dose GHRP and GHRH is viewed as being permissive in nature. Additionally, an effect of GHRP on SRIF (somatostatin) release or action does not explain its hypothalamic action (11). Collectively, results support that the GH releasing action of GHRP is much more closely related to the action of GHRH than to SRIF. To explain the GHRP hypothalamic action at low dosages *in vivo,* we have postulated that GHRP acts on the hypothalamus to release U-factor (Unknown Factor) which, together with endogenous GHRH, releases GH by a direct pituitary action of both peptides (10,11).

II. ISOLATION

Considerable indirect evidence has accumulated to reasonably consider that the putative GHRP-like hormone exists and may be involved in the physiological regulation of GH secretion. From our efforts to isolate the putative GHRP-like hormone from porcine hypothalami over a 3 year period, unpublished studies support its possible existence. Partially purified fractions have been considered to have GHRP activity because of the following results. Fractions that did not contain GHRH released GH by a direct pituitary action, *in vitro*. Cyclic AMP was not increased by these fractions. The GHRP competitive antagonist, HisDTrpDLysTrpDPheLysNH$_2$ (DLys^3-GHRP-6), but not the GHRH competitive antagonist, [DArg^{2}Ala^{8,9,15}]GHRH 1-29NH$_2$, inhibited the GH releasing activity of GHRP-2. Adding the fractions together with GHRH additively released GH *in vitro*. One surprising result was a greater release of GH when the fractions were combined with a maximal dose of GHRP. These fractions were obtained after considerable purification of extracts from about 100,000 porcine hypothalami. The sequential purification scheme involved preparative and semipreparative column separation as well as different types of analytical chromatography. We were unable to detect GH releasing activity in the fractions from the final stages of purification. This we believe was due to the very small amount of peptide in the fractions and possibly because the peptide, in such low concentrations, may bind or stick to the glassware.

The empirical synthetic peptidyl approach utilized in the development of the GHRPs *de novo* in 1976 is again being used to reveal the putative natural GHRP-like hormone *de novo*, prior to its isolation and complete amino acid sequence determination. The inability to incorporate unnatural D sterioisomeric amino acid residues into the putative all L-amino acid natural hormone was a major restriction that has been difficult to overcome. The hybrid synthetic GH releasing peptides which consist of only L amino acid residues have been derived from structure-activity data accumulated on the 4 different types of GHRPs, $DTrp^2$, $DTrp^3$, $DTrp^{2,3}$ and $DTrp^2$, $LTrp^4$-GHRP that we have previously developed (10). Although the first 3 types were only active *in vitro* but not *in vivo* in our earlier studies between 1976 and 1980, we have subsequently developed highly active synthetic peptides of each of these 3 different chemical types *in vivo*. Our GHRP structure activity database has been considerably expanded by this achievement and this additional information is being utilized in the development of hybrid all L amino acid peptide molecules which eventually may be partially or closely related to the amino acid sequence and activity of the putative natural GHRP-like hormone. Noteworthy is that the results of our isolation studies as well as the high activity and small molecular size of our unnatural GHRPs are both compatible with the extrapolation that the natural GHRP-like hormone is small in size, i.e., 5-15 amino acids. Notably the hypophysiotropic tripeptide, TRH, was indeed a surprise! Highly active synthetic peptides with 3-5 amino acid extensions at the N or C terminus of GHRP-1 or GHRP-2 are also compatible with the possibility that the natural hormone may be longer than a hexapeptide. From our GHRP studies, certain amino acids which have been found to be favorable in one of the 4 different chemical types of GHRPs were not infrequently unfavorable in one of the other types. Nevertheless, generally favorable aromatic amino acids were Trp and Phe. Overall, a rather broad and flexible database is available for the *de novo* discovery of the natural GHRP-like hormone independent of its isolation.

Part of the basic strategy being pursued is the gradual "dismantling" of unnatural GHRPs by substituting L amino acids and removing D amino acid residues in the 4 different types of synthetic GHRPs. Furthermore, certain L-amino acids have been selected because of mass spectrometry results obtained by Hunt et al from our active porcine hypothalamic fractions (unpublished). Even though these fractions were obtained after extensive purification they were still definitely impure and contained a number of small peptides in very low quantities. Nevertheless, certain L amino acids have been selected from these latter isolation results as well as our synthetic peptide studies. Also, favorable small sequences of L-amino acids from a number of different active synthetic peptides have been incorporated into composite hybrid all L-amino acid peptide molecules.

III. PHYSIOLOGY VERSUS PHARMACOLOGY

Specific differences become apparent when GHRP-2 or a GHRP-like hormone is considered in the role of a possible physiological regulator or a pharmacological agent (6,10). Indirect evidence in humans indicates low and high dose GHRP-2 may have similar but different mechanisms of action involved in the release of GH.

Low dose GHRP-2 GH responses have been obtained in the same normal young men and women with iv bolus 0.03 or 0.1 μg/kg GHRP-2 and GHRP-2 + 1.0

μg/kg GHRH (Figures 1 and 2). Even these very low dosages of ~2-7 μg GHRP-2 administered together with a high maximal dose of GHRH synergistically released GH (6,13). The high dose of GHRH administered with GHRP-2 obviates the possibility that low dose GHRP-2 produces synergism by releasing endogenous GHRH.

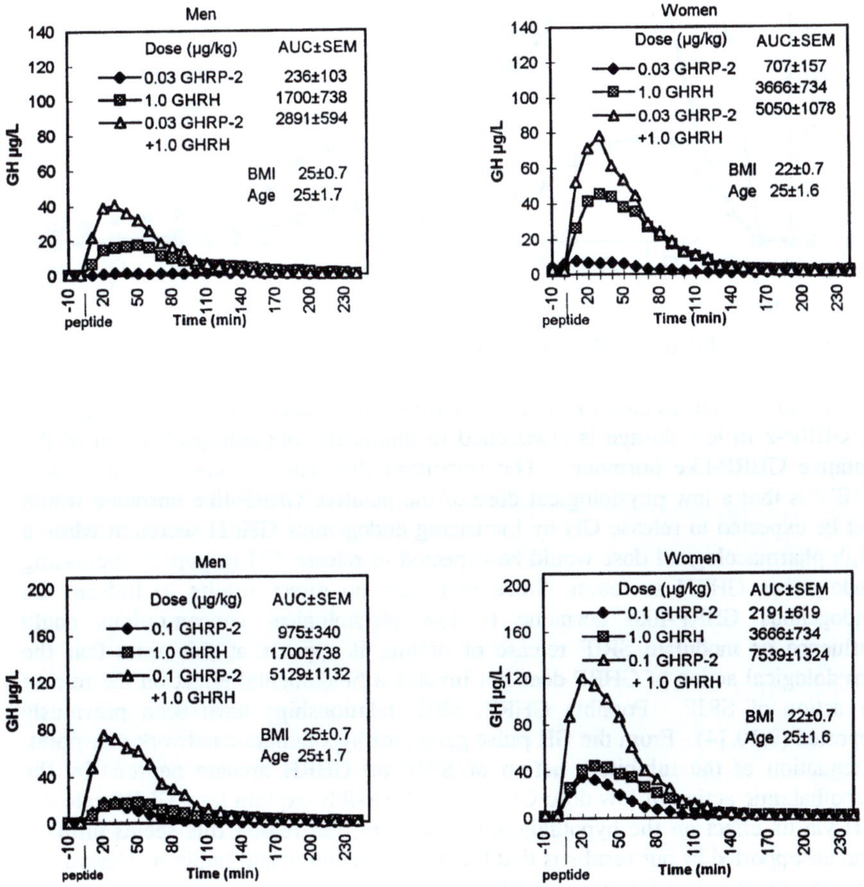

Figures 1 and 2: Synergism in Normal Young Men and Women. n=9 (men); n=7 (women).

High dose GHRP-2 GH responses, as demonstrated in Figure 3, also have been obtained in normal young men (13). In this study, 10 μg/kg sc GHRP-2 released the same amount of GH as 1+1 μg/kg iv GHRP-2+GHRH. The high dose GHRP-2 administered alone has been interpreted to mimic the synergistic release of GH induced by GHRP-2+GHRH by increasing endogenous GHRH.

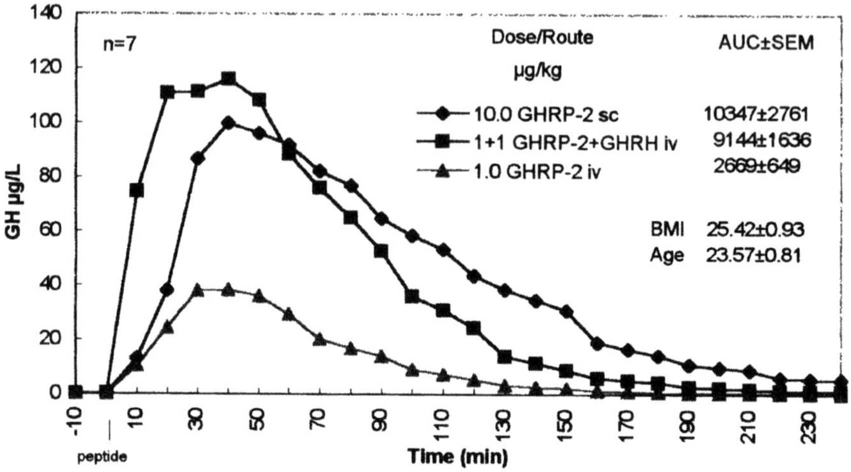

Figure 3: High Dose GHRP-2 in Normal Young Men

Several instructive conclusions are derived from the studies recorded in Figures 1-3. GHRP-2 in low dosage is envisioned to mimic the physiological action of the putative GHRP-like hormone. The important difference of low and high dose GHRP is that a low physiological dose of the putative GHRP-like hormone would not be expected to release GH by increasing endogenous GHRH secretion while a high pharmacological dose would be expected to release GH in part by increasing endogenous GHRH secretion. Since there are no direct results to indicate an endogenous GHRP-like hormone in low physiological concentrations could influence or modulate SRIF release or action, it appears at this time that the physiological action of GHRP does not involve a fundamental effect on the release or action of SRIF. Possible GHRP, SRIF relationships have been previously reported (9,10,14). From the GH pulse generator hypothalamic network viewpoint, attenuation of the inhibitory action of SRIF on GHRH arcuate neurons by the hypothalamic action of low dose GHRP could possibly explain how GHRP releases GH via its effect on the hypothalamus. However, the reason this seems unlikely and unsupported by our results is that this GHRP action would imply it releases GH because it releases endogenous GHRH.

Since GHRP acts on both the hypothalamus and pituitary to release GH, a relevant question is raised of whether the main action of a putative GHRP-like hormone would be primarily on the hypothalamus or pituitary. Currently the hypothalamus seems to be our choice (15).

The putative physiological GHRP-like hormone is envisioned to act on the hypothalamus to release U-factor (unknown factor) and U-factor+GHRH would act on the pituitary to release GH. Presumably, GHRP and U-factor would be a helper or co-factor in the release of GH by GHRH. The reason GHRH may need a helper would be to attenuate the strong inhibitory pituitary action of SRIF on GH release. With a helper, the GHRH action on GH release would be less dependent on a coordinated increased release of GHRH and a concomitant decreased release of endogenous SRIF from the hypothalamus into portal blood. Thus this would result in GHRH being more of a primary regulator of GH secretion. One proof of the neuroregulatory conceptual model of GH secretion would be the eventual identification of a putative hypothalamic GHRP-like hormone as well as hypothalamic U-factor and the determination of interrelated patterns of secretion of the GHRP-like hormone, U-factor, GHRH and SRIF into the hypothalamic-hypophyseal portal system. If the above conceptual model is correct, increased amounts of GHRP-like hormone, U-factor and GHRH would be secreted together with only minor changes in the secretion of SRIF. This model minimizes but does not eliminate the pituitary action of the GHRP-like hormone in the physiological regulation of GH release (16).

Many valuable and important details about the GHRP hypothalamic receptor location and action are now being revealed by neurochemical and electrophysiological approaches (5,17-22). These studies are gradually being directed to understand and describe a possible hypothalamic pulse generator network which may be involved in the regulation of the physiological pulsatile secretion of GH (5,9,15,21,22). Major focus is on the arcuate nucleus and a possible interaction of GHRH, SRIF and the putative GHRP-like hormone as well as the feedback effects of GH and IGF-I. Undoubtedly, another basic increment of understanding will evolve from availability of the GHRP-like hormone as well as subsequent neurochemical identification and location of the neurons which produce this putative hormone.

Collaborative studies with Van den Berghe, de Zegher and Veldhuis indicate continuous administration of GHRP-2 to normal young men as well as elderly subjects on the Intensive Care Unit induces an increase in the amplitude but not frequency of spontaneous pulsatile GH secretion and raises serum IGF-I levels (23). From current studies in stabilized, critically ill patients, low serum IGF-I levels were positively correlated with GH secretory burst amplitude and basal GH secretion. In these studies, both basal and pulsatile GH secretion were moderately increased by continuous infusion of 1 μg/kg/h GHRH, substantially increased by 1 μg/kg/h GHRP-2 and strikingly increased by GHRH+GHRP-2 at these dosages. GHRP-2 alone or combined with GHRH elicited a robust rise in circulating IGF-I levels within 24h without altering serum cortisol levels. These results suggest a special relationship exists between increased GH production via an increase of pulsatile GH secretion and elevation of serum IGF-I and IGFBP-3 levels. In the future, these same types of relationships as well as the metabolic effects on serum glucose, lipids, and insulin levels during continuous administration of GHRP-2 will need to be determined in normal older men and women.

The pharmacological action of GHRP includes all of the same presumed physiological actions plus an additional action of releasing endogenous GHRH. In principle, when endogenous GHRH secretion is decreased, GHRP in large pharmacological dosages may become a valuable agent for increasing its secretion. In both children and adults, GH secretion is frequently decreased because of a hypothalamic disorder which some investigators believe is due to a decreased secretion of endogenous GHRH. Thus, determination of whether chronic administration of GHRP in high dosages can influence the endogenous release of GHRH will need to be elucidated. As the GHRP mechanism of action and the pathophysiology of decreased GH secretion in children and adults become better understood, including their similarities and differences, the use of GHRP as a pharmacological agent will become increasingly better established on a more rational basis (24-27).

Acute studies with GHRP and GHRH in normal older men and women with decreased GH secretion have led to tentative conclusions about the pathophysiology of decreased GH secretion (Tables 2-4).

Table 2: GH Responses in Elderly Men and Women. All GH responses slightly decreased indicating a possible mild deficiency of the putative endogenous GHRP hormone.

				GHRH	GHRP-2	G+G μg/kg	GHRP-2	IGF-I μg/L
	INITIALS	AGE/SEX	BMI	1.0	0.1	0.1+1.0	1.0	n=4
						GH PEAK μg/L		
1	GJR	59/M	25.5	9.0	5.8	31.6	46.6	200±6.9
2	BOD	72/F	27.7	10.6	15.6	50.8	43.5	121±4.5
3	N-G	60/M	24.2	14.5	6.3	57.2	47.4	185±3.7
4	JRB	64/M	23.8	14.8	2.3	24.7	22.8	112±5.5
5	RAG	64/M	25.3	24.6	1.4	28.7	45.9	139±5.1
	MEAN	63.8	25.3	14.7	6.3	38.6	41.2	151
	±SEM	±2.3	±0.7	±2.7	±2.5	±6.5	±4.7	±17
						GH AUC μg/L•4h		
1	GJR	59/M	25.5	721	286	1816	2597	200±6.9
2	BOD	72/F	27.7	531	687	2732	2307	121±4.5
3	N-G	60/M	24.2	1548	486	3502	3107	185±3.7
4	JRB	64/M	23.8	1199	273	1425	1173	112±5.5
5	RAG	64/M	25.3	1058	148	1506	2948	139±5.1
	MEAN	63.8	25.3	1011	376	2196	2426	151
	±SEM	±2.3	±0.7	±179	±95	±402	±344	±17

Table 3: GH Responses in Elderly Men and Women. Low GH response to maximal dose GHRH (mean peak=4.2) and increase of this response by low dose (0.1 µg/kg) GHRP-2.

	INITIALS	AGE/SEX	BMI	GHRH µg/kg 1.0	GHRP-2 µg/kg 0.1	G+G µg/kg 0.1+1.0	GHRP-2 µg/kg 1.0	IGF-I µg/L n=4
				GH PEAK µg/L				
1	MAB	70/F	27.2	1.9	0.5	7.3	12.8	102±0.8
2	JDB	73/M	26.3	3.0	0.7	14.0	20.9	104±4.9
3	DCR	75/F	28.2	2.1	2.3	13.7	----	100±5.2
4	ENB	66/F	21.4	2.7	0.7	44.1	47.6	85±10
5	RPL	66/M	24.7	3.3	1.9	24.4	34.3	119±4.3
6	M-L	63/M	27.5	4.3	0.7	13.4	25.0	122±5.3
7	RED	65/M	29.1	4.8	4.6	17.2	15.6	134±6.4
8	EES	60/M	29.0	4.8	9.3	52.1	66.5	59±3.3
9	WAB	69/M	25.8	6.2	4.2	33.0	37.2	135±5.5
10	B-G	70/M	30.0	6.3	12.4	17.9	37.6	185±8.0
11	JMJ	69/M	24.3	6.7	0.8	27.3	30.2	182±8.3
	MEAN	67.8	26.7	4.2	3.5	24.0	32.8	121
	±SEM	±1.3	±0.2	±0.5	±1.2	±4.2	±5.1	±11
				GH AUC µg/L•4h				
1	MAB	70/F	27.2	198	96	548	553	102±0.8
2	JDB	73/M	26.3	314	81	803	1025	104±4.9
3	DCR	75/F	28.2	127	175	584	----	100±5.2
4	ENB	66/F	21.4	183	221	2436	2540	85±10
5	RPL	66/M	24.7	368	156	1757	2476	119±4.3
6	M-L	63/M	27.5	388	45	731	1583	122±5.3
7	RED	65/M	29.1	432	223	1149	831	134±6.4
8	EES	60/M	29.0	286	593	3379	4103	59±3.3
9	WAB	69/M	25.8	428	185	1275	2290	135±5.5
10	B-G	70/M	30.0	475	526	778	1765	185±8.0
11	JMJ	69/M	24.3	668	86	1978	1732	182±8.3
	MEAN	67.8	26.7	352	217	1402	1890	121
	±SEM	±1.3	±0.2	±46	±54	±272	±327	±11

Table 4: GH Responses in Elderly Men and Women. All GH responses markedly decreased indicating a possible severe deficiency of the putative endogenous GHRP hormone.

	INITIALS	AGE/SEX	BMI	GHRH	GHRP-2	G+G	GHRP-2	IGF-I
						µg/kg		µg/L
				1.0	0.1	0.1+1.0	1.0	n=4
						GH PEAK µg/L		
1	GAS	61/M	30.5	1.2	0.7	8.6	11.2	130±2.5
2	JWM	71/M	21.7	2.2	2.6	8.0	18.5	127±7.1
3	WEE	63/M	27.7	3.1	1.1	7.2	9.5	149±7.9
4	PEH	74/F	26.1	6.7	0.5	4.7	18.1	62±4.5
	MEAN	67.3	26.5	3.3	1.2	7.1	14.3	117
	±SEM	±3.1	±1.8	±1.2	±0.4	±0.9	±2.3	±18
						GH AUC µg/L•4h		
1	GAS	61/M	30.5	107	64	454	584	130±2.5
2	JWM	71/M	21.7	223	204	387	950	127±7.1
3	WEE	63/M	27.7	290	84	402	432	149±7.9
4	PEH	74/F	26.1	430	60	321	858	62±4.5
	MEAN	67.3	26.5	263	103	391	706	117
	±SEM	±3.1	±1.8	±67	±34	±27	±120	±18

Evidence and reasons for these conclusions are the following. Of the 20 normal older subjects studied, GH responses to the maximal 1 µg/kg iv bolus dosage of GHRH was mildly (Table 2) or markedly impaired (Tables 3 and 4). Particularly noteworthy are GHRP-2 results with and without GHRH which demonstrate that the impaired pituitary GH response to GHRH is not the result of pituitary GH depletion nor an irreversible pituitary action of GHRH. Noteworthy is that these results are compatible with the hypothesis of a basic disorder of decreased release of the putative GHRP-like hormone and not with decreased release of endogenous GHRH. Although the impaired GH response to GHRH implies a primary pituitary disorder, the GH release induced by GHRP indicates a secondary pituitary disorder.

Since such a low dose of 0.1 µg/kg GHRP administered together with a high dose of 1.0 µg/kg GHRH very substantially reverses the impaired pituitary GH response to GHRH, it has been hypothesized that the decreased secretion of GH in some normal older subjects is due to a primary hypothalamic disorder, i.e., deficiency of the putative GHRP-like hormone. Furthermore, it is postulated that this reversal of the impaired pituitary GH response of GHRH is via the hypothalamic rather than the pituitary action of GHRP and involves the release of U-factor. Additionally, the substantial GH response of GHRP-2 alone at the higher 1 µg/kg iv bolus dosage again indicates endogenous GHRH is being secreted because in the absence of endogenous GHRH GHRP releases only a small amount of GH *in vivo*. The GHRH GH responses recorded in Table 3 indicate that even though physiological amounts of GHRH were probably being secreted in these subjects as indicated by the GH response to 1 µg/kg GHRP-2 alone, the pituitary is refractory to GHRH. From the results in Tables 2 and 4 the type of pathophysiology of decreased GH secretion is less readily apparent; however, it may be more a matter of degree rather than type. Subjects in Table 4 had low GH responses to high dose GHRP as well as high dose GHRH. In one subject the GH response to iv bolus 1 µg/kg GHRH+GHRP was still markedly impaired (not recorded). Whether this is due primarily to a decreased pituitary content of GH or more collectively to a combination of disorders, i.e., decreased GHRP-like hormone and GHRH secretion as well as in part to excess secretion of SRIF will require future studies.

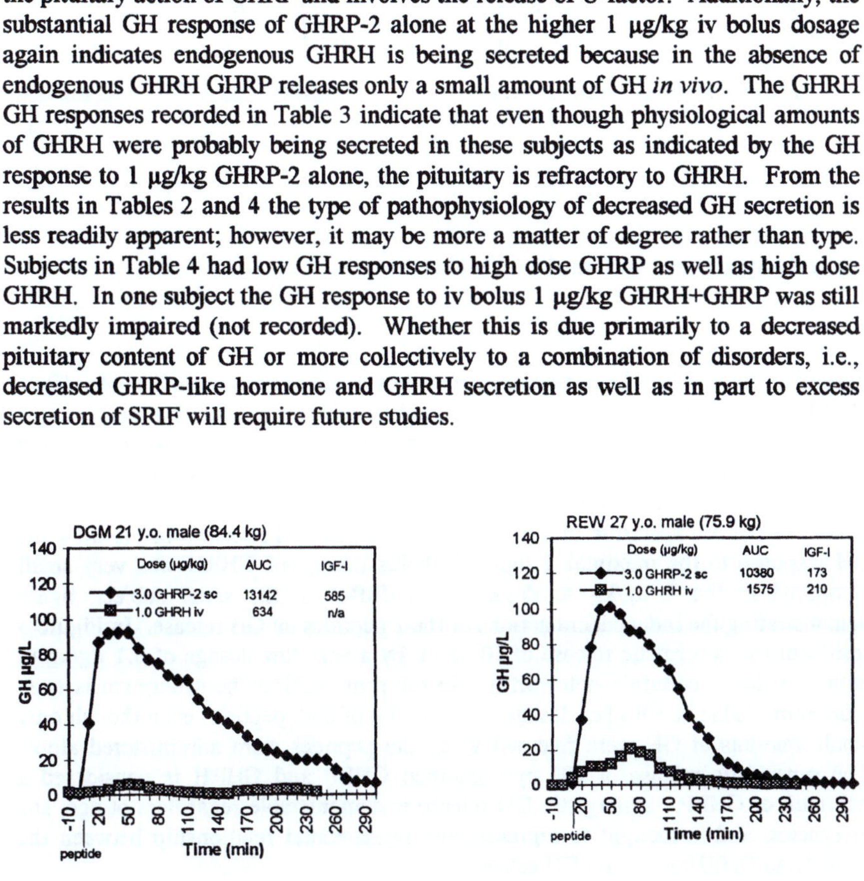

Figure 4: Massive GH Release in Normal Young Men

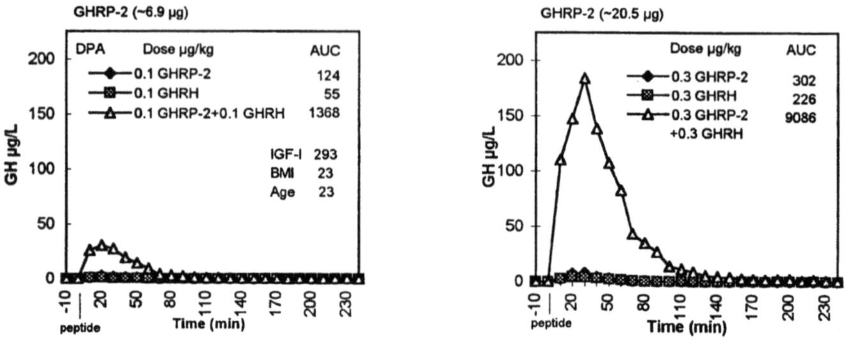

Figure 5: Synergism in a Normal Young Man

Results in Figures 4 and 5 demonstrate new, novel and at times unbelievable effects of GHRP that sometimes have been observed in select individual normal younger men and women. These unusual results appear to reflect new dimensions and mechanisms of GH release not previously appreciated. Once they are understood, it is envisioned that they will impart new insight into the regulation of GH secretion as well as the full spectrum of the effect of GHRP on GH release. In Figure 4 the GH response to the maximal 1 µg/kg iv bolus dosage of GHRH was very small compared to the 3 µg/kg sc dosage of GHRP-2 in the same subject, again demonstrating the independent actions of the 2 peptides on GH release. In Figure 5 results of the synergistic release of GH even by a very low dosage of 0.1 µg/kg of both peptides is certainly noteworthy. In the same subject the synergism is even more remarkable at a higher dosage of 0.3 µg/kg of both peptides even though only small amounts of GH were released when the peptides were administered alone. The synergistic release of GH by combined GHRP and GHRH is considered a hallmark of GHRP. Synergistic GH release is demonstrable regardless of age, sex or species, and is thought to represent the fundamental relationship between the GHRP and GHRH action on GH release.

IV. PITUITARY TUMORS

A number of basic and clinical studies indirectly suggest that the putative GHRP-like hormone may be involved in the function and possibly the pathophysiology of pituitary tumors in humans. At times, GHRP dramatically releases GH in patients with acromegaly (28). Also, GHRP releases GH *in vitro* from pituitary tumor cells of these patients sometimes even more consistently than GHRH (29). *In vitro* studies of Adams et al (30) on human pituitary somatotrophinoma cells reveal that

GHRP-2 and GHRP-6 stimulate the phosphoinositol (PI) intracellular signal pathway and GH release regardless of whether the cells are positive or negative for the adenyl cyclase-activating gsp oncogene. The protein kinase C inhibitor, phoretin, but not the GHRH antagonist $NAcTyr^1DArg^2GHRH$ 1-29NH_2, inhibits GH release from these tumor cells. GHRP-2 and GHRP-6 stimulate cAMP production in somatotrophinoma cells with but not without constitutively activated adenyl cyclase activity. Additionally, GHRP-2 and GHRP-6 potentiate the stimulation of cAMP production induced by GHRH in tumors without gsp oncogenes. Evidence for cross-talk between the PI and cAMP transduction pathways has evolved from these studies.

In supplemental studies on human pituitary tumors Adams et al demonstrated GHRP/GHS receptor mRNA by the reverse transcription polymerase chain reaction (RT-PCR) approach (31). This receptor mRNA was demonstrated in 6 of 6 human pituitary somatotrophinomas removed from patients with acromegalia but PI hydrolysis and GH release were induced in only 4 of these tumor cells. Additionally, in GH rat pituitary tumor cells, mRNA for the GHRP-GHS receptor was demonstrated but the cells were not responsive to PI hydrolysis or GH release (31). These results again indicate the presence of expression of this receptor in pituitary tumor cells of both humans and rats and thus absence of the GHRP receptor may not be the reason GHRP does not stimulate PI hydrolysis or GH release in some tumor cells.

The *in vitro* studies of Adams et al also revealed expression of the GHRP/GHS pituitary receptor in 3 human pituitary prolactinoma cells and in 2 of these tumors GHRP-2 increased PI hydrolysis (31). Utilizing a semi-quantitative RT-PCR approach, Korbonits et al (32) recently reported expression of the GHRP/GHS receptor in normal pituitary cells as well as several different types of tumors. For example, the GHRP/GHS receptor was not only expressed in human pituitary tumors of somatotrophs (3 of 3), prolactotrophs (3 of 3), corticotrophs (1 of 1), and nonfunctional pituitary adenoma (NFPA) cells (3 of 3), but the relative number of receptors in the tumor cells versus normal cells revealed overexpression in all of the somatotroph tumors, 2 of 3 prolactotrophs tumors, and 2 of the NFPAs. Besides the GHRP GH release in patients with acromegaly, Ghigo et al (33) reported that GHRP also released more ACTH than CRH in patients with Cushing's syndrome which presumably was due to a pituitary tumor.

V. CONCLUSION

It is now well established that GHRPs act on both the pituitary and hypothalamus to release GH *in vivo*. Somewhat different than established for the known hypothalamic hypophysiotrophic hormones that act on the pituitary to release hormones, GHRPs act on both anatomic sites in a highly dependent and complementary way to release GH. Fundamentally important in considering and elucidating the action of GHRP is that it acts on different receptors and signal transduction pathways than GHRH.

The effects and molecular actions of GHRP are much better understood on the pituitary than on the hypothalamus. The relative importance of the effects of GHRP on the pituitary and the hypothalamus still vary among different investigators. In part, this probably arises because of the many variables that influence the effects of GHRP such as dose, route, duration and frequency of administration, pharmacokinetics of the particular GHRP/GH secretagogue as well as species, experimental design, age and sex.

Without endogenous GHRH, the *in vivo* effect of GHRP on GH release in animals and humans is markedly less. In studies of normal older subjects with decreased GH secretion, many subjects have had impaired GH responses to GHRH that are reversed when a low dose of GHRP is administered together with GHRH. Even though GHRP has been demonstrated to activate GHRH neurons in the arcuate nucleus of the hypothalamus, to release GH into portal blood and to be dependent on endogenous GHRH, release of GH by low dose GHRP *in vivo* is considered to be passively related to GHRH and thus does not release GH because GHRP releases GHRH via its hypothalamic action. In contrast, a high dose of GHRP releases GHRH from the hypothalamus, and thus endogenous GHRH is considered to be actively released rather than passively involved in the GH response of GHRP. Also, the effect of GHRP on the release or action of SRIF does not appear to explain how it releases GH. What seems particularly important is the hypothesis that the hypothalamic action of GHRP involves the release of a still unidentified factor (U-factor or unknown factor) that acts on the pituitary together with GHRH and GHRP to release GH.

A final point is that it is hypothesized that low dose GHRP reflects the physiological action of the putative GHRP-like hormone while high dose GHRP reflects the pharmacological action of GHRP. We are now attempting to isolate the putative U-factor as well as the GHRP-like hormone from the hypothalamus. U-factor is thought to primarily mediate and reflect the physiological effects and it is the release of endogenous GHRH in combination with U-factor and GHRP that is thought to primarily mediate and reflect the pharmacological effects of GHRP via the unique and novel synergistic release of GH.

ACKNOWLEDGMENT

This work was supported in part by National Institutes of Health grant 5 M01RR05096 (GCRC). Special appreciation is expressed to Dr. Granda-Ayala, to the technicians and the fellows of the Endocrinology and Metabolism Section of the Department of Medicine, and to Robin Alexander for typing the manuscript.

REFERENCES

1. Zadik Z, Chalew SA, McCarter RJ, Meistas M, Kowarski AA. The influence of age on the 24-hour integrated concentration of growth hormone in normal individuals. J Clin Endocrinol Metab 60:513, 1985.
2. Ho KY, Evans WS, Blizzard RM, Veldhuis JD, Merriam GR, Salojlik E, Furlanetto R, Rogol AD, Kaiser DL, Thorner MO. Effects of sex and age on the

24-hour profile of growth hormone secretion in man: importance of endogenous estradiol concentrations. J Clin Endocrinol Metab 64:51, 1987.

3. Smith RG, Cheng K, Schoen WR, Pong SS, Hickey G, Jacks T, Butler B, Chan WWS, Chaung LYP, Judith F, Taylor J, Wyvratt MJ, Fisher MH. A novel nonpetidyl growth hormone secretagogue. Science 260:1640-1643, 1993.

4. Howard AD, Feighner, SD, Cully DF, Arena JP, Liberator PA, Rosenblum CI, Hamelin M, Hreniuk DL, Palyha OC, Anderson J, Paress PS, Diaz C, Chou M, Liu KK, McKee KK, Pong SS, Chaung LY, Elbrecht A, Dashkevicz M, Heavens R, Rigby M, Sirinathsinghji DJS, Dean DC, Melillo DG, Patchett AA, Nargund AA, Griffin PR, DeMartino JA, Gupta SK, Schaeffer JM, Smith RG, Van der Ploeg LHT. A receptor in pitutiary and hypothalamus that functions in growth hormone release. Science 273:974-977, 1996.

5. Bennett PA, Thomas GB, Howard AD, Van der Ploeg LHT, Smith RG, Robinson ICAF. Expression and regulation of the growth hormone secretagogue-receptor (GHS-R) gene in normal and dwarf rats. 79th Annual Meeting of the Endocrine Society, Minneapolis (Abstract) p. 127, 1997.

6. Bowers CY. GHRP+GHRH synergistic release of GH: scope and implication. 2nd International Symposium on Growth Hormone Secretagogues, Tampa, February 1997. In: Growth Hormone Secretagogues, edited by B. Bercu, R. Walker, Marcel Dekker, Inc, New York (in press).

7. Bowers CY, Reynolds GH, Kaf-Alghazal A, Albustamy A, Almasri A, Lopez A, Granda-Ayala R. Indirect evidence for the existence of an endogenous putative GHRP hormone. 79th Annual Meeting of the Endocrine Society, Minneapolis (abstract) p. 333, 1997.

8. Bowers CY. GHRP historical perspective: basic and clinical. In: Human Growth Hormone: Basic and Clinical Research, Contemporary Endocrinology, edited by R. Smith, M. Conn, Humana Press, New York (in press).

9. Smith RG, Pong SS, Hickey G, Jacks T, Cheng K, Leonard R. Modulation of pulsatile GH release through a novel receptor in hypothalamus and pituitary gland. Recent Prog. Horm. Res. 51:261-286, 1996.

10. Bowers CY, Veeraragavan K, Sethumadhavan K. Atypical growth hormone releasing peptides. In: Growth Hormone II: basic and clinical aspects, edited by B. Bercu, R. Walker, Springer-Verlag, New York. p. 203-222, 1994.

11. Bowers CY, Sartor AO, Reynolds GA, Badger TM. On the action of the growth hormone releasing hexapeptide GHRP. Endocrinology 128:2027-2035, 1991.

12. Pandya N, Barkan AL, DeMott-Friberg R, Bowers CY, Jaffe, CA. GHRP-6 requires GHRH for maximal GH stimulation. 79th Annual Meeting of the Endocrine Society, Minneapolis (abstract) p. 92, 1997.

13. Bowers CY, Reynolds GA, Field G, Arkieh A, Kaf-Alghazal A, Krishnamurthi L, Granda-Ayala R. Independency, dependency and synergism of GHRP-2 and GHRH on GH release in humans. 10th International Congress of Endocrinology (abstract), San Francisco, p. 770, 1996.

14. Fairhall KM, Mynett A, Thomas GB, Robinson ICAF. Central and peripheral effects of peptide and nonpeptide GH secretagogues on GH release *in*

vivo. In: Growth Hormone Secretagogues, edited by B. Bercu, R. Waler, Springer-Verlag, New York, p. 219-236, 1996.
15. Robinson ICAF. Regulation of growth hormone output: the GRF signal. In: Growth Hormone II: basic and clinical aspects, edited by B. Bercu, R. Walker, Springer-Verlag, New York, p. 47-65, 1994.
16. Bowers CY. GH releasing peptides (GHRPs). In: Handbook of Physiology, edited by J. Kostyo, Oxford University Press (in press).
17. Dickson Sl, Leng G, Robinson ICAF. Systemic administration of growth hormone releasing peptide activates hypothalamic arcuate neurons. Neurosci. 53(2):303-306, 1993.
18. Dickson SL, Leng G, Dyball REJ, Smith RG. Central actions of peptide and non-peptide growth hormone secretagogues in the rat. Neuroendocrinol 61:36-43, 1995.
19. Dickson SL, Doutrelant-Viltart O, McKenzie DN, Dyball REJ. Growth hormone releasing peptide activates rat arcuate neurons recorded *in vitro*. 10th International Congress of Endocrinology (abstract) San Francisco, p. 286, 1996.
20. Dickson Sl, Doutrelant-Viltart O, Leng G. Growth hormoen (GH) deficiency dw/dw rats and lit/lit mice show increased Fos expression in the hypothalamic arcuate nucleus following systemic injection of GH-releasing peptide (GHRP-6). J Endocrinol 146:519-526, 1995.
21. Dickson Sl, Luckman SM. Induction of c-*fos* messenger ribonucleic acid in nueropeptide Y and growth hormone (GH)-releasing factor neurons in the rat arcuate nucleus following systemic injection of the GH secretagogue, GH-releasing peptide-6). Endocrinol 138:771-777, 1997.
22. Kamegai J, Hasegawa O, Minami S, Sugihara H, Wakabayashi I. The growth hormone-releasing peptide KP-102 induces c-*fos* expression in the arcuate nucleus. Mol Brain Res 39:153-159, 1996.
23. Van den Berghe G, de Zegher F, Veldhuis JD, Wouters P, Awouter M, Verbruggen W, Schetz M, Verwaest C, Lauwers P, Bouillon R, Bowers CY. The somatotropic axis in critical illness: effect of continuous GHRH and GHRP-2 infusion. J Clin Endocrinol Metab 82:590-599, 1997.
24. Laron Z, Bowers CY, Hirsch D, Almonte AS, Pelz M, Keret R, Gil-Ad, I. Growth hormone-releasing activity of growth hormone releasing peptide-1 (a synthetic heptapeptide) in children and adolescents. Acta. Endocrinol 129:424-426, 1993.
25. Pihoker C, Middleton R, Reynolds GA, Bowers CY, Badger TM. Diagnostic studies with intravenous and intranasal growth hormone releasing peptide-2 in children of short stature. J Clinical Endocrinol Metab 80:2987-2992, 1995.
26. Mericq V, Cassorla F, Garcia H, Avila A, Bowers CY, Merriam G. Growth hormone responses to growth hormone releasing peptide (GHRP) and to growth hormone releasing hormone (GHRH) in growth hormone deficient children (GHD). J Clin Endocrinol Metab 80:1681-1684, 1995.
27. Tuilpakov, An, Bulatov AA, Peterkova VA, Elizarova GP, Volevodz NN, Bowers CY. Growth hormone (GH) releasing effects of synthetic peptdie GHRP-2

and GH releasing hormone (GH 1-29NH_2) in children with GH insufficiency and idiopathic short stature. Metabolism 44:1199-1204, 1995.

28. Alster DK, Bowers CY, Jaffe CA, Ho PJ, Barkan AL. The GH response to GHRP (His-DTrp-Ala-Trp-DPhe-Lys-NH_2) GHRH and TRH in acromegaly. J Clin Endocrinol Metab 77:842-845, 1993.

29. Renner U, Brockmeier S, Strasburger CJ, Lange M, Schopohl J, Muller OA, Werder KV, Stalla GK. Growth hormone (GH) releasing peptide stimulation of GH release from human somatotroph adenoma cells: interaction with GH-releasing hormone, thyrotropin-releasing hormone, and octreotide. J Clin Endocrinol Metab 78:1090-1096, 1994.

30. Adams EF, Bowers CY. Protein kinase c-dependent growth hormone releasing peptides stimulate cAMP production by human pitutiary somatotrophinomas expressing GSP oncogenes: evidence for cross-talk between transduction pathways. Mol Endocrinol 10:432-438, 1996.

31. Adams EF, Huang B, Buchfelder M, Howard A, Smith RG, Bowers CY, Fahlbusch R. Presence of growth hormone secretagogue-receptor (GHS-R) mRNA in human pituitary tumors and rat GH_3 cells. J Clin Endocrinol Metab (in press).

32. Korbonits, M, Aylwin SJB, Trainer PJ, Burrin JM, Besser GM, Grossman AB. Growth hormone secretagogue-receptor mRNA is expressed in human pitutiary adenoma tissue. 79th Annual Meeting of the Endocrine Society (abstract) Minneapolis, p. 128, 1997.

33. Ghigo E, Arvat E, Ramunni J, Colao A, Gianotti L, Deghenghi R, Lombardi G, Camanni F. Adrenocorticotropin and cortisol-releasing effect of hexarelin, a synthetic growth hormone-releasing peptide, in normal subjects and patients with Cushing's syndrome. J Clin Endocrinol Metab 82:2439-2444, 1997.

INDEX

Page numbers in *italic* indicate figures. Page numbers followed by "t" indicate tables.

9 780792 384786

MIX
Papier aus verantwortungsvollen Quellen
Paper from responsible sources
FSC® C105338

If you have any concerns about our products,
you can contact us on
ProductSafety@springernature.com

In case Publisher is established outside the EU,
the EU authorized representative is:
Springer Nature Customer Service Center GmbH
Europaplatz 3, 69115 Heidelberg, Germany

Printed by Libri Plureos GmbH
in Hamburg, Germany